Definitions
and
Doctrine
of the
Military Art

Past and Present

Definitions
and
Doctrine
of the
Military Art

Past and Present

John I. Alger

Thomas E. Griess
Series Editor

DEPARTMENT OF HISTORY
UNITED STATES MILITARY ACADEMY
WEST POINT, NEW YORK

AVERY PUBLISHING GROUP INC.
Wayne, New Jersey

Series Editor, Thomas E. Griess
In-House Editor, Joanne Abrams
Editorial Staff, Diana Puglisi and Jacqueline Balla
Cover design by Martin Hochberg
Photograph Sources, U.S. Army, U.S. Navy, and U.S. Air Force

All original artwork by
Edward J. Krasnoborski

Library of Congress Cataloging in Publication Data

Alger, John I.
  Definitions and Doctrine of the Military Art.

  (The West Point military history series)
  Bibliography: p.
  Includes index.
  1. Military art and science--Terminology.
2. Military history--Terminology. 3. Military
art and science--History. I. Title. II. Series.
U104.A54 1985   355'.0014   85-18673
ISBN 0-89529-309-9
ISBN 0-89529-275-0 (pbk.)

10 9 8 7 6 5 4 3 2 1

Printed in the United States of America

Contents

Illustrations, ix
Acknowledgements, xi
Foreword, xiii
Introduction, xv

1   Fundamental Concepts in the Language of the Military Profession, 1
2   Notes on Offensive, Defensive, and Retrograde Operations, 15
3   From the Ancients to the Gunpowder Revolution, 29
4   Gunpowder, Renaissance, and Reason, 41
5   The Age of Napoleon, 59
6   From Napoleon Through the American Civil War, 73
7   To the Great War, the First World War, 93
8   From the Great War to the Second World War, 123
9   The Age of Small Wars and Nuclear Deterrence, 167
10   Toward A More Perfect Understanding, 193

Index, 203

To Mom and Dad
and to true friends near and far
especially to
Chuck, Bob, Terry, Mike, and Ralph
who ennobled themselves by serving their country
and who in death ennoble us.

Illustrations

Chapter 1

The Threads of Continuity, 4
Standard Terminology, 11

Chapter 2

Distribution of Forces, 16
Classic Pursuit, 17
The Offensive Phase of a Classic Defensive-Offensive, 18
Penetration, 18
The Oblique Order, 19
Envelopments, 19
Classic Envelopments, 20
Turning Movement, 20
Classic Turning Movement, 21
Central Position Creates Interior Lines, 22
Concentration of Forces, 22
Defense Sector Areas, 23
Division Area Defense, 24
Division Mobile Defense, 24
Delay on Successive Positions, 25
Delay on Alternate Positions, 25
A Division River Crossing, 26
Classic Deliberate River Crossing, 26

Chapter 3

Chart: From the Ancients to Gunpowder, 30
Crossbow and Arming Device, 36
Onager, 36
Ballista, 37
Trebuchet, 37

Chapter 4

Chart: The Age of Gunpowder, Renaissance, and Reason, 42
The Swiss Phalanx, 45
Pole Arms: Bill, Ceremonial Halberds, and Spontoon, 47
The Matchlock, 48
Sixteenth and Seventeenth Century Matchlock Musket—
    Caliber .80, 48
Circa 1750 "Brown Bess" Flintlock Musket and Bayonet—
    Caliber .75, 49
Vasi or *Pot de Fer,* 49

Bombard Mounted on a Swivel With Elevating Capability, 50
Fifteenth Century Breech-Loader, 50
Leather Gun, 1630, 52
General Characteristics of Guns, Howitzers, and Mortars, 52
Profile of an Eighteenth Century Permanent Fortification, 54

Chapter 5

Chart: Some Key Events During the Napoleonic Era, 60
Chart: Wars of the Republic and Empire and the Career of
    Napoleon Bonaparte, 62
French Musket—Model 1763—Caliber .70 (Charleville), 65
Eighteenth Century Field Gun, 65
Eighteenth Century Field Gun Attached to Limber, 67
The *Ordre Mixte,* 69

Chapter 6

Chart: The Half Century After the Defeat of Napoleon, 74
Chart: Principal Campaigns, Battles, and Command Changes
    of the American Civil War, 76
Hall Rifle—Model 1819—Caliber .52, 81
Dreyse Needle Gun—Model 1854—Caliber .60 (Carbine
    Shown), 82
A Sectional View of a Needle Gun Cartridge, 82
A Sectional View of a Minié Ball, 82
U.S. Special—Model 1861—Rifle-Musket—Caliber .58, 83
Sharps Carbine—New Model—1863—Caliber .52, 83
Spencer Carbine and Rifle—Caliber .50-.56, 84
Henry Rifle—Caliber .44, 84
Napoleon Gun Howitzer—U.S. Model 1857, 84
Parrott Rifle—U.S. Model 1861, 85
9-Inch Dahlgren Gun on a Marsilly Carriage, 86
A Typical Attack Formation Used by a Division During the
    Civil War, 90

Chapter 7

Chart: From the American Civil War Through the Great
    War, 94
Chart: Principal Campaigns, Battles, and Command Changes
    of the Great War, 96
German Mauser—Model 1871—11-mm, 103
United States Rifle—Model 1892—Caliber .30, 103

German Mauser–Model 1898–7.92 -mm,103
United States Rifle–Model 1903–Caliber .30(Springfield), 104
Gatling Gun–Model 1862–Caliber .58, 104
Maxim Heavy Machinegun–Model 1893, 105
German Maxim–Model 08-15–7.92 -mm,105
Browning Machinegun–M1917A1–Caliber .30, 105
French Light Machinegun–Model 1915–8 -mm, 106
Lewis Machinegun–Model 1915, 106
German MP–Model 18-1, 107
French 75-mm Gun–Model 1897, 107
The Paris Gun, 108
Sopwith Camel, 110
Fokker D-VII, 110
SPAD 13, 110
Fokker DR-1, 111
Gotha G-V, 111
Handley-Page 0/400, 111
British Tank, Mark IV, 112
British Medium Tank, Mark A Whippet, 112
French Tank, St. Chamond, 112
Submarine–USS *Holland,* 114
Battleship–HMS *Dreadnought,* 115
Factors in the Formulation of the Concept of Operations Ashore, 119
Factors in the Selection of Landing Areas, 120
Schematic of the Landing Area, 121

Chapter 8

Chart: Between the World Wars, 124
Chart: Principal Events of the Second World War, 126
United States Rifle M-1–Garand–Caliber .30, 130
Browning Automatic Rifle–Model 1918–Caliber .30, 131
Browning Light Machinegun–M1919A6–Caliber .30, 131
British Bren Gun-Mk I–Caliber .303, 131
German MG-42 Machinegun–7.92 -mm,131
Japanese Type II Light Machinegun–6.5 -mm,132
United States 105-mm Howitzer–Model 101A1, 133
United States Gun Motor Carriage M7, 133
GermanV-2 Ballistic Missile, 134
German V-1 Flying Bomb, 134
Hiroshima "Little Boy", 136
Nagasaki "Fat Man", 136
Mitsubishi A6M Type Zero, 138
Curtis P-40, 138
North American P-51 Mustang, 139
Republic P-47 Thunderbolt, 139
Messerschmitt Bf-109, 140
Vickers Supermarine Spitfire, 140
Focke-Wulfe Fw-190A, 141
Messerschmitt Me-262 Schwalbe, 141
Yakovlev Yak-3, 142
Boeing B-17F Flying Fortress, 142
Consolidated B-24H Liberator, 142
Boeing B-29 Superfortress, 143
Douglass SBD Dauntless, 144
Avro Lancaster Mk I, 144
Junkers Ju-87 Stuka, 145
Mitsubishi G4M Betty, 145

Gliders: American Waco CG-4A and German DFS-230A, 146
Glider: British Horsa With U.S. Markings, 146
A "Funny": United States M4 (Sherman) With Flailing Attachment, 147
United States Medium Tanks, M3, 148
United States Medium Tank, M4–General Sherman, 148
United States Medium Tank, M26–General Pershing, 148
British Medium Tank, Mk II–Matilda, 149
British Medium Tank, Mk III–Crusader, 149
British Heavy Tank, Mk IV–Churchill, 150
United States Light Tank, M5A1 (37-mm Gun)–Stuart, 150
French Light Tank, *Char Léger* 1935R, 150
Soviet Medium Tank, T-34, 151
German Light Tank, PzKw I Model B, 151
German Medium Tank, PzKw III, 152
German Heavy Tank, PzKw V–Panther, 152
German Heavy Tank, PzKw VI–Tiger, 153
Fleet Submarine, 154
Battleship– USS *Washington,* 155
Destroyer–USS *Willard Keith,* 155
Landing Ship Tank (LST), 157
Graphic Representations of Control Measures, 160

Chapter 9

Chart: The Age of Small Wars and Nuclear Deterrence, 168
United States Rifle M-14–7.62 -mm,173
United States Rifle M-16–5.56 -mm,173
Soviet Kalashnikov Assault Rifle AK-47–7.62 -mm,173
Soviet Simonev SKS–7.62 -mm,173
Soviet SGM Heavy Machinegun–7.62 -mm,174
M-79 Grenade Launcher–40-mm, 174
Redeye Guided Missile, 174
M-220 (TOW), Ground-Mounted Missile, 175
M-72A (LAW)–66-mm HEAT Rocket, 175
Pershing II Missile, 176
MLRS Launcher, 176
Chaparral, 177
Sergeant York Division Air Defense Gun, 177
Patriot Missile, 178
North American F-86 Sabre, 179
MiG-15 With U.S. Markings, 180
General Dynamics FB-111 With Wings Extended, 180
Convair B-36D, 181
Boeing B-52 Stratofortress, 181
CH-21 Shawnee, 182
AH-1 Cobra, 182
CH-47 Chinook, 182
UH-60 Blackhawk, 183
AH-64 Apache, 183
British Medium Tank–Centurion, 184
United States Main Battle Tank, M60A1, 184
Soviet Medium Tank, T-62, 184
United States Main Battle Tank–M1 Abrams, 185
M-113 Armored Personnel Carrier, 186
M2 Bradley Fighting Vehicle, 186
Aircraft Carrier and Destroyer, 187

Chapter 10

Chart: The Threads of Continuity Revisited, 195

x

Acknowledgements

The many diverse tasks associated with the creation of this book were accomplished under the able direction of Brigadier General (Retired) Thomas E. Griess. His ideas and efforts guided the author and contributed immeasurably to the text's preparation, from inception to publication. Respected as a professor, a leader, and an example by all who know him, he is also richly deserving of the sobriquet "editor-in-chief." For his gracious sharing of wisdom gained during long and faithful years of service to man and nation, I also thank Brigadier General (Retired) Edwin V. Sutherland, whose remarks on leaving West Point are quoted at length in Chapter 10. I gratefully acknowledge the inspiration and encouragement that were provided by my mentor, Professor Peter Paret, and by other distinguished professors of history whom I have had the pleasure of knowing—notably Don Horward, John Keegan, Gordon Craig, and Gordon Wright. Sincere and warm thanks are extended to my colleagues in the Department of History of the United States Military Academy, who unselfishly shared their encouragement, time, effort, and expertise, and to colleagues at the United States Army School of Advanced Military Studies and the National War College, whose thoughts, writings, and students will have a positive impact on the United States Army for years to come. To Mr. Edward J. Krasnoborski, whose cartographic and drafting genius are evident throughout, go accolades and appreciation. Several agencies at the Military Academy deserve special mention: I thank the staff of the Library; members of the Department of Foreign Languages; the Photo Laboratory; and the "Gold Coats," the consultation element of the Dean's Academic Automation Division, whose tireless efforts assisted greatly with the initial computer indexing of the manuscript. To special friends and family, especially the kids (Chip, Kelly, Scott, Jason, and Andrew), I express my deepest appreciation for their support and love. For her help in preparing this second edition and for her support in countless ways, I am eternally grateful to Dottie.

John I. Alger

Arlington, Virginia
April 1985

Foreword

Doctrinal writings extracted from field manuals have been available for the use of students enrolled in the United States Military Academy's course in the History of the Military Art since 1943, when a booklet entitled *Notes on Combat Maneuvers* was published. Successive reference texts followed, each generally being more detailed than its predecessor. These texts served several purposes: they provided a lexicon of terms used in the course; they provided a brief and ready guide to current military doctrine; and they provided other information essential to the study of the military art. In recent editions of such texts, historical examples and definitions that no longer appeared in contemporary United States Army field manuals were included, along with current extracts from doctrinal sources. This is also true of the present volume, in which a chronological organization has been adopted to emphasize the changes that have affected definitions and doctrine in the period extending from early Greek civilization to the present. Doctrinal concepts are presented in every chapter, but long discussions of current doctrine are confined to Chapters 1 and 2.

It is hoped that this volume will provide a basic reference work that will be particularly helpful to students beginning their in-depth study of the History of the Military Art. The chronological orientation of the book facilitates its use as an introduction to the study of a given period of military history, as a factual companion to the study of a given period, or as a review of a given period. The chapters dealing largely with current operational doctrine will further the reader's understanding both of today's military profession and of past experiences. The comprehensive index allows the work to serve as a dictionary of terms and doctrine encountered throughout the study of the military institutions of the past and present. One caveat is in order regarding doctrinal concepts and definitions in *current* use. As the military services make doctrinal changes—a fairly frequent occurrence—a reference text such as this one becomes outdated in part. This is difficult to avoid without an annual revision, an expensive and impractical task.

The Department of History at the United States Military Academy and a number of West Point graduates are indebted to John I. Alger for his detailed research and careful writing of *Definitions and Doctrine of the Military Art*. A student of military theory and doctrine, Alger prepared both the first edition of the text in 1979 and the present edition. This new edition contains revisions related to the organization of material and the coverage of some doctrinal matters. As editor, I have attempted to clarify certain passages for the general reader, explained purely military terminology, and tried to improve the evenness of the narrative. The editor is grateful for the advice and suggestions that were tendered by Rudy Shur and Joanne Abrams of Avery Publishing Group, Inc. Their assistance was timely and helpful. Ms. Abrams immeasurably improved the narrative through her painstaking editing, corrections of lapses in syntax, and penetrating questions related to clarity of expression.

Thomas E. Griess
Series Editor

Introduction

In the literature of military history, modern concepts are often used in describing events long past. Conversely, some long-established concepts that no longer appear in modern doctrine are well suited for use in discussions of current events. The language of military history and the language of the military profession are, therefore, a blend of terms and doctrine, old and new. Accordingly, a comprehensive reference work for the study of the history of the military art needs to include concepts and definitions from both the past and present.

The 10 chapters of this volume contain definitions or discussions of over 2,500 terms and concepts that are a part of the military vocabulary. Some of these terms can be found in current doctrine and some cannot, but all are important to the understanding of the military past and present.

In order to ease the task of locating definitions in the text, words are boldfaced at the point where each word or concept is most explicitly defined. The page numbers in the index that are boldfaced indicate the pages on which the indexed terms are most explicitly defined.

The terms and concepts defined in this book are derived from predilections based on nearly a quarter of a century in military uniform, beginning with the blues of an Air Force ROTC cadet in September 1960; from the repeated perusal of doctrinal sources, pertinent histories, and reference works; and from the words and concepts used in the History of the Military Art Course at the United States Military Academy and the West Point Military History Series. No single volume could begin to define all the terms encountered in a career of military service or a lifetime of reading military history, but the material included in this book will surely get the student started and, on occasion, inform even the most widely read and practiced members and followers of the military profession.

The terms and concepts presented in this book are organized into seven chapters, each of which is concerned with a major period of military history. Before delving into the terms associated with the seven historical periods, however, fundamentals that will facilitate the study of every period are examined. Hence, the first chapter deals generally with "The Language of the Military Profession," and the second chapter briefly deals with "Notes on Offensive, Defensive, and Retrograde Operations." The next seven chapters are devoted to military history, from the time of the ancient Greeks through the wars of the very recent past. The final chapter reviews some of the fundamental concepts and presents a few considerations on future warfare and some thoughts for future military leaders.

Each of the seven chronologically-ordered chapters is divided into six sections that are respectively entitled "Major Themes of the Age," "Participants in the Profession of Arms," "Principal Organizations," "The Technology of the Age," "Operations," and a brief "Selected Bibliography." A hasty perusal of these chapters will reveal that many concepts and terms are used anachronistically—that is, in periods when those concepts and terms did not exist, or, possibly, when they meant something else or connoted something different. To be totally accurate in describing the ancient Greeks, for example, it would be necessary to use some words precisely as they were used then—in Greek. Since Greek is not a common second language among English-speaking students of the military art, modern English terms and concepts, and especially recent and current doctrinal terms, are frequently used when discussing the military institutions of the Greeks. Terms for items that are tangible and distinct tend not to change. For example, the *gladius,* a sword used by the Romans, is still known as a *gladius.* Similarly, the word describing the Roman organizational unit, the legion, is still in use. In the sections of the chronologically-ordered chapters that deal with organizations and technology, the terms tend to be distinct and applicable to a specific period. Concepts that are less concrete and specific, however, often are best defined by modern terms; hence, when discussing such concepts as command, tactics, operations, and strategy, the author has relied heavily upon current doctrinal terminology. Although a chronological progression occurs throughout the central chapters of the book, the reader should be alert to the fact that recent and current terms are frequently defined in chapters that deal primarily with earlier periods of warfare. The narrative, however, attempts to alert the reader to the doctrine and terms that are used anachronistically.

Because meaning is sometimes lost or nearly impossible to grasp in a tongue different from the original, some foreign terms have been left untranslated. Whenever foreign words are used in this text, they are italicized, but even this convention is difficult to adhere to strictly, because words of one language sometimes are assimilated into another language without change in form. For example, a coup in French is precisely the same as a coup in English. Hence, it is not italicized. A word that is set in a bold italic typeface is a foreign word that is being defined. Titles of books and names of ships also are italicized. Italics that appear in quoted materials are unaltered from the original.

The diversity and sometimes anachronistic appearance of the defined terms suggest that an encyclopedia might have been created in lieu of a narrative. A readable reference, complemented by a useful index, however, provides advantages not offered by encyclopedias. The narrative format lends itself particularly well to the introduction and review of the military aspects of a given period. Individual chapters and individual sections within the chapters can be read independently. Together, the chapters and sections provide historical background; independently, they provide topical background. Throughout, they emphasize implements and institutions rather than people and events in an effort to provide the support that is essential in the student's quest to understand better how wars have been fought and how warfare has evolved.

As concerns capitalization, spelling, and hyphenation, it is virtually impossible to use a consistent style in regard to the nomenclature of military hardware, and even the spelling of some military terms, because of the variety of styles used in different countries, different services, and different time periods. Moreover, familiar names for items often bear no resemblance to official names, and official names often consist of model designations that themselves create confusion. For example, an M-60 in the United States Army can be a main battle tank or a machinegun, and an M-1 can be a rifle, a carbine, a mortar, a howitzer, a cannon, a bayonet, or one of a number of other items. Even slight modifications in design change the military nomenclature. The inclusion of hyphens and the use of upper- and lower-case letters in titles also vary unendingly. Aircraft, weapons and, vehicle nomenclature is especially confusing, but throughout this volume an attempt has been made to establish some general rules: popular names have been used in conjunction with official designations; manufacturers have been listed with aircraft; and relatively insignificant suffixes to basic model designations have been ignored.

Statistics, also, create problems. Specific statistical characteristics of items, such as speeds of aircraft, ranges of weapons, or armaments of tanks, vary on occasion from model to model. Also, for security reasons, figures on current and some recent weapons systems may have been intentionally understated in public sources. However, statistics are needed to provide a basis for comparison. For those who desire more explicit detail, other references, many of which are cited in the bibliographical sections of the chapters, should be consulted.

Definitions and Doctrine of the Military Art is not a history book, for history involves far more analysis and interpretation than is offered here. Nor does this book claim to be a definitive source of current doctrine, for even though current doctrine addresses some "constants" in warfare (see Chapter 1), doctrine is forever a living catalog of concepts. Doctrinal words and concepts change in response to both the myriad changes that are synonymous with life and the ideas espoused by dedicated and determined military thinkers everywhere. This reference is merely a starting point—a primer that seeks to be a concise and useable introductory reference. For some, it may contribute to a long and rewarding association with the language and life of the military profession.

Fundamental Concepts in the Language of the Military Profession

1

Every profession has a language of its own—a language that must be familiar to all who wish to understand the profession. Because the vocabulary of the language changes from age to age, all professionals must understand the basic vocabulary of their own day and of previous periods if they are to succeed in the critical task of communicating with others in their chosen field. The profession of arms is no exception in this regard.

War and Peace

Because war and the deterrence of war are human activities that continue to persist, military and naval force is necessary, and leadership in war has become the special field of the professional officer. The definition of war must therefore be thoroughly understood by every student of the profession of arms. **War**[1] is the condition in the life of a political group (an alliance, a state, a nation, or an organized faction) in which violence and destruction of considerable duration or magnitude are directed against a rival political group that is powerful enough to make the outcome of the conflict uncertain for a time. The object of war is to impose the will of one group upon a rival. When the objectives of groups are in accord, war's opposite, **peace**, prevails among those groups. When the will of one group does not threaten the existence or vital interests of a rival group, compromise through diplomacy, or **peaceful coexistence**, usually results. When rival groups are unable to resolve vital issues through diplomacy, war, which is sometimes called **diplomacy by force**, often results.

Levels of War

War and peace are not clearly delimited conditions. Rather, they form a spectrum of conditions between absolutes that rarely exist. Absolute peace is the extreme condition of harmony between a group and all its like groups. In times of absolute peace, diplomacy is sufficient to solve every difference. Relative peace is the condition in which groups compete economically, culturally, or politically; alliances develop along ideological lines; and the threat of or potential for war is an important element in diplomacy. When relative peace exists among major powers, **cold war** will often exist as well. This situation is characterized by hostile propaganda, international boycotts, seizures of property and personnel, subversion, border clashes, sabotage, and assassinations. In a cold war there is no overt armed conflict between the major powers, but the role of the military profession is nevertheless significant. In addition to the **show of force,** defined as an exhibition intended to demonstrate military might to a potential belligerent, the military forces function as a deterrent against hot war or general war, and may even be involved in a hot war with lesser powers. A **hot war**—or simply a war, as it is commonly called—occurs when the territory of one group is occupied by a rival, or when naval or air elements attack a rival's territory. An **invasion,** or an entering of a rival's territory by a hostile armed force, is often the initial act of violence in a hot war. States and nations engaged in hot war are referred to as **belligerents.** Belligerents are protected by and subject to the laws of war—even though enforcement of the laws of war is not always possible. When the belligerents in a hot war are recognized states or nations, **international war** results.

Hot wars can be divided into limited wars and general wars. When the term **limited war** is used to describe wars such as those fought between European states in the late seventeenth and early eighteenth centuries, it refers to a war in which the principal objectives of the societies involved differ from the objectives of the combatants that presumably represent the societies. For example, in the late seventeenth and early eighteenth centuries, the great majority of the people

1

represented by opposing armies were principally concerned with economic, religious, and cultural issues, while the armies were principally concerned with the outcome of the war. The second definition of limited war is more contemporary and refers to wars in which conscious restraints are placed on the use of available and militarily significant weapons, methods, manpower, time, or geographical area by at least one of the principal belligerents. For example, the United States, one of the principal belligerents in the Korean War, neither used nuclear weapons nor extended the war beyond Korea. Hence, for the United States, the Korean War was a limited war. The term does not imply that the war is limited for all participants. Furthermore, it does not imply a small war in regard to the number of participants and casualties or the extent of the geographical area. Finally, it does not imply that the results are necessarily of minor consequence at a national or international level.

Major powers can fight limited wars, but when national survival is at stake, a general war ensues. A **general war** is a war in which the survival of a major power is in jeopardy and the societies involved share common interests with the combatant forces that represent them. A **total war** is a general war in which there is an involvement not only of the rival combatants, but of all the resources of the opposing societies. A Soviet definition reveals the Russian view of total war:

> TOTAL 'NAYA VOYNA (foreign) (total war)—An all-embracing imperialist war, waged by all manner of means, not only against enemy armed forces, but against the entire population of a nation, with a view to its complete destruction. Characteristic of total war are the methods by which it is waged, namely, the most perfidious and the most brutal methods, inhuman with respect to the world's population.[2]

An **absolute war** is a general war that has no limits. It is the most extreme form of war, and, like absolute peace, will probably never exist. On a practical level, the magnitude of destruction that is possible would threaten all civilization and the very existence of man. The nineteenth century Prussian general, Carl von Clausewitz, the dean of modern military theorists, postulated in his work, *On War,* that absolute war cannot exist in reality because "countless minor incidents—the kind you can never really foresee—combine to lower the general level of performance."[3] He explained that just as a machine is made up of many parts, an army is made up of many parts, and that just as a machine cannot be 100 percent efficient because of friction between the parts, an army cannot fight an absolute war because of friction. Friction—which is created by "countless minor incidents" such

as the weather, mud, the enemy, fear, fortune, and fate—limits war.

Nuclear war, which is any war that is fought with nuclear weapons, can be further described as being limited or general. A **limited nuclear war** is a nuclear war in which restraints are placed on such factors as weapon size, region of use, or number of weapons used. A **general nuclear war, strategic nuclear war,** or **strategic exchange** is a nuclear war in which major powers strike each other with nuclear weapons.

Revolutionary Warfare

Within the various levels of war discussed above, many forms of war, or warfare, can exist.* **Revolutionary warfare** is the combination of political action and violence that is directed against the population, either by a competitor who wishes to establish legitimacy in order to seize governmental power or by an incumbent who wishes to maintain legitimacy in order to retain power. **Insurgency** refers to the actions and violence of the competitor in revolutionary warfare. **Counterinsurgency** refers to the actions and violence of the incumbent in revolutionary warfare.

Another form of warfare that is often encountered in the study of revolutionary warfare is internal warfare. **Internal warfare** is a conflict that results from unsatisfied dissenters who have gone beyond their attempts to secure evolutionary change or reform in the exercise of governmental power by seeking the destruction and replacement of the government. Internal warfare includes such subcategories as putsches; coups, or *coup d' états*; revolutions; revolts; jacqueries; insurrections; rebellions; and civil war. A **putsch** is a secretly plotted and suddenly executed attempt to overthrow an existing government. A **coup,** or *coup d' état,* is a violent act, generally of short duration, by an elite movement that attacks the power base of an incumbent government. It seeks to replace one elite with another, and is often referred to as a "revolution from above." A **revolution** is a mass movement that attempts to remove the government by replacing the decision-making element of governmental power with a popularly based government. A revolution is often referred to

---

*Distinctions are sometimes made between "war" and "warfare," but it is not possible to both be precise and stay within the limits of general usage with these two words. In some reference works, "war" refers to the periods of violence, while other authors refer to "warfare" as the periods of violence. Similarly, some authors use the term "history of war" to describe the study of military organizations in peace and war, while other authors use the term "history of warfare" for the same purpose. Be aware that although distinctions are sometimes made (for example, revolutionary warfare and revolutionary war connote different phenomena), "war" and "warfare" are often used interchangeably as they are in this text.

as "revolution from below." A **revolt** is a revolution of relatively short duration. **Jacqueries, insurrections,** and **rebellions** are mass movements aimed at the resolution of a specific local grievance, and not at the governing power. A jacquerie generally refers to a rebellion by peasants. An insurrection is generally conducted by a very weak internal group. A **civil war** is a combined mass and elite movement in which one group attempts to wrest control of a region or state from its existing government. A **revolutionary war** is a covert civil war in which the elites remain hidden; hence, the conflict appears to be a revolution. A revolutionary war can occur either with or without the support of outside nations or states.

Other Forms of Conflict

Since wars must, by definition, involve rivals of sufficiently equal strength to render the outcome uncertain for a time, many conflicts between armed forces occur that do not fit comfortably in the spectrum between peace and war. For example, when one group is so much stronger than a rival that the outcome of an armed conflict is never seriously in doubt, neither war nor peace prevails, according to the definitions above. Such a condition occurs when a powerful nation's military force is directed against a primitive people. These activities are referred to as **pacifications** when the conflict occurs within the political domain of the greater power, as **expeditions** when the conflict occurs within the empire of a power, or simply as **explorations** when the conflict occurs in an area that is neither claimed nor defended by a major power. Another example of an armed conflict that is not a war occurs when a great power fights a much smaller state. Such conflicts are called **interventions** when the conflict arises over any issue, exclusive of issues arising from acts directed against the great power. When a great power responds to an act committed against its government, its people, or its properties, the conflict is referred to as a **reprisal**. A **massacre** occurs when a considerable number of combatants or noncombatants are killed under circumstances of cruelty or atrocity. Massacres can occur either within the context of war or outside the context of war. A **liberation** occurs when violent means are used to free the people of a region or state from the control of an oppressive government and to grant them a greater degree of participation in government and a greater degree of individual, humanitarian rights.

The Nature of War

Leadership in war is the special province of the professional soldier. Even though he is sometimes called upon to serve in conflicts that lie outside the formal definition of war, he is charged to know the heavy responsibility that rests with him in war and in his preparation for war. The Prussian military philosopher, Carl von Clausewitz, observed:

> War is no pastime; it is no mere joy in daring and winning, no place for irresponsible enthusiasts. It is a serious means to a serious end, and all its colorful resemblance to a game of chance, all the vicissitudes of passion, courage, imagination, and enthusiasm it includes are merely its special characteristics.[4]

War is not a blind struggle between mobs lacking guidance and coherence. To the contrary, it is a conflict of well organized masses, moving in concert with other masses and acting under the impulse of a single will and purpose. Each mass seeks to impose its will on other masses that are capably led, comparably equipped, and possessing unknown resolve.[5]

The Threads of Continuity[6]

One of the most obvious aspects of military history is that warfare changes. Each war has been different from the preceding one. Sometimes the changes have been small, involving, for example, minor improvements in individual weapons; sometimes the changes have been major, involving entirely new systems of tactics and strategy. The rate of change has often been gradual, although at times it has been abrupt and revolutionary. Soldiers, leaders, and generals have always had to adapt to these changes, often under the pressure of command in battle or supreme command in a major war. Failure to recognize the impact of the changes, often because of an obsession with imagined similarities, has resulted in loss of life, loss of wars, and the defeat of armies and nations. On the other hand, there are brilliant examples of professional soldiers who have recognized the impact of changes, reacted forcefully and in time, and thus changed the course of battles, campaigns, and sometimes history. In the hope of joining the latter category, rather than the former, professional soldiers study the process of change in history. The student's problem is to discover a way to analyze the process of change in warfare systematically. In the past, there have been common factors that either were a part of the military profession or affected that profession. By focusing on these factors in different ages, the student can perceive the changes that have occurred more clearly, and can thus understand the meaning of the past more perfectly. These factors can be called the "threads of continuity." While the threads

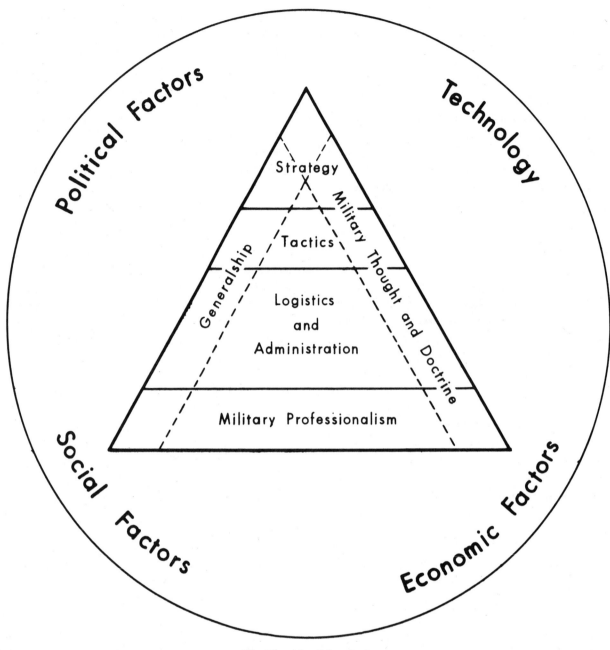

The Threads of Continuity

have no inherent worth, they can provide students with a way of obtaining information, and serve as a lens through which events can be examined and placed in perspective. The military past can be envisioned as a carpet that is woven from strands representing the threads. The carpet is a complex one, showing man's activities, ideas, and discoveries. Moreover, it is in a constant state of subtle change. The importance of individual threads vary from one era to another—that is, the strands in the tapestry of the military past fluctuate in size as their importance to the tapestry as a whole undergoes change. The threads do not change in importance at the same rate.

Taken together, the threads further the student's understanding of the past.

The Internal Threads

The threads of continuity that are entirely, or almost entirely, a part of the military profession are military professionalism, tactics, strategy, logistics and administration, military theory and doctrine, and generalship.

Military Professionalism

A **profession** is an occupation or calling that requires specialized knowledge of a given field of human activity and long and intensive training. A profession maintains high standards of achievement and conduct through force of education or concerted opinion, commits its members to continued study, and has the rendering of a public service as its prime purpose. **Military professionalism** encompasses activities practiced by those whose goal is the preservation of peace through deterrence or, should deterrence fail, the restoration of peace through combat. Military professionals are experts in the management of violence, and are characterized by their dedication to improvement and their sense of responsibility to subordinates and the state. Attitude distinguishes the professional members of the military services from those who are not professionals. Those who seek to create or strive to perfect a profession of arms are military professionals. Those who think about or practice the conduct of war solely for personal glory or material gain are not military professionals. Mercenaries and pirates are of the latter category.

Tactics

The second thread of continuity that is strictly a part of the military profession is tactics. **Tactics** is the planning, training, and control of the ordered arrangements (formations) used by military organizations when engagement between opposing forces is imminent or underway. The word tactics is derived from the Greek *taktos,* which means ordered or arranged. In effect, tactics, which includes the use of supporting weapons, is the art of fighting battles. In the nineteenth century, the term was further refined by adding the adjectives "grand" and "minor." While neither refinement is any longer used, **grand tactics** was the tactics of large organizations, and **minor tactics** was the tactics of small organizations or of organizations consisting entirely of one arm (infantry, cavalry, or artillery). In recent years, the United States Army has added to its doctrine the **operational level of war** to distinguish the theory of larger unit operations from that of smaller units. It is defined as the activity concerned with using available military resources to attain strategic ends in a theater of war. As the link between tactics and strategy, it governs the manner in which operations are designed to meet strategic ends and the way in which campaigns are conducted. The term **applied tactics** refers to the application of tactical doctrine to practical problems.

Strategy

The third internal thread of continuity, strategy, no longer belongs entirely to the military professional, for today's military leaders generally work closely with civilian officials in the field of strategy. The term is derived from the Greek *strategos,* which is the art or skill of the general. This definition of the ancient Greeks remains useful in understanding modern definitions of the term. Until the late eighteenth and early nineteenth centuries, the specific tasks of generals differed little from the tasks of subordinate commanders or politicians, and no specific term was used to describe the art or skill of the general. Political and military leadership of a group was often vested in the same individual, and the resources of small unit leaders on the battlefield differed little from the resources of the general in overall command.

By the late eighteenth century, the existence of a resource available to higher ranking leaders was recognized and given the name **stratagem:** a ruse or a trick that gives an advantage to one side in battle or war. By the early nineteenth century, strategy referred to the use of resources or the tasks of war that were particular to the high-ranking officer. It was defined as the preparation for war that took place on the map,[7] or the use of battles to win campaigns.[8] Since the modern appearance of the term, no precise definition has received universal acceptance. Yet the term continues to be widely used, and is among the principal concepts employed to examine and describe the evolution of the profession of arms. Although other thoughtful definitions exist, here we will define **strategy** as the planning for, coordination of, and concerted use of the multiple means and resources available to an alliance, a nation, a political group, or a commander, for the purpose of gaining an advantage over a rival. Strategy allows the achievement of adopted goals in war or peace, and if it is to be successful, those goals must be clearly defined and attainable. However, because conditions in war and peace are constantly changing, strategy must be modified as it is being executed, and at times even the goals of strategy must be altered.

Although tactics, strategy, and the operational level of war are not clearly separable activities, distinctions between their characteristics can be made. Tactics (the planning, training, and control of ordered arrangements of troops just before and during battle) is the primary concern of battalion, company, and platoon officers, because they immediately control the dispositions of the fighting forces. Strategy (the planning for and concerting of various means and resources available to gain an advantage over a rival) lies primarily within the province of national leaders and commanders of divisions, corps, and armies, because they control the multitude of available means and resources. The operational level encompasses the planning of campaigns within a given theater of operations. Tactics is generally prescribed by regulations, field manuals, or convention. Strategy and the operational level are dependent upon the conditions of a given time and are rarely prescribed by regulation or convention. The operational level is guided by the tenets of **AirLand Battle**, United States doctrine that calls for the integration and coordination of air and land combat resources against close and distant targets.

The adjectives "tactical" and "strategic" are also distinguishable. Tactical refers to aspects of battle where the ordered arrangements are employed, while strategic refers to matters or goals that are the concerns of national leaders and division, corps, or army commanders. In contrast to tactical operations, strategic operations have a long-range effect on opposing forces rather than an immediate one.

Strategy, like tactics, is further refined by restrictive modifiers. For example, **grand strategy** is the strategy of a nation or an alliance. The goal of grand strategy is the attainment of the political objective of a war. Grand strategy is formulated by heads of state and their principal civilian and military advisers. Grand strategy is more accurately called **national strategy** when the goals of a single nation are of primary consideration. A third refinement or level of strategy is **military strategy**, in which the means and resources are those of the armed forces of a nation and the goal of strategy is the securing of objectives consistent with national policy through the application or threat of force. Although military strategy can be formulated by military commanders at all levels, commanders below general officer rank are rarely involved in strategy that affects national policy. A fourth level of strategy, **campaign strategy,** is the strategy of a commander of a force of considerable size that is acting independently. Its immediate goal is generally the occupation of territory or the defeat of all or a significant part of the enemy armed forces; its long-term goal is the support of political goals.

The four levels of strategy can be illustrated by examining Anglo-American experiences in World War II. Grand strategy was the responsibility of Prime Minister Winston Churchill and President Franklin D. Roosevelt, their civilian advisers, and the British and American chiefs of staff. Two of the goals agreed upon were: (1) Germany would be defeated before the principal effort would be directed against Japan, and (2) the German surrender would be "unconditional." During World War II, the national strategy of both Great Britain and the United States was nearly synonymous with grand strategy, but the British goal to remain in the war might be more accurately described as a national strategy. The national goal was supported by Churchill, the Cabinet, British military leaders, and the Parliament. Military strategy was determined primarily by the Anglo-American Combined Chiefs of Staff, who established goals and allocated resources

to various theaters of operations. Based on their assigned goals and allocated resources, theater commanders, like General Dwight D. Eisenhower in Europe and Admiral Chester W. Nimitz in the Central Pacific, formulated campaign strategy at the operational level.

In addition to the levels of strategy, strategy can be further defined by its purpose. Two of the latter forms, the strategy of annihilation and the strategy of exhaustion, were referred to extensively by the late nineteenth century theorist, Hans Delbruck. He claimed that the **strategy of annihilation** seeks the complete destruction of the enemy army, and that the operations of Alexander the Great, Caesar, and Napoleon best illustrate such a strategy. In contrast, the **strategy of exhaustion** seeks to gradually destroy the enemy's will and capacity to resist. Delbruck labeled Pericles, Gustavus Adolphus, and Frederick the Great as strategists of exhaustion. A third form of strategy, the **strategy of attrition**, sometimes is loosely used as a synonym for the strategy of exhaustion. However, attrition more accurately connotes that the enemy force is slowly being destroyed. Exhaustion seeks to erode the will and resources of the enemy; attrition erodes the force itself.

Military literature refers to still other descriptive forms of strategy. For example, an army that reacts to, rather than dominates, the activities of its enemy follows a **strategy of survival**. One strategy retains the name of the man who is credited with having devised it. During the Second Punic War (218–201 B.C.), the Roman general, Fabius, strongly believed that combat against Hannibal's Carthaginians would result in a battle of annihilation for Rome. (In a **battle of annihilation,** one side is physically unable to continue the fight.) Fabius therefore chose to conduct a campaign of delays in an effort to exhaust the Carthaginians. This strategy, which intentionally avoids battle for fear of its outcome, is referred to as **Fabian strategy,** the **Fabian way of war,** or a **strategy of evasion.**

Logistics and Administration

The fourth thread of continuity, logistics and administration, is much like strategy in the sense that even though most of its functions are a part of the profession of arms, many functions are dependent upon and interact closely with civilian-controlled activities. In addition to having this similarity to strategy, logistics and administration provide many of the resources that strategy puts to work. **Logistics** is the provision, movement, and maintenance of all services and resources necessary to sustain military forces. Most of these services and resources originate in the civilian sector; hence, logistics involves the nation's economic capacity and the closely related capability of the nation to support its military forces. It includes the design, development, acquisition, storage, movement, distribution, maintenance, evacuation, and disposal of materiel; the movement, evacuation, and hospitalization of personnel; the acquisition or construction, maintenance, operation, and disposition of facilities; the acquisition of civilian labor; and the acquisition or provision of services such as baths, laundry, libraries, and recreation. **Administration** is the management of all services and resources necessary to sustain military forces. Since administration applies to the management of the functions of logistics, it is inseparable from logistics.

Military Theory and Doctrine

Military theory is the body of ideas that concern war, especially those concerning the organization for, training for, and fighting war. **Doctrine** is the *accepted* body of ideas concerning war. The acceptance of ideas can be the result of either long usage or official sanction by the appropriate military authorities of a particular service, nation, or political group. Those men who have thought deeply about war or whose thoughts about war have influenced considerable numbers of soldiers are known as military theorists. After examination and acceptance by highly experienced professionals, theory becomes doctrine. The battlefield leader can employ this accepted theory, or doctrine, with a reasonable assurance of positive results. Doctrine does not, however, alleviate the requirement for sound judgment, for the best solutions to every critical problem are not always found in doctrine. Doctrine in modern armies is generally disseminated through manuals, regulations, circulars, and handbooks that prescribe standardized procedures and organizations.

Generalship

Generalship is the art of command at high levels. It involves strategy, tactics, and logistics, and is heavily laced with administration—the management of all available logistical resources. It also involves military theory and doctrine, and connotes a deep understanding of the conduct, aims, and qualities of members of the military profession; a high degree of personal courage; complete dedication to the profession; and an acute awareness of the value of morale and esprit.

The External Threads

In addition to the important role played by those factors within the military profession, there are external threads of continuity that exercise considerable influence on the preparation for and the practice of war. The most significant of these external threads are political factors, social factors, economic factors, and technology.

Political Factors

Those ideas and actions of governments or organized groups that affect the preparation for war are termed **political factors**. Political factors determine the composition and strength of military organizations, often establish the goals and policies for which wars are fought, and affect the way in which wars are fought. Until the middle of the nineteenth century, the political chiefs, or heads of state or government, were usually the commanders of the military forces as well. Alexander the Great, Julius Caesar, Gustavus Adolphus, Frederick the Great, and Napoleon are prime examples of such leaders. In each of these cases, political policy and military goals were nearly synonymous. However, in modern democratic societies, such as Great Britain and the United States, national security policy lies more in the domain of civilian leaders. Regardless of the conditions, political factors maintain a major influence upon the military profession. In modern democratic societies, political factors have a double function. At one level, they involve the activities of the military profession that influence legislative and administrative decisions regarding national security. At another level, they involve the consequences of military actions on the international balance of power and the behavior of foreign states.[9]

Social Factors

The activities of or ideas emanating from human groups and group relationships that affect the preparation for and conduct of war are **social factors**. These factors involve such diverse concepts as popular attitudes, roles of religious institutions, levels of education, functions of educational institutions, psychological warfare, reactions to and actions of mass media, interracial and minority rights questions, combat psychology, standards of morality and justice, and—ultimately—the will of a people to fight.[10] In total war, social factors can be as important as terrain objectives or the destruction of the military forces in the field.

Economic Factors

Those activities and ideas that influence the production,

distribution, and consumption of the material resources of the state are **economic factors.** Different types of economies —for example, capitalist, Communist, laissez faire, industrial, agrarian, commercial, subsistence, or common market—affect warfare differently. Economic war, which takes such forms as blockade or boycott, can occur both in times of war and in peacetime.

The interrelationship of political, economic, and social factors is complex, especially in modern societies, and a complete understanding based on the detailed study of one alone is not possible. Together, these factors provide the foundation of national power. Without them, there would be no armies. Without their influence, there would be no wars.

Technology

While political, social, and economic factors provide the foundation of power, technology often provides the limits to power. **Technology** is the use of knowledge to create or improve practical objects or methods. In modern times, technology has become the application of science to war. Within the military profession, it leads to progress in such important areas as transportation, weapons, communications, construction, food production, metallurgy, and medicine. Technology has an undeniable influence on strategy, tactics, logistics, military theory and doctrine, and generalship. When a nation is technologically superior to its enemy, its probability of success in military endeavors is greatly improved.

The ten threads of continuity discussed above do not provide an infallible means for learning about every aspect and innuendo of the military past. Rather, they form a conceptual framework on which the student can reconstruct a general outline of the tapestry of the military past. The full meaning and magnitude of that tapestry can be appreciated only after long study or long years of military service and significant contribution to the profession of arms.

The Principles of War

Many theorists have tried to glean the essential components of success in war from a study of the campaigns of the greatest commanders. Physical components, such as the size of armies and the relative lethality of weapons, were known to be important. In addition, intangible components, such as methods, training, morale, and leadership, were recognized as potentially decisive factors that often brought victory to the physically inferior adversary. Among those who thought deeply about the components of success in war was Antoine-

Henri Jomini, the Swiss theorist whose study of Frederick the Great and whose personal experiences with and study of Napoleon gave him considerable insight into the art of war.[11]

Throughout his writings, Jomini stressed the importance of unchanging principles that influence the conduct of war. For years, the existence of such principles was questioned by many theorists, but in the aftermath of World War I, both the American and British Armies included definitive lists of principles in their regulations for the conduct of war. These lists—made up of a series of from 8 to 11 brief titles that represent the most critical of the nonphysical elements regulating the conduct of war—have been called the **principles of war.** Critics of the lists argued that the circumstances that a commander might face in war were so diverse that these principles could be neither applicable in every situation nor useful in many situations, and might even be detrimental in certain situations. Other critics claimed that the accepted principles were in some cases contradictory, that some principles were redundant, and that certain important principles had been omitted from the official lists. The British list differed from the United States list, and by 1930, criticism was strong enough to cause the lists—and even the term "principles of war"—to be deleted from the doctrine of both nations. After World War II, however, new lists of principles of war appeared in British and United States doctrine. The principles were again referred to by title, but fuller explanations of their meaning and application were included. These two lists differed both from the previous doctrinal lists and from each other. (*See Table on page* 9.) Not only have the titles differed, but as British and United States doctrines have been revised since the late 1940s, the explanations of each principle have been modified. Other nations have adopted different titles and explanations, and still other nations have steadfastly denied that a brief list of fundamental principles is valid or meaningful.

As with any given compendium of principles formulated after much reflection and study, each individual principle can be used as a criterion for the evaluation of campaigns and battles. When used for this purpose, however, it must be remembered that many commanders lived and led armies before the acceptance of such lists or served when applicable doctrine did not present such lists. These professionals may have been guided by valid considerations that were not included in a given list of principles of war or even contradicted principles that have appeared in such lists. For example, the principle of unity of command was violated by the Duke of Marlborough and Prince Louis, the Margrave of Baden, in the Campaign of 1704 when they agreed that each would command on alternate days; the value of the coalition was greater than the concept known to the twentieth century

British (1920)	United States (1921)	British (1948)	United States (1949)
Maintenance of the Objective	Objective	Selection and Maintenance of the Aim	Objective
Offensive Action	Offensive	Offensive Action	The Offensive
Surprise	Surprise	Surprise	Surprise
Concentration	Mass	Concentration of Force	Mass
Economy of Force	Economy of Force	Economy of Effort	Economy of Force
Security	Security	Security	Security
Cooperation	Cooperation	Cooperation	Unity of Command
Mobility	Movement	Flexibility	Maneuver
—	Simplicity	—	Simplicity
—	—	Maintenance of Morale	—
—	—	Administration	—

British and American Principles of War

American Army as the principle of unity of command. Neither Marlborough nor Louis had a doctrinal list of principles to guide or direct him—even though most, if not all, of the concepts now embodied as principles were known to experienced leaders throughout the centuries. But Marlborough and Louis wielded a much simpler administrative and tactical system than those that prevail in the twentieth century, enabling command to be alternated more easily and effectively than would be possible today. The concept of unity of command, therefore, probably meant far less at that time than it does now.

In addition to the value of the lists of principles of war as a tool of analysis, and in spite of the risk of imposing a standard on a commander justly ignorant of the standard, the principles provide further benefits. They give names to complex concepts; and hence facilitate discussions of important considerations in the study, planning, and conduct of war. They are also a ready source of important and often decisive considerations for the planners and commanders of military operations. The principles, however, must never be thought of as a substitute for sound, rational thinking. Rather, they should be viewed as a stimulus and a point of departure for discussions between military professionals.

In 1978, the list of principles of war that appeared in United States Army doctrine from 1949 to 1976 reappeared in a new manual entitled *The Army*. In August 1982, the principles of war were again listed in the successor to the manuals in which they had appeared from 1949 to 1976.

The principles of war that follow have been entitled in accordance with the Army manuals that have presented these listings since 1949. Through the years, the explanation of each principle has changed, and there is good reason to believe that the place of the principles in doctrine and their interpretation will continue to change in the future. The explanations given below provide a concise definition and a brief discussion of each concept. Although these descriptions are not necessarily verbatim accounts of the most recent Army doctrine, they are true to the spirit of the doctrine. Even if these principles are again altered or expunged from future doctrine, the current list will serve the purpose that led to its exposition by providing a useful tool for the serious student of war and delineating significant concepts in the history of military thought.

The order in which the principles are listed below suggests their relative importance, but that order has not always been followed in the various doctrinal publications in which they have appeared. In explaining some of the principles, imperatives are used that suggest an authoritativeness that prescribes rather than describes. The explanations of most of the principles, however, tend to be more descriptive than dogmatic. The introduction to some lists states that "depending on the circumstances" the separate principles "may tend to reinforce . . . or to be in conflict" with one another and that "the emphasis on any particular principle or group of principles will vary with the situation." They are not inviolable, nor are they a panacea. Rather, they are offered in the hope of being a boon to the study and consideration of past military operations, and as a guide to the study of and reflection on future military operations.[12]

Objective

Direct every military operation toward a clearly defined, decisive, and attainable objective. The ultimate military objective of war is the defeat of the enemy's armed forces. Correspondingly, each operation must contribute to the ultimate objective. Intermediate objectives must directly, quickly, and economically contribute to the purpose of the ultimate objective. The selection of objectives is based on consideration of the mission, the means and time available, the enemy, and the operational area. *Every commander must understand and clearly define his objective and consider each contemplated action in light thereof.*

Offensive

Seize, retain, and exploit the initiative. Offensive action is necessary to achieve decisive results and to maintain freedom of action. It permits the commander to exercise initiative and impose his will on the enemy, to set the terms and select the place of battle, to exploit enemy weaknesses and rapidly changing situations, and to react to unexpected developments. The defensive may be forced on the commander as a temporary expedient while awaiting an opportunity for offensive action, or may be adopted deliberately for the purpose of economizing forces on a front where a decision is not sought. Even on the defensive, the commander must seek opportunities to seize the initiative and achieve decisive results through offensive action. The defense must be active, not passive.

Mass

Concentrate combat power at the decisive place and time. Superiority results from the proper combination of the elements of combat power. Proper application of this principle, in conjunction with other principles of war, may permit numerically inferior forces to achieve decisive combat superiority at the point of decision.

Economy of Force

Allocate minimum essential combat power to secondary efforts. This principle is the corollary of the principle of mass. Minimum essential means must be employed at points other than that of the main effort. Economy of force requires the acceptance of prudent risks in selected areas to achieve superiority at the point of decision, and places a premium on flexibility of thought and action. Economy-of-force missions may require limited attack, defense, cover and deception, or retrograde actions.

Maneuver

Place the enemy in a position of disadvantage through the flexible application of combat power. Maneuver is an essential ingredient of combat power. It contributes materially to the exploitation of success, the preservation of freedom of action, and the reduction of vulnerability. The object of maneuver is to concentrate (or disperse) forces in a manner that will place the enemy in a position of disadvantage, and thus achieve results that would otherwise be more costly in men and materiel.

Unity of Command

For every objective, insure unity of effort under one responsible commander. The decisive application of full combat power requires unity of command. Unity of command results in unity of effort by coordinating the actions of all forces and directing them toward a common goal. While coordination may be achieved through cooperation, it is best achieved by vesting a single commander with the requisite authority.

Security

Never permit the enemy to acquire an advantage. Security is essential to the preservation of combat power. Security results from the measures taken by a command to protect itself from espionage, observation, sabotage, harassment, or surprise. It is a condition that results from the establishment and maintenance of protective measures against hostile acts or influences. Since risk is inherent in war, application of the principle of security does not imply undue caution or the avoidance of calculated risk.

Surprise

Strike the enemy at a time and/or place and in a manner for which he is unprepared. Surprise can decisively shift the balance of combat power. With surprise, success out of proportion to the effort expended may be obtained. Surprise results from striking an enemy at a time and/or place and in a manner for which he is unprepared. It is not essential that the enemy be taken unaware, but only that he becomes aware too late to react effectively. Factors contributing to surprise include speed, cover and deception, application of unexpected combat power, effective intelligence, variations of tactics and methods of operation, and operations security (OPSEC). OPSEC consists of signals and electronic security, physical security, and counterintelligence to deny enemy forces knowledge or forewarning of intent.

Simplicity

Prepare clear, uncomplicated plans and clear, concise orders to insure thorough understanding. Simplicity contributes to successful operations by reducing the possibility of misunderstanding and confusion. Other factors being equal, the simplest plan executed promptly is preferable to the complex plan executed later.

Rudiments of Military Organization

The Anatomy of an Army

The term **army** refers to an organization of individuals who have a common purpose to serve. In a strictly military sense, an army is composed of the organized forces that represent a significant political group and have as their principal task the conduct of war on land. The United States Army is a complex organization. **Combat units** or **combat arms,** such as the infantry, armor, and artillery, do the fighting. Units whose primary function is to render support to other organizations rather than to engage in combat, such as the Finance Corps, the Quartermaster Corps, and the Medical Service Corps, are called **combat service support units.** Units that function at times as combat units and at times as combat service support units are **combat support units.** Examples of such units are the Military Police Corps, the Corps of Engineers, and the Signal Corps.

No army can function or even exist without reinforcement from combat support and combat service support units. **Support** refers to those activities that aid, protect, complement, create, administer, manage, and sustain other forces. Supporting activities are directed toward the improvement of the effectiveness of the command. Combat units themselves are often called upon to support other units in the command. A combat unit that can come to the aid of another in a timely manner is said to be in **mutual support** or **mutually supporting.** As war has become more complex, supporting activities have played an increasingly critical role in the profession of arms.

When a unit of any size or type is confronted by an enemy force, that portion of the unit nearest to the enemy is called the **front.** If there is no confrontation with an enemy force or if the location of the enemy force is not known, the direction that a unit is facing is called the **front.** A **rank** refers to a line of soldiers standing side by side; a **file** is a line of soldiers standing front to rear. The left or right portion of the unit is termed the **flank.** The left flank is the portion on the left as you face the front. The portion between the flanks is the **center.** The **rear** of a unit is that portion farthest from the front.

When units are designated by a block, as in the figure on page 11, the outline of the block represents the geographical limits of concentration of the subunits of the organization. Blue figures generally represent friendly forces, but in a

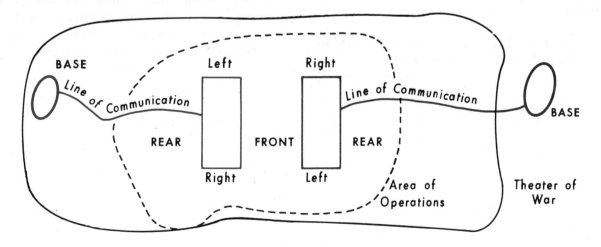

Standard Terminology

historical context, they often represent the force that is the major object of the study. Sometimes the choice is arbitrary. Red represents those forces opposing the blue.

The Army Environment

A **theater of war** comprises all those regions in which combat is likely to occur. In wars of considerable magnitude, there are often several designated theaters. For example, in literature of the American Civil War, the terms Eastern Theater and Western Theater are often used; in World War II texts, the terms European Theater and Pacific Theater are frequently encountered. In more modern usage, a theater is a geographical area outside of the continental United States for which a senior commander has been assigned military responsibility.

A **base** refers to a geographical area from which an army obtains its reinforcements and resources, and from which an army initiates its actions when it takes the offensive. If the activities and movements of combat and combat support units originate in an area far removed from the combat service support units, the area is referred to as the **base of operations.** In such a case, the area in which the combat service support units are based is referred to as the **base of supply.** For example, the base for the Normandy invasion in June 1944 was England. An **area of operations** is a geographical region within a theater of war wherein either offensive or defensive military operations are expected to occur.* **Lines of communication,** also known simply as **LOCs,** are those air, water, and land routes that connect an operating military force with its base of operations. The term **SLOC (sea line of communication)** and **ALOC (air line of communication)** are also used.

The Activities of an Army

In the course of providing services for modern societies, armies become involved in certain functionally defined activities. **Garrison duty** describes the condition of units that are assigned to an area—generally for a significant period of time —for the purpose of maintaining equipment and facilities and conducting small-unit and small-scale training. **Garrison troops** are those that rarely leave the garrison environment. **Field duty,** which contrasts with garrison duty, describes the activities of units conducting or simulating operations essential to the conduct of war. When war appears imminent, ar-

---

*Some dictionaries treat "theater of operations" and "area of operations" as synonymous terms. The same may be said of "theater of war" and "area of war." The distinctions made in this text, however, are widely accepted.

mies, as well as societies, often mobilize. To **mobilize** is to prepare for war by assembling and organizing the military resources and, at times, the societal and economic resources of a nation or other political group. When an offensive is undertaken in war, armies conduct campaigns. A **campaign** comprises a series of operations designed to accomplish certain military objectives, or a series of operations that occur within a defined period of time. Until the present century, a campaign generally lasted less than one calendar year—that is, the period from an army's departure from its garrisons in the spring until its return in the late fall.

During a campaign, an army conducts offensive, defensive, or retrograde operations. An **offensive operation** is one in which an army is attacking, ready to attack, or simply moving toward the enemy or his expected location. The purposes of the offensive are to destroy enemy resources (including his will to continue the fight), to secure important terrain, and to gain information about the enemy. No significant movement toward or away from the enemy is contemplated in the course of a **defensive operation.** The defensive also serves to preserve forces for future offensive operations or for offensives in other areas, to gain time to refit and reinforce, to retain important terrain, or to force the enemy to mass so that he is more vulnerable to friendly firepower. The **retrograde,** or **retrograde operation,** describes the action of an army that is moving away from the enemy. The retrograde is undertaken when there are insufficient forces to attack or defend successfully, when units can be employed to better advantage in other locations, or when the purpose of a campaign has been achieved.

In the course of offensive, defensive, and retrograde operations, armies are called upon to maneuver, to fight battles, and occasionally to besiege enemy positions. **Maneuver** refers to movement intended to place an army, units, troops, ships, materiel, or fire in a better position with respect to the enemy. Maneuver is generally directed against the flanks or rear of the enemy, and often aims at posing a threat to or disrupting the enemy's line of communication. A **battle** is a violent and prolonged confrontation of opposing military organizations. Significant casualties generally accrue to one or both sides in a battle. Violent confrontations of lesser magnitude are called **engagements,** a term that connotes an accidental meeting when information about the enemy is being sought, or **actions,** which generally are sharp, but brief, encounters. In addition to maneuver and battle as means of successfully achieving the objectives of war, armies are also called upon to **besiege** or to **lay siege,** that is, to place an army or part of an army around a fortified place, which is generally heavily and strongly defended, in order to compel its surrender through negotiation, exhaustion, or military action. Sieges rarely last less than a few weeks, and, depending on the strength and size of the fortified place, can continue for months before the outcome is resolved. When the forces conducting a siege are needed elsewhere, are defeated, or voluntarily leave the area of the besieged position, perhaps because of negotiations, the siege is said to be **lifted** or **raised.**

The Purpose of an Army

Every major political organization establishes objectives that, if attained, guarantee the survival of the organization, the protection of its members from external harm, and recognition of the organization by like organizations. Different organizations rarely have precisely the same objectives. When objectives conflict, political organizations must seek to resolve their differences. Policy translates the objectives into actions. Policy also determines the size and composition of military forces, the resources with which the forces are to be supported, the manner in which the forces may be employed, and the boundaries within which the forces must operate. One fundamental tenet of American policy is that objectives should be obtained through peaceful measures; military forces exist primarily as a deterrent to war. Should deterrence fail, United States military forces must be capable of gaining political objectives through the use of force. For example, they may destroy enemy forces, secure and hold important geographic objectives, or compel the enemy to submit to terms—regardless of the level of conflict. While reflecting on the seriousness of these means to the political end, every military professional should recognize that his particular goal is to develop the combat power of American forces to the maximum extent possible in order to insure the successful accomplishment of assigned tasks.

The Art and Science of War

In studying the military past, students will encounter some courses of instruction and some references that claim to be about military science, while other courses and references, seemingly covering similar material, will claim to be works on military art. In fact, many theorists have debated this question since science first shed its light on the Dark Ages and, as a result, a variety of explanations and conclusions have been offered. When the theory and doctrine of war are being examined, some think of war in scientific terms—that is, war represents a discipline that requires the systematic study of theories that have been validated through application. In advocating a science of war, some theorists further insist that theory and doctrine are based upon unchanging principles that have been derived from the scrutiny of past wars. Those who maintain that the conduct of war is essentially an art claim that each commander is called upon to express himself creatively and uniquely; theory and doctrine lose the quality of science and become guides to artful execution. Of course, whether the conduct of war is an art or a science also depends on accepted definitions. If the art of war refers to the trade or skill needed in war, then few could successfully argue that war is not an art. If the science of war means that unbreakable laws established by hypothesis, test, observation, comparison, and conclusion regulate the outcome of battle, then few could argue successfully that war is a science. When unique situations and relatively constant changes are the object of study, a belief in an art of war prevails. When standardization and attention to the details of established procedures are the object of study, a belief in the science of war prevails. Regardless of the perspective or emphasis, these two dimensions of war exist. It is of far greater importance, however, to recognize that, whether art or science, military topics demand intense and rigorous study by all members of the profession.

Selected Bibliography

A practical bibliography intended for anyone who might have an interest in current United States Army doctrine should begin with a reference to Department of the Army Pamphlet 310-1, *Consolidated Index of Army Publications and Blank Forms.* This reference work—it is no longer a pamphlet since it is available only on microfiche—lists all current Army publications, and is updated quarterly. Its many lists include, for example, the numbers, titles, dates, changes, and proponent (sponsoring) agencies of all field manuals (FMs), training circulars (TCs), Army Training Programs (ATPs), Tables of Organization and Equipment (TOEs), and Tables of Allowances (TAs).

Army doctrine is primarily contained in the publications that are cataloged in Department of the Army Pamphlet 310-1, which lists over 400 field manuals. These field manuals constitute the primary reference work on doctrine for land warfare. The capstone manual on United States Army doctrine is FM 100-5, *Operations.* Since 1939, when the numbering of field manuals was adopted by the Army, FM 100-5 has been the manual to which all doctrine is subordinate and the manual that all others must complement. Until the 1968 revision of FM 100-5, the title *Field Service Regulations* was given to this manual. In fact, *Field Service Regulations,* which was first prepared in 1904, was the capstone manual of United States Army doctrine long before the numbering system was initiated. Prior to the publication of *Field Service Regulations,* doctrine was promulgated through general orders and regulations, including such works as the *United States Army Regulations of 1861,* and through commercially published books that included the phrase "as authorized by the War Department" on their title pages. An example of the latter is Lieutenant Reed's *Elements of Military Science and Tactics,* which was used in the 1880s and 1890s. The doctrine of other nations is found in publications similar to those of the United States, and in many nations, such as Great Britain, France, and Japan, *Field Service Regulations* serve as the principal manual. Some Soviet doctrine is available to English readers through translations of a series called "The Officers' Library." English titles in this series include A.A. Sidorenko's *The Offensive* and Radziyevskiy's *Dictionary of Basic Military Terms: A Soviet View.*

Other contemporary doctrinal sources of a very general nature include the first chapter of Maurice Matloff's *American Military History,* the introduction to David Chandler's *The Art of Warfare on Land,* and John Quick's *Dictionary of Weapons and Military Terms.*

Notes

[1] Unless quotation marks are used, definitions have not been taken verbatim from any single source. Rather, each definition is consistent with widely accepted usages of the word. The author has frequently consulted and heavily relied upon definitions appearing in such general works as *Webster's International Dictionary,* various editions of the *Encyclopaedia Britannica,* and different editions of the *Dictionary of Army Terms* and *Field Service Regulations.*

[2] *Dictionary of Basic Military Terms: A Soviet View,* trans. by Secretary of State Department, Ottawa, Canada (Washington, D.C., United States Air Force [originally published in Moscow in 1965]), p. 223.

[3] Carl von Clausewitz, *On War,* trans. by Michael Howard and Peter Paret (Princeton, 1976), p. 119.

[4] *Ibid.,* p. 86.

[5] Adapted from G.F.R. Henderson, "War," *Encyclopaedia Britannica,* 1911, XXVIII, 305.

[6] The "threads of continuity" is a concept that was developed by Brigadier General (Ret.) Thomas E. Griess for use in the History of the Military Art course at the United States Military Academy at West Point. The second section of this chapter relies heavily on his essay, "The Threads of Continuity," which introduced this concept to students in the course prior to its inclusion in the first edition of *Definitions and Doctrine of the Military Art.* This section also appears in General Griess' "Introduction to the West Point Military History Series," in the first volume of the series, *Ancient and Medieval Warfare.*

[7] Antoine-Henri Jomini, *The Art of War,* trans. by G. H. Mendell and W. P. Craighill (Westport, CT, 1971), p. 69.

[8] Clausewitz, *On War,* p. 177.

[9] Morris Janowitz, *The Professional Soldier: A Social and Political Portrait* (New York, 1971), p. 12.

[10] See Harry G. Summers, Jr., *On Strategy: The Vietnam War in Context* (Carlisle Barracks, PA: Strategic Studies Institute, 1981) for an excellent appraisal of the role of popular will in the Vietnam War.

[11] See John I. Alger, *The Quest for Victory: The History of the Principles of War* (Westport, CT 1982).

[12] The Principles of War are from FM 100-1, *The Army,* September 29 1978, and FM 100-5, *Operations,* August 20, 1982.

Notes on Offensive, 2

Defensive and

Retrograde Operations

One means of learning the language of the military profession is to start at the beginning of recorded history and analyze the developing concepts of the profession, especially those that strongly influenced later terms and doctrine. While this method may seem logical, the historian soon discovers that in discussing ancient civilizations, he must use modern language and concepts in order to communicate with his readers. Accordingly, even ancient military history is laced with modern concepts. The timeliness of the concepts, however, transcends the chronology of sample selection and emphasizes their importance to an understanding of military operations of any period. Thus, in the following commentary on offensive, defensive, and retrograde operations, concepts are drawn from different periods of the military past, always with the view of selecting the most illustrative examples. These three types of operations deserve the complete coverage that a chapter provides because they are elemental to most military activities. During combat, armies engage in actions that involve one or more of these operations. In peacetime, armies are trained and equipped in preparation for these operations.

Offensive Operations

The Designation of Forces

In modern doctrine, the forces that participate in offensive operations have been designated by titles that help to explain the function that each part of the force performs. Although these descriptive titles (such as skirmish, main attack, secondary attack, and supporting attack) are used today, they have at times not appeared in contemporary doctrinal sources. Past commanders often designated a main attack in the **operation order**—the instructions that set forth the situation, the mission, the plan of action, and those details that insure

the coordinated, efficient, and effective execution necessary to obtain maximal performance from the command. In other eras, the commander merely assigned missions to subordinate elements. One element of the force could, however, be favored by being given more artillery or airpower in support, or by enjoying the advantages of having the reserve positioned where it could most readily benefit that element.

The **reserve** is any force that is not committed at the start of an attack and is available for the commander to commit at the decisive time and place—that is, the time and place that will exploit success and insure the accomplishment of the mission. A force that receives the benefit of the favorable positioning of the reserve, the bulk of the air support, and the artillery assets is the **main attack**, even though the words "main attack" might not be used in the operation order. The **supporting attack** is that portion of the attacking force that contributes to the success of the main attack by controlling terrain, destroying enemy forces, or deceiving the enemy. In the United States doctrine of the 1960s, the term "supporting attack" replaced the term **secondary attack**, which had been officially defined as "any attack whose importance is secondary to that of the main effort" and is "characterized by lack of depth, reduction of reserves to the minimum, maximum firepower in the attacking echelon, wide zones of action, and usually limited objectives."[1] During the conduct of an attack, a commander must be prepared to change his initial concept. It is conceivable that the reserve could be committed in the zone of the secondary or supporting attack and that, together, these united forces would become the main attack. Often, the determination of which forces compose the main attack and which compose the supporting attack can be made only in retrospect. Nevertheless, they remain useful terms for not only the historian of war, but also the practitioner.

Other terms used to designate forces involved in offensive operations are skirmishers, security forces, and strategic reserve forces. **Skirmishers** are those soldiers who precede the main force in order to discover, interrupt, confuse, or delay enemy forces. Skirmishers can be part of a **security force**, which consists of all those elements that provide protection against surprise, observation, or interference by the enemy. A **screening force** is a security force that lacks sufficient combat power to risk voluntary engagement with any but the very smallest of enemy units. When a force is not **deployed for combat** (anticipating imminent engagement by a major enemy force), its security forces closest to the front are often called the **van, vanguard,** or **advance guard**. Those security forces in the rear are called the **rear, rearguard,** or **rear security forces**. The bulk of forces, the **main body**, is comprised of those found between the van and rear. When the force deploys for combat, the extremities of the main body are called the **wings. Mobile reserve forces** are those forces in the rear that are prepared for immediate commitment and that are almost always **mounted,** on horses or vehicles. **Strategic reserve forces** are those forces that are withheld from commitment until required to influence a battle that could decide the outcome of an individual campaign or the entire war. Modern airborne forces are often part of a strategic reserve.

The Purpose of Offensive Operations

Even though national policy might dictate that military forces initially adopt a defensive posture, it is virtually impossible to conceive of victory without offensive operations. To obtain decisive results, a force that is outnumbered must attack or outflank the enemy in order to destroy his support elements, his command and control apparatus, and, eventually, his combat elements. The offensive may also be undertaken to secure key terrain, to deceive and divert the enemy, or to learn more about enemy dispositions and intentions.

Categories of Offensive Operations—Current and Historical

Offensive operations have been categorized in many ways, but recent United States Army doctrine has focused on the following five: the movement to contact, hasty attack, deliberate attack, exploitation, and pursuit. The **movement to contact** is an operation with the purpose of finding and engaging the enemy. The force conducting a movement to contact, or an **advance to contact**, should be a highly mobile, well balanced force prepared to accomplish its mission well forward of the main body. When opposing forces are moving to contact and engage before either can adequately plan to attack or defend, a **meeting engagement** or **encounter battle** occurs. When one force waits in a carefully prepared position for another force, moves along a road or any other commonly traveled route, and then attacks, the encounter is called an **ambuscade.** The term **ambush** is also used to describe such a surprise attack, but ambush is properly a verb whose nounal form is ambuscade.

The **hasty attack** is a planned attack made without pause in the forward momentum of the force upon initial contact with the enemy. If momentum is lost, or if the commander decides to take time to develop the situation more carefully because he faces a strong enemy force in well prepared defensive positions, he conducts a deliberate attack. A **deliberate attack** is characterized by greater knowledge of enemy positions, more extensive preparation, greater volumes of more effectively

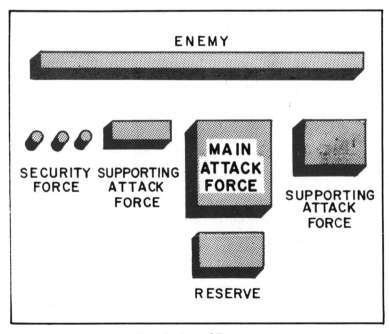

Distribution of Forces

delivered supporting fires, more extensive deception, and other measures beyond the scope of those possible in a hasty attack.

The **exploitation** is an operation undertaken to capitalize on the success of an attack. Previously uncommitted forces are used to strike for deep objectives, seize command and control facilities, sever escape routes, destroy reserves, and deny the enemy a chance to reorganize.

The **pursuit** is used to intercept and annihilate a retreating enemy. A pursuit should be conducted when the enemy has lost his ability to operate effectively and attempts to flee. Pursuit requires great energy; the resolution to press on despite fatigue, dwindling supplies, or the approach of darkness; and the proper coordination of all resources, especially highly mobile forces. In addition, commanders must be positioned well forward to provide impetus, encouragement, and timely, proper judgments. In a successful pursuit, the **direct pressure force** maintains contact with the enemy main force while an **encircling force** or **blocking force** cuts the enemy line of retreat.

In addition to the categories of offensive operations just enumerated, many others have been and still are widely used, including coordinated attack, attack of an organized position, reconnaissance in force, diversion, demonstration feint, raid, defensive-offensive, counteroffensive, and counterattack. The **coordinated attack** is a carefully planned and executed operation in which the various elements of a command are employed in a way that maximizes their benefit to the command as a whole. Chronologically speaking, before its designation as a coordinated attack, such an operation was referred to as an **attack of an organized position**. The term **deliberate attack**, which is still in use, describes similar activities. **Reconnaissance in force**, or **reconnoitering in force**, describes operations intended to test the enemy's strength and disposition. **Diversions** are operations intended to draw the attention and forces of an enemy from the area of a major operation. There are two types of diversions: demonstrations and feints. **Demonstrations** seek to divert the enemy without engagement, while **feints** seek to divert the enemy by the taking of a shallow objective. Although **raids** also divert the enemy, they are primarily sudden attacks intended to destroy resources or disrupt lines of communication. In raids, there is no intention of holding the attacked position. The **defensive-offensive** is an operation wherein a commander intentionally takes the defensive in order to fix and exhaust the enemy before launching an offensive. Classical defensive-offensives occurred at Cannae (216 B.C.) during the Second Punic War, at Austerlitz (1805), and at Cowpens (1781) in the American Revolution. If a force has unintentionally been on the defensive for a period of weeks, months, or years, its adoption of the offensive is called a **counteroffensive.** Counteroffensives were used after the German attack through the Ardennes in December 1944, and in Vietnam after the Tet Offensive of 1968. A **counterattack** is an offensive action that occurs immediately after an enemy attack has been halted.

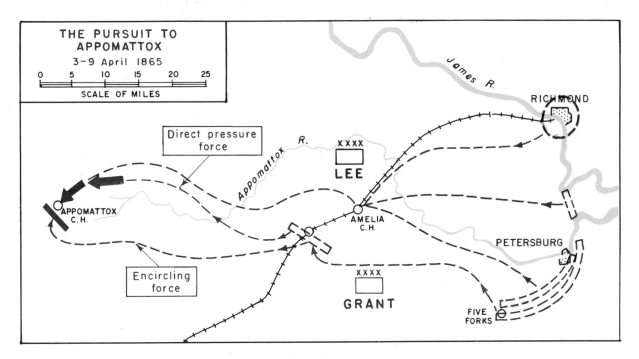

Classic Pursuit

Forms of Maneuver

In the course of conducting offensive operations, an attacking force is required to maneuver to gain an advantage over the enemy. It may attack frontally, or it may strike at the enemy's flank or rear. In a third case, the attacker uses maneuver to force the enemy from his original position to another position in which he can still be attacked.

In the period of the Greek city-states, most battles consisted of **frontal attacks,** in which phalanx moved against phalanx, hacking and slashing until one side was pushed or driven from the field of battle. The frontal attack strikes the enemy all along his front and seeks to destroy or overrun him in his position. In some cases, frontal attacks result in **penetrations**, a form of maneuver wherein an attacking force destroys or overruns a portion of the enemy's defensive position. After being thus divided, each portion of the enemy forces is subject to **defeat in detail** (the defeat of one part of a force before it can be reinforced by the other part of the force). When penetrations develop from frontal assaults or from other forms of maneuver, they are called **penetrations of opportunity.** Such a penetration was achieved by a portion of Frederick the Great's forces in the Battle of Prague in 1757; hence, the penetration of opportunity is also known as the **Prague maneuver.** Successful penetrations require the concentration of superior combat power at the point selected as the focus of the attack. Penetrations should be considered when a wealth of fire support is available, when the enemy is overextended, or when his flanks are unassailable. The penetration consists of first rupturing the enemy line, then widen-

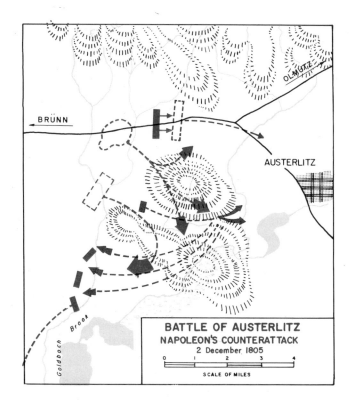

The Offensive Phase of a Classic Defensive-Offensive

ing the gap and protecting its shoulders, and finally securing objectives well to the rear of the unbroken portion of the enemy line. Penetrations, which are used during both hasty attacks and deliberate attacks, often lead to the exploitation.

Another form of maneuver directed against the enemy front is called the **oblique order.** Used by Epaminondas of

Penetration

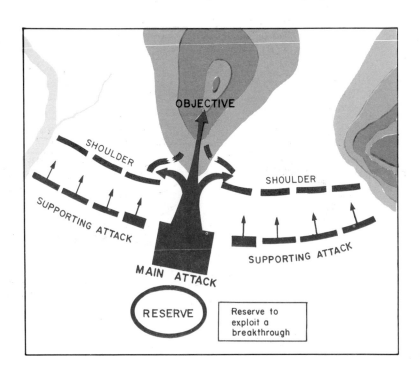

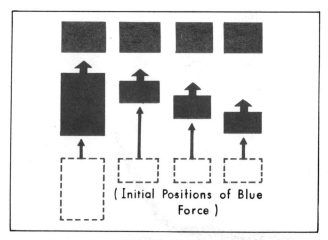

The Oblique Order

Thebes in the Battle of Leuctra (371 B.C.), it consists of an advance by the weighted wing of a force, followed by the advance of an adjacent portion of the line and, in turn, succeeding adjacent portions. By thus striking the opposing front in **echelon,** the advancing force prevents the opposing commander from shifting the uncommitted portions of his line for fear of exposing a flank to the advancing forces. Furthermore, once the wing of the advancing force defeats the subunit to its immediate front, it can threaten and maneuver against the flank of the next portion of the enemy force as it is being struck from the front. The oblique order is thus able to defeat the enemy force in detail, and even though it requires highly skilled troops and a talented commander, it is a means by which numerically inferior troops can succeed in the face

of overwhelming odds. For example, at the Battle of Leuthen in 1757, Frederick the Great, who was outnumbered 70,000 to 36,000, used the oblique order to win a decisive victory over an Austrian army.

Maneuvers that are directed against the enemy's flanks or rear are called **flanking movements** or **envelopments.** In an envelopment, the maneuvering force may either pass around one or both of the enemy's flanks in order to strike him in the flank or rear, or advance frontally from a position opposite the enemy flank (from a **flanking position**) to strike the enemy's flank or rear. When the maneuvering force passes around both of the enemy's flanks, a **double envelopment** results. A classic example of the envelopment of a single flank occurred in the Battle of Nashville on December 15, 1864; a classic double envelopment occurred at Cannae in 216 B.C., when Hannibal decisively defeated the Romans.

When a flank can be enveloped, it is called an **assailable flank,** or a **tactical flank.** When a force has positioned its flank against an obstruction, such as an unfordable river or an impassable swamp, and has thus protected itself from assault by significant numbers, that flank is unassailable, and the force is said to have **refused a flank.** In the course of an envelopment, a supporting attack called a **holding attack,** is nearly always necessary to **fix the enemy**—that is, to occupy him so that he cannot maneuver. The supporting attack can also deceive the enemy regarding the form of maneuver being employed. Since the advent of airborne forces and helicopters, a **vertical envelopment**—a maneuver against the flank or rear of an enemy force from above—is also possible.

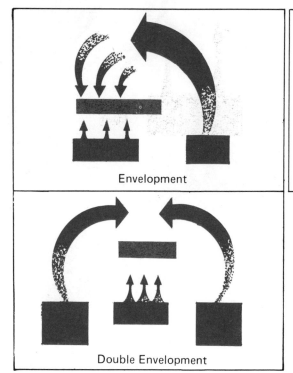

Envelopment

Double Envelopment

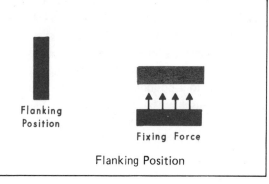

Flanking Position

Envelopments

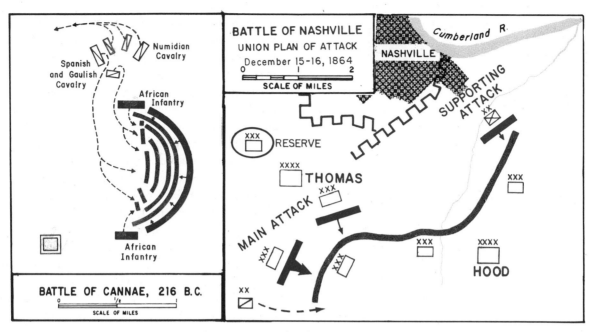

Classic Envelopments

The third form of maneuver consists of feigned withdrawals and turning movements—maneuvers that seek to draw the enemy from an area that is well defended, and perhaps unassailable, to an area of the maneuvering force's choosing. The movement of both forces occurs before decisive engagement is undertaken. In the **feigned withdrawal,** a force seeks to induce its opponent to advance from a strong position "by tempting him with an apparent flight and the prospect of an easy victory."[2] The feigned withdrawal is a part of the defensive-offensive operation, and is also closely associated with the double envelopment. Cannae is thus a classic example of the feigned withdrawal, the defensive-offensive, and the double envelopment.

In the **turning movement,** a force attempts to induce its opponent to move from a strong position by advancing beyond the opponent's flanks to a position that makes the enemy's situation untenable—usually by threatening his line of communication. Since this maneuver, like the envelopment, requires movement around the enemy flanks, it is sometimes called a **strategic envelopment,** or, in the parlance of modern football, an **end run.** However, the distinction between the envelopment and the turning movement is important. In the former, there is an intention to hold the enemy in his position while striking him in the flank or rear. The enemy may be forced to **change his front** (face his unit in a new direction), but he does not leave the immediate area that he occupied when the maneuver began. In the turning movement, the enemy may be fixed by a supporting attack as in the envelopment, but the element that is turning him forces him to abandon both his position and the area that he occupied when the maneuver began. Also, in the turning movement, the

maneuvering force and fixing force will probably be out of **supporting distance,** the distance that separated forces must not exceed if defeat by a determined enemy of each in detail is to be avoided. When a commander feels the risks involved in turning the enemy are justified, he generally seeks to turn the enemy's **strategic flank,** the flank that when turned drives the

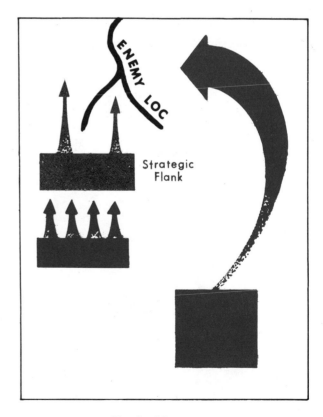

Turning Movement

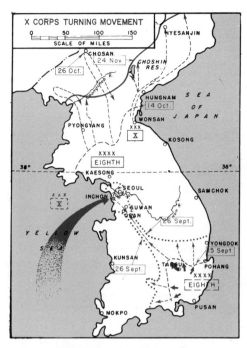

Classic Turning Movement

enemy away from his line of communication. A classic turning movement was made in Korea in September 1950 when the United States X Corps turned the North Korean position at Pusan. Another term that is nearly synonymous with turning movement is **indirect approach.** This term is used widely in British circles because, in the present century, the theory of indirect approach has been featured in the writings of British theorist-historian Sir Basil H. Liddell Hart, and British political leader Sir Winston Churchill. Adherents of the indirect approach maintain that by striking the enemy where he is weakest and where he least expects to be struck, greater results in proportion to losses can be achieved.

If an enemy force is turned on both flanks, an encirclement may result. Encirclements also occur when the forces conducting a double envelopment meet in the rear of the opposing force. In an **encirclement,** every section of the perimeter must be occupied, preferably simultaneously. The intended result of an encirclement is the capture of large units. When encircled units attempt to maneuver, they soon recognize the futility of continued resistance. Encirclements were often referred to as **pincers movements** in World War II.

Other Operational Concepts

Three further concepts are essential to an understanding of offensive operations: the piecemeal attack, interior lines, and concentration on or off the battlefield.

A **piecemeal attack** is an attack in which the various units

of a force are employed as they become available, or an attack in which the timing breaks down and forces are committed in an uncoordinated manner. Piecemeal attacks should be deliberately undertaken only when time is of such importance that no other means of commitment will succeed, or when the commander has superiority of combat power throughout the piecemeal commitment of his units.

Interior lines is a concept that was prevalent in the military theory and doctrine of the nineteenth century. It describes the condition of a force that can reinforce or concentrate its separated units faster than the opposing force can reinforce or concentrate. A force has interior lines when it is in a **central position** relative to the enemy—unless the enemy can move laterally so much faster than the force with a central position that it can concentrate or reinforce faster. In the latter case, the force without a central position is said to have interior lines as a result of **superior lateral communications.** For example, because of his central position during the Seven Years' War, Frederick the Great was able to concentrate against one of the allied armies opposing him while using economy-of-force measures to prevent attacks on his rear by other allied armies. In more modern times, railroads have often provided one side with the advantage of interior lines due to superior lateral communications. Although the concept of interior lines is generally applied at the campaign or theater level, it can have meaning on the battlefield. When terrain, training, skill of the commander, communications, or mobility enable one force to reinforce along the **lines of engagement** faster than its enemy, the force is said to have **tactical interior lines.** A force that does not have interior lines is said to operate on **exterior lines.**

A military force is concentrated when all its units are within supporting distance of one another. If subordinate units are brought within supporting distance before arriving at the field of battle, they have **concentrated off the battlefield.** If they are brought within supporting distance as the battle is beginning or after the battle had begun, they have **concentrated on the battlefield.** Napoleon generally concentrated off the battlefield. Successful examples of concentrations on the battlefield include the Prussian defeat of the Austrians at Königgrätz in 1866 and the German defeat of the Russians at Tannenberg in 1914. When units concentrate on the battlefield, they risk defeat in detail.

Defensive Operations

Purpose

Defensive operations employ all available means and

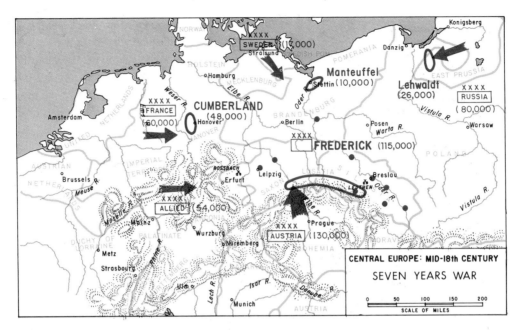

Central Position Creates Interior Lines

methods to prevent, resist, or destroy an enemy attack. A force may assume the defensive for one or more of several purposes: to cause an enemy attack to fail; to gain time; to control essential terrain; to wear down enemy forces as a prelude to offensive operations; to retain tactical, strategic, or political objectives; or to economize forces in one area so as to allow for concentration elsewhere. Any effort employed to free troops for use in another area is called an **economy-of-force measure**.

In defensive operations, the defender seeks to seize and retain a degree of initiative by selecting the area of battle, by forcing the enemy to react in conformity with the defensive plan, and by exploiting enemy weakness and error. Defensive operations may be necessitated by an inability to attack, or they may be deliberately undertaken (in combination with deception) to destroy the enemy. The shift from offense to defense, or vice versa, may occur rapidly and with considerable frequency. Elements of a command may defend,

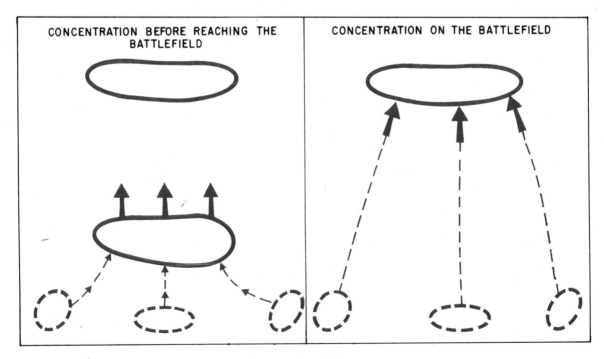

Concentration of Forces

delay, attack, feint, or deliver fire as part of the defense. An offensive attitude is necessary so that the defender is always ready to seize opportunities to destroy the enemy. Psychological preparation of troops and strong leadership in the defense are essential to maintain high morale, alertness, and an aggressive attitude. Troops must be made to understand that an effective defense is an opportunity to defeat the enemy.

Defensive Areas

In the defense, the battlefield is normally organized into three areas: the covering force area, the main battle area, and the rear or reserve area. The **covering force area** is made up of the **outer line of defenses,** or **outposts,** which includes the **general outpost** (security forces controlled by the division commander), the **combat outpost** (security forces controlled by the brigade commander), observation posts, **listening posts** (used when visibility is limited, and especially to detect night attacks), and patrols. The covering forces seek to halt the **spearhead,** or **lead elements,** of the enemy advance. Sometimes the covering force is able to launch a **pre-emptive attack,** or a **spoiling attack,** which occurs as the enemy is organizing for his attack. The security forces may succeed in forcing the enemy's security forces to withdraw after a brief **fire fight** (exchange of small arms fire), and thus **uncover,** or expose, the enemy's main force to observation and attack. If the enemy main force is on its **approach march,** or in **march column** (that is, not deployed for combat), the security force may be able to **carry a flank,** or **roll up a flank,** of the enemy main force. However, it is unlikely that a security force could **rout,** or decisively defeat, the main body of an enemy attack, since attacking forces must generally outnumber defending forces by a margin of at least 3 to 1 to be successful.

The **main battle area,** or **forward defense area,** is the area in which the decisive battle is fought. It consists of the **main line of defense,** which is made up of the principal combat elements of the defending unit. According to modern doctrine, the **setpiece battle,** or **pitched battle,** is an anticipated battle that is to be fought in the main battle area. When the troops on both sides fight from standing, uncovered positions, the setpiece battle is referred to as an **open battle,** or **battle in the open field.**

The **forces of the rear** or **reserve area** are the primary means by which the defender regains the initiative. Retention of a relatively large reserve, consistent with the requirement for forces in other defensive areas, permits offensive action both within and forward of the main battle area. The combat power allocated to the reserve includes fire as well as maneuver elements. When nuclear fires are authorized, the reserve can concentrate overwhelming combat power quickly in a given area. In addition, the reserve provides flexibility, and may be used to reinforce units or occupy positions; insure retention of key terrain; assist in disengagement of units; **relieve,** or replace, forward units; extend flanks; provide security against infiltration and airborne or air-landed attack; and conduct operations against irregular forces. When a relief occurs, the units involved must undergo a **passage of lines,** which refers to one unit's passage through the positions or lines of another unit. A passage of lines can result in a forward movement, a rearward movement, or a **relief in place.**

In a defensive position, the **frontline** (that is, the **forward line of troops,** or **FLOT**) can be anywhere in the covering force area or the main battle area. An occasional **probe**—a small-scale attack intended to test the strength of the enemy's line—or a **sortie**—a sudden advance intended to break or harass the enemy—is used to determine the location of the enemy's front line.

Forms of Defense

Because doctrinal sources have presented a variety of forms of defense throughout the years, many terms that are no longer in use will be encountered when studying the past. The discussion that follows deals with the most important and widely used terms of the past and the present.

The **active defense** capitalizes on the use of ground mobility and massed firepower to engage the enemy from a series of battle positions that are **deployed in depth** (that is, arrayed

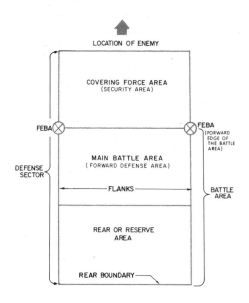

Defense Sector Areas

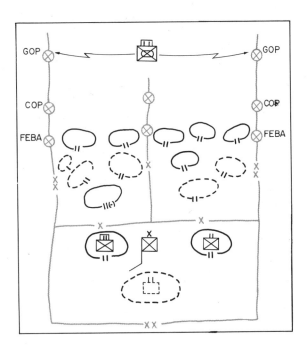

Division Area Defense

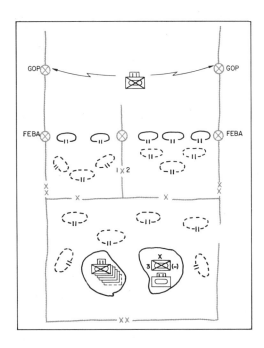

Division Mobile Defense

throughout the battle area). It envisions rapid movement between and within battle positions. In the **position defense,** formerly called the **zone defense,** units are placed either on line or in depth, where their weapons will be most effective against the primary threat. Zone defenses often provided **defense in depth** by utilizing successive main battle areas. A position defense can become a **static defense,** which is a defense void of offensive activities. **Strongpoint defense,** or **point defense,** is utilized to prevent a catastrophic penetration by armor or other highly mobile forces; it is essentially a strongly defended battle position that affords all-round protection. In German literature, such a defense is called the **Schwerpunkt** (military strongpoint) **defense.** A **perimeter defense,** also called the **cordon defense, all-around defense, box defense,** or **laager defense,** is organized when a unit must hold critical terrain in areas where there are no adjacent units. The line occupied by units in a perimeter, cordon, or laager defense is called the **perimeter.**

The area defense and the mobile defense were important defensive forms used for a time in the period after World War II. In the **area defense,** the primary purpose of the defending force was to hold specific terrain, and the defending commander allocated the majority of his available combat power to the forward defense area. In the **mobile defense,** the principal objective was to destroy any attacking force, and the priority of combat power was allocated to the reserve, which formed a **mobile strike force** (a force composed of mobile units whose mission was to attack on short notice to destroy enemy forces).

When defensive positions are prepared on high ground, they should be placed along the **military crest,** which is the line on the forward slope of a hill from which maximum observation of the slope can be obtained, rather than the **geographical crest,** which is the highest point on the high ground. In addition, defensive positions should not be traced so as to create **salients,** which are bulges into the enemy line, unless an objective point located within the salient is worth more than the increased number of men and weapons required to hold the salient.

Retrograde Operations

Armies of every era have conducted **retrograde operations,** a term that refers to any movement of a unit to the rear, or away from the enemy. In some sources, retrograde operations are treated as a form of defensive operation, which illustrates the fact that the operations of forces in the field are likely to be a blend of offensive, defensive, and retrograde forms. Like the terms used to describe the offensive and the defensive, those used to identify forms of retrograde operations are numerous, and include retreat, retirement, withdrawal, and delay. Only the latter three terms, however, can be found in recent United States Army doctrine.

A **retreat** is simply a movement away from the enemy. However, the term generally implies that the movement is forced by the enemy and is characterized by a high degree of disorder. The **line of retreat** is the route planned or followed

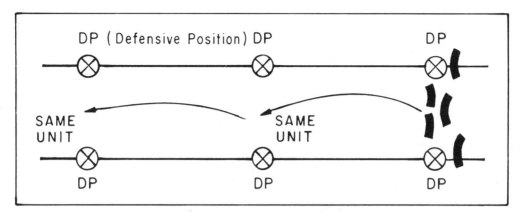

Delay on Successive Positions

by a retreating force. **A retirement** occurs when a force that is not **engaged,** or not **in contact,** chooses to move away from the enemy. Retirements are generally orderly and well planned. **Withdrawals** occur when a force in contact with the enemy chooses to **break contact,** or **disengage,** by moving to the rear. Disengagement is a difficult task, because if a unit waits too long to disengage, it risks destruction. In some earlier Army manuals, withdrawals were further subclassified as daylight withdrawals and withdrawals not under pressure, also known as **deliberate withdrawals** or **voluntary withdrawals.** In most withdrawals, the commander designates a **covering force,** which is a mobile force strong enough to impede the advance of the enemy. A covering force often creates and occupies a series of strong positions, known as **covering posts.** One of the most demanding of all ground combat operations is the **delay,** which is a retrograde operation in which a force conducts any or all types of combat operations in order to gain time for something else to happen—for example, time for reinforcements to arrive, time for forces to concentrate elsewhere, or time for other forces to withdraw. A delaying force generally seeks to maintain contact with the enemy force and to use all available means to inflict maximum casualties on the enemy. A delaying action can be accomplished from a single defensive position, successive positions, alternate positions, or a combination of these positions.

Retrograde operations are conducted when there are insufficient forces to attack or defend successfully, when a unit can be better employed elsewhere, when the strength of the enemy necessitates the use of such an operation, or when the purpose of an operation or campaign has been achieved. In large commands, many types of retrograde operations can occur simultaneously, and some units and individuals may be involved in offensive and defensive operations in the course of a larger unit's general retrograde movement. Retirements are frequently preceded by withdrawals, and both retirements and withdrawals may employ a covering force whose mission is to delay.

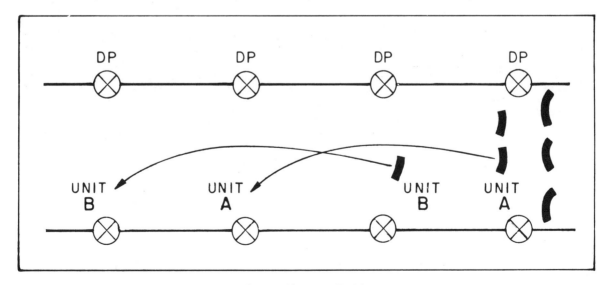

Delay on Alternate Positions

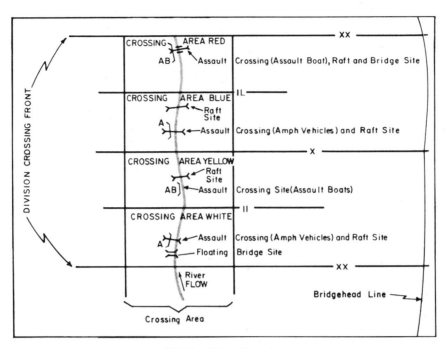

A Division River Crossing

Special Operations— River Crossings

Certain military operations that are peculiar to neither the offense nor the defense and that require special troops, equipment, or techniques are called **special operations.** An example of a special operation, and one that has played a prominent role in the military profession from earliest times to the pres-

ent, is the **river crossing operation,** an operation whose purpose is to move "a force across a river obstacle as rapidly and as efficiently as possible." [3]The **crossing force** is the entire force involved in the operation, the entire distance along the river in the zone of the crossing force is the **crossing front.** **Crossing areas** are designated within the crossing front to facilitate the flow of troops and materiel. **Crossing sites,** where bridges, rafts, amphibious vehicles, fords, or boats are

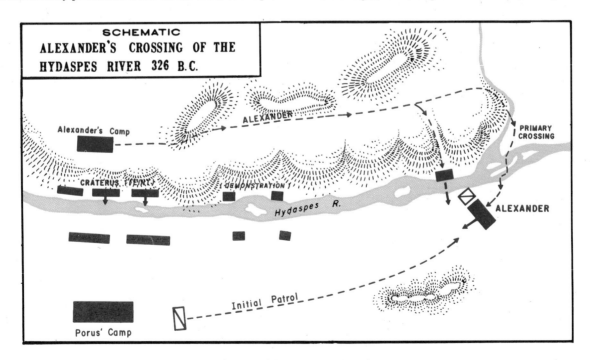

Classic Deliberate River Crossing

located, are designated within the crossing areas. The objective of the crossing force is the **bridgehead line,** an imaginary line that generally contains defensible terrain and that encompasses the **bridgehead,** an area large enough to accommodate and facilitate the maneuver of the crossing force once the crossing is completed. Although it may require some detailed planning, a **hasty river crossing** is an operation that is conducted as a continuation of the attack. It is characterized by speed, surprise, a minimal loss of momentum at the river, and a minimal concentration of personnel and materiel. A **deliberate river crossing** occurs when a hasty crossing is unfeasible (such as when enemy defenses are very strong or when the river obstacle is severe) and is characterized by detailed planning, a deliberate buildup of personnel and materiel, a loss of momentum at the river line, the use of deception, and centralized planning and control. Alexander the Great's crossing of the Hydaspes in 326 B.C. fits the description of the deliberate river crossing.

Selected Bibliography

Operational concepts are treated in many military history texts, but the most detailed explanations are found in the library of both current and past official publications. The editions of the United States Army FM 100-5 published since 1941—a test volume appeared in 1939—and the *Field Service Regulations,* which dates back to 1904, are invaluable starting points for an in-depth study of the American doctrine of the twentieth century. Further information on doctrinal sources can be found in the Bibliography section of Chapter 1.

Notes

[1]United States Army, Special Regulation 320-5-1, *Dictionary of United States Army Terms,* November 1953.

[2]David Chandler, *The Art of Warfare on Land* (London, 1974), p. 16.

[3]United States Army, FM 31-60, *River Crossing Operations,* March 1972, pp. 1-3.

From the Ancients to the Gunpowder Revolution

3

During the period extending from the creation of the early Greek city-states to the emergence of gunpowder as a significant influence upon the conduct of war in Western civilization, significant changes occurred in the nature and attitude of soldiers, the fighting organizations, the technology of war, and the means that armies used to seek victory in battles and wars. The Greek city-states date from the eighth or seventh century before the birth of Christ, while the significant use of gunpowder commenced perhaps in the fourteenth century, when hand-held cannon were used, and certainly by the mid-fifteenth century, when the walls of Constantinople were beaten down by Turkish artillery. More than 2,000 years are included in this age, but in spite of this vast time span, generalizations can be made that not only describe the art of war in that era, but also provide the foundation necessary for an understanding of more modern times.

Major Themes of the Age

In the area of military professionalism, which includes the nature and attitude of the soldier, several generalizations can be made. In the early Greek city-states, the soldier fought because it was a duty shared by all citizens. In Carthaginian society, however, accumulated wealth was used to purchase soldiers from less affluent and more militaristic societies, while in ninth and tenth century Europe, loyalty and obligations to land owners brought soldiers together to serve in poorly organized armies. The motivations and attitudes of soldiers thus differed significantly from society to society and from age to age.

Generalship is another area that provides interesting and significant contrasts with respect to the transition from the Greek city-state to the politically fragmented societies of the Middle Ages. In early Greece, leadership positions were sometimes elective and sometimes hereditary. During the height of the Roman Empire, military leaders had to gain political influence before they could raise and command military forces. By the time of the Middle Ages, the large landholders (the dukes and princes) often provided leadership for troops in the field because of feudal obligations. The leader's place during the battle also changed considerably. In the early Greek phalanxes, the leader often positioned himself in the center or at the extreme right of the front rank. Alexander was often described as being at the head of his charging cavalry, and Hannibal is said to have been at the center of the intentionally weakened part of his line at the Battle of Cannae. By the time of Caesar, however, the leader was often removed from the brunt of the battle. The trend toward removing senior military commanders from the critical point of engagement during the conduct of battle has continued to the present.*

As a political theme, the creation and subsequent eclipse of one extensive empire after another dominated the ancient period. In the fifth century B.C., the Persians attempted to extend their empire to include the Greek city-states. In the fourth century B.C., Alexander conquered the world as known to him in a campaign of just 11 years duration. In the first and second centuries A.D., the Roman Empire included all the shores washed by the Mediterranean and all of modern Portugal, Spain, France, and England. In the sixth century A.D., Justinian extended the Byzantine Empire to all the lands of the central and eastern Mediterranean, as well as the coast of modern Spain. In the early ninth century A.D., Charlemagne's empire encompassed all of modern France,

---

*The helicopter has perhaps altered this trend—especially in the realm of low-intensity warfare. The size of forces employed, the terrain, the type of operation, and communications capabilities are all considerations in determining the place of senior commanders in battle. During recent Arab-Israeli Wars, especially in 1967 and 1973, Israeli tank commanders in the grade of colonel and general often took the lead in advances across the desert terrain of the Middle East.

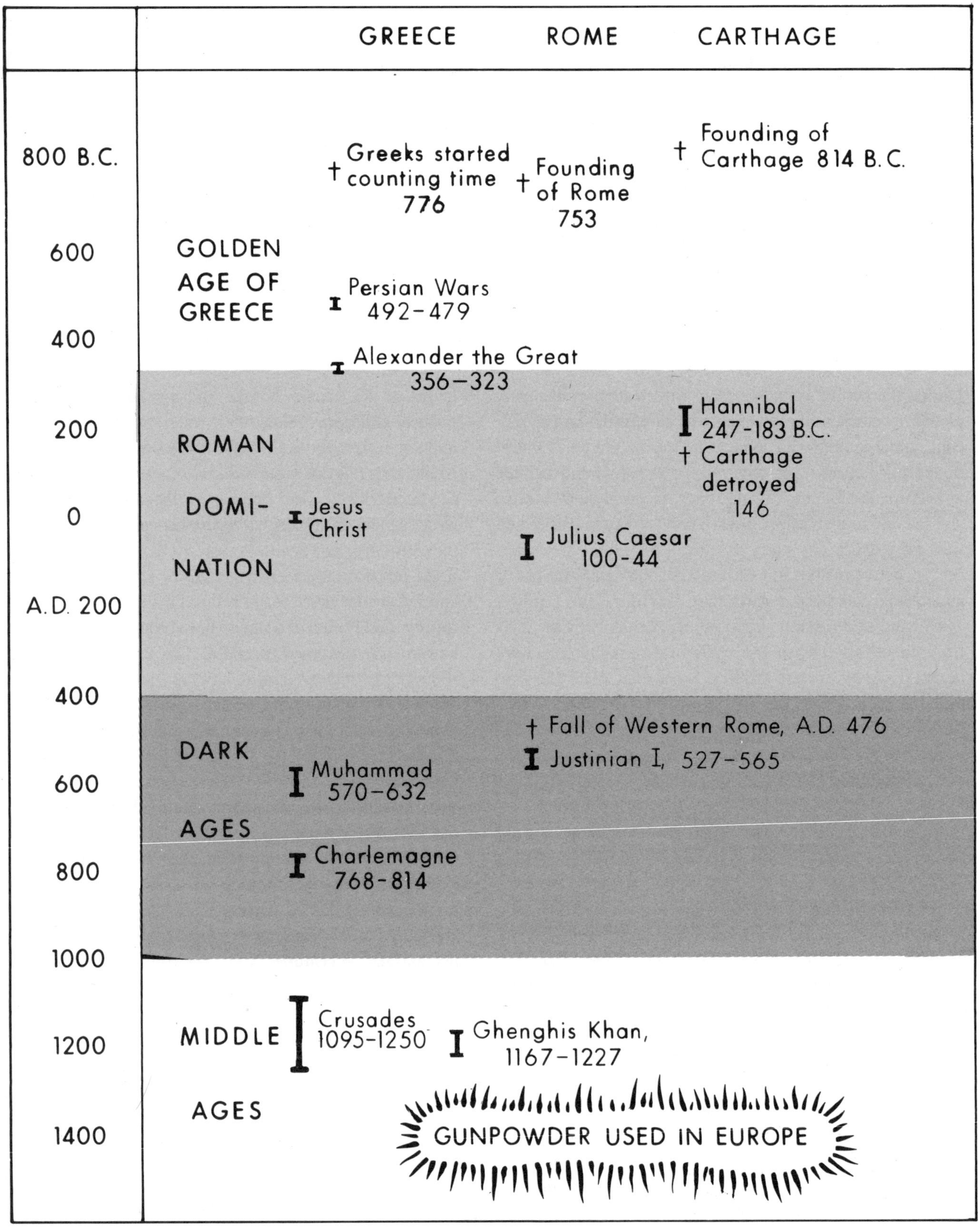

	GREECE	ROME	CARTHAGE

800 B.C.
                                                                       Founding of

† Greeks started   † Founding   † Carthage 814 B.C.
counting time   of Rome
776   **753**

600   GOLDEN

AGE OF

GREECE   ⊥ Persian Wars
492–479

400
⊥ Alexander the Great
356–323

200   ROMAN   **I** Hannibal
247–183 B.C.
† Carthage
DOMI-   ⊥ Jesus   detroyed
Christ   146

0   **I** Julius Caesar
100–44
NATION

A.D. 200

400

† Fall of Western Rome, A.D. 476
DARK   **I** Justinian I, 527–565

I Muhammad
600   570–632

AGES

I Charlemagne
800   768–814

1000

MIDDLE **I** Crusades   **I** Ghenghis Khan,
1095–1250   1167–1227

AGES

GUNPOWDER USED IN EUROPE
1400

From the Ancients to Gunpowder

Belgium, the Netherlands, and Switzerland; most of Germany, Austria and Italy; and parts of Spain. After Charlemagne, political fragmentation characterized most of European civilization until the emergence of modern states in the seventeenth, eighteenth, and nineteenth centuries.

A final dominant theme of the latter portion of the period from the ancient Greeks to the gunpowder revolution involves one of the principal social factors that affect warfare—religion. The influence of religion upon daily life generally increased during the period, commencing with the persecution of Christians by Roman authority and continuing through the late Middle Ages, when religious institutions controlled the wealth, the land, the courts, the accumulation of knowledge, and even the thoughts of men. The shibboleth of Charlemagne's army was "Baptism or Death"; the fundamental purpose of the Crusades was to restore access to the Holy Land for Christian pilgrims. The dominance of the Church was guaranteed by the point of the sword.

Participants in the Profession of Arms

Many names and titles have been bestowed upon those who participate in the profession of arms. Some of the terms distinguish those whose service is a full-time occupation from those whose service is provided only in situations of need. Other terms describe those whose roles are determined by their social and economic backgrounds. Still other terms identify positions in the hierarchy of authority.

The first of the fundamental distinctions used to categorize members of the military profession involves the amount of time that an individual devotes to the profession. In most simple societies, members bear arms only in times of emergency or during relatively brief periods of drill and training. In politically advanced societies where such an obligation exists between the individual and state, the individuals are called **citizen-soldiers,** and the force to which they belong is called a **militia,** or a **citizen militia.** In other societies, military service requirements are fulfilled by individuals whose sole function is to serve in the profession of arms. These individuals together form a **standing army,** and in modern societies they are generally referred to as **regulars.**

Most citizen-soldiers serve because of an obligation to their society. Within the regular or standing army, however, the factor that primarily influences an individual to serve in the profession of arms creates a second means of distinguishing between members of the military. When a soldier's prime motivation to serve is material gain—which might be money, title, or another form of remuneration—he is referred to as a

mercenary. Generally, a mercenary serves a society or political group to which he does not belong. Mercenaries who were recruited from provinces and fought for the Roman Empire in return for the promise of Roman citizenship at the conclusion of 25 years of service were called *auxilia.* During the reign of Constantine, mercenaries who received land in return for their defense of the Roman Empire were called *limitanei.* Mercenaries who served the Danish kings and early English kings during the Middle Ages were called **housecarls.** Individuals who held land from the English kings in the Middle Ages and who paid for the land by performing military service were called *thegns* or **thanes.** In modern times, the mercenary who combines his thirst for money and recognition with a sincere desire to fight is called a **soldier of fortune.** Additional examples of the motivational distinctions made within the profession of arms will be encountered during the study of other eras.

On the continent of Europe, feudalism encompassed complex social, political, and economic relationships between members of the society. Part of the system involved the feudal landowner's exaction of military service from his subjects in return for their use of land and for his guarantee of their protection from other influential individuals of unlike temperament. The chief warrior in this arrangement was the **knight,** a mercenary who fought to fulfill the obligation to a higher landowner and who could expect wealth and social advancement in return for service faithfully performed. The knight, or **man-at-arms,** as he was often called, was highly skilled, well equipped with weapons and armor, and generally mounted on a stout charger, since the weight and unwieldliness of his armor acted to his disadvantage when he was on foot. In origin, the knight was a mere cavalryman; but by the twelfth century he had become a member of a hereditary caste—except in England, where elaborate pageantry was associated with each knighting ceremony. The horse and its training were such an indispensible part of knighthood that the French word *chevalier,* the Spanish word *caballero,* and the German word *Ritter* all translate as horseman or knight. Chivalry became associated with knighthood through the Germanic concepts of family and loyalty, and the Church went beyond these ideals by stressing Christian virtues to the mounted warrior caste. Charity to the poor was encouraged, and the strain of idealized love of fair and noble women was derived from the important role of the Virgin Mary in the medieval church.

> The church commanded the knight to defend it and to turn his sword against the infidel ceaselessly and without mercy. [The knight] fought in certainty of

heavenly reward to which he would be borne by the angels, as was sometimes represented on his tomb. The perfect knight was like St. Louis who attended mass every day. The knight was exhorted to avoid all base actions and to love truth about everything, to defend the right and avenge injustice, and to be courteous to all and humble in all things.[1]

These ideals lay at the foundation of the unwritten modern code that states what the professional officer should be.

Members of the profession of arms may also be distinguished in accordance with their level of social, economic, or political acceptance—acceptance that was sometimes related to demonstrated military prowess. In the Theban state in the fourth century B.C., the **Sacred Band** was a group of elite warriors who were highly trained and who pledged to stand by their partners until death. In the Macedonian armies led by Philip and Alexander the Great, the **Companions** was the elite force assigned the most difficult tasks; the elite infantry recruited from the nobility were the *hypaspists.* The Roman field army of the fourth century A.D. consisted of three types of units, and eligibility for each of the units was based on the individual's social standing. The *palatians* were from the highest class. Originally, their task was to protect the palaces of the Empire. The *comitatenses* and *pseudocomitatenses* were mobile reserves from the lower ranks of society. Justinian's army of the sixth century A.D. consisted of *numeri* (regular troops recruited from the Empire), *foederati* (mercenary troops from areas outside the Empire), and *bucellarii* (troops whose loyalty was pledged to their leaders, whom they served as bodyguards). In the Middle Ages, knights tended to represent a more established class than did the poorly armed and generally unarmored common foot soldiers, who were drawn almost exclusively from the peasantry. In more modern societies, elite groups continue to exist, but admission to the elite, especially in democratic societies, has tended to be based on merit rather than social or economic background.

A fourth means of categorizing soldiers involves differentiation according to function. An **infantryman** is a soldier who fights primarily on foot. A **cavalryman** fights primarily mounted. A **light infantryman** is a foot soldier who carries the minimally essential equipment, and hence is more mobile than most other infantrymen. In contrast, a **heavy infantryman** or **heavy cavalryman** carries heavier equipment, and therefore was once less mobile but more powerful than other soldiers. In the present century, armored fighting vehicles have been developed, greatly increasing both the mobility and the firepower of the heavy infantryman and cavalryman.

In specific ages and armies, specialized terms were used to describe fighters with common purposes. The *hoplite* was the heavy infantryman, citizen-soldier of Greece; he was often accompanied by archers, slingers, and javelin men. The *peltasts* were highly trained light infantry, armed with a small round shield. The terms *hastati, principes,* and *triarii* were used to distinguish Roman soldiers according to their relative positions within their fighting formations. A **crusader** was anyone who traveled and usually fought for the purpose of establishing or re-establishing the Church as the dominant institution of the age.

Among the most significant terms applied to soldiers are terms of **rank**—those terms that identify the levels of authority that exist within the profession. In the most primitive of societies, rank consisted simply of the leader and the led, and in the citizen armies of ancient Greece, further distinctions were slight. The commander-in-chief was the *polemarch,* and the tiny army of *hoplites* was commanded by a *strategos,* or general. When fighting units increased in size, intermediate commanders were introduced. One of the most significant of these was the *centurion,* an officer who commanded a Roman force composed of about 100 men. Originally, the *centurion* was chosen from the ranks at the start of each campaign. In the second century B.C., however, *centurions* complained that they should not have to be reselected each time. They felt that once selected they should remain in that position, or higher positions, throughout their military careers. The institutionalization of this belief was an important step in the establishment of the professional officer corps.

Principal Organizations

Before discussing the principal organizations of the ancient and medieval periods, it is useful to establish a standard against which the size of the units under discussion can be measured. Three units are frequently used to provide such a standard: the company, the regiment, and the squadron. The origins of the term regiment are found in the late Middle Ages when units larger than the **company,** an administrative unit and later a tactical unit consisting of from 100 to 300 troops, were permanently organized. These larger units were called regiments because they were under the regimen, or rule, of a single individual. Generally, a **regiment** has been an administrative and tactical unit that is made up of anywhere from 1,000 to 3,000 troops, while the **squadron** has been a unit made up of cavalry forces numbering between 100 and 300. The term is derived from the Italian word *squadra,* or square, because the earliest bodies of horsemen were arranged in square formations. The size of the early squadron was determined by the fact that a commander's voice could be heard along a rank of about 50 mounted men and to a depth of from 3 to 6 files. Hence the original size varied from 150 to 300 mounted troops. Current usage restricts the application of the term to cavalry units, aviation units, and certain naval organizations.

Naval organizations differ considerably from army organizations of the same name. A **naval squadron,** for example, consists of two or more divisions of ships, of two or more Navy divisions, or **flights,** of aircraft. A **naval division** is made up of an unspecified number of ships or naval aircraft, but all are of similar type and are grouped for operational and administrative command. Other naval organizations frequently encountered are the **fleet,** an organization of ships, aircraft, marine forces, and shore-based fleet activities all under a single commander; the **armada,** a fleet of armed ships, or simply a large massed formation of armed ships; and the **flotilla,** an organization of smaller ships.

The traditional formation of the early Greeks was called the *phalanx.* The term has remained in general use and has been applied to any unit of foot soldiers formed in close, deep ranks and files. Although effective as an instrument of shock action, the Greek phalanx was unwieldy in the attack and vulnerable in the flank and rear. On the defensive, however, it was a powerful formation. The best troops in the phalanx were positioned in the front and rear of the formation. In modification of the early Greek phalanx, the Macedonians under Alexander the Great grouped their territorial infantry battalions together to create a form of a phalanx. Each battalion, called a *taxis,* consisted of 1,536 infantrymen. Each *taxis* consisted of six *syntagma.*

Throughout the ancient period, tactical organizations remained compact. The early Roman army was organized into *legio,* or **legions,** which consisted of 3,000 citizens organized into three subunits of about 1,000 each. The soldiers of the legions, called **legionnaires,** fought in a compact formation resembling the early Greek phalanxes; because of the similarity, the legion was further defined as the **phalangeal legion** by the fourth century B.C. Further reforms occurred between 300 B.C. and 100 B.C., when smaller subunits, called **maniples,** consisting of 60 or 120 men each, were created within the legion. The **manipular legion** was made up of 30 maniples. The final major organization effected by the Romans was the **cohortal legion.** Its development is attributed to Gaius Marius, who felt that the maniples were too small to be effective against the Barbarian forces aligned against Rome. The cohortal legion was made up of 10 **cohorts,** each consisting of 600 heavy infantrymen, who could form in six ranks of 100, three ranks of 200, or ten ranks of 60. The 10 cohorts within the legion could be arranged at the discretion of the commander, but two or three ranks of cohorts were generally used.

In the Byzantine Army of the late sixth century A.D., the basic unit of both infantry and cavalry was called the *band.* Each *band* consisted of about 400 men, and two or three *bands* composed a unit, roughly of regimental size, called the *drunge.* Three *drunges* formed the largest military organization of the period, the *turma.* These three military units persisted as the dominant military units for over 600 years.

The feudal armies of Charlemagne and of later periods contributed little of significance to the history of military organizations. Foot soldiers preparing for imminent battle were arranged in what has been called an **array** or, perhaps more accurately, a horde or mob. Cavalry forces were slightly better organized, generally being formed into three great masses called **battles.** When battles met, a disastrous **melee,** —a confused intermingling of opposing forces—usually followed. The conflict ended when one side realized that it was physically or psychologically overmatched.

Tactics—the management and control of ordered arrangements of troops—scarcely existed in the twelfth through fifteenth centuries; strategy, logistics, military thought and doctrine, and even military professionalism were rarely evident. In the field of military art, the Dark Ages prevailed.

The Technology of the Age

Technology, which is the application of knowledge to create or improve practical objects or methods, changed slowly during the period extending from the predominance of the ancient Greeks to the significant use of gunpowder. Nevertheless, some sophisticated objects, and certainly a wide variety of objects, were developed.

Items useful to the military professional can be offensive or defensive in nature. **Weapons,** which constitute the first group of items, are those objects that improve the fighting ability of individuals or units. Weapons are further broken down into two major subgroups: first, the **striking weapons,** which remain with the soldier when they are in use; and second, the **missile weapons,** which hurl a projectile or are themselves thrown. Striking weapons can be further classified as **thrusting, cutting,** or **bludgeoning weapons,** but these distinctions are often difficult, as a single weapon can be used in more than one way. Regardless, all striking weapons depend on **shock**—the simultaneous physical contact of the weapon and both adversaries in a fight. Defensive items, which afford protection from the weapons of the enemy, range from metal foot coverings to the Maginot Line (the complex system of permanent fortifications that the French built in the 1920s and 1930s to protect their eastern frontier from German incursion). Other objects that have had a profound effect on the military profession include naval items (such as ships and mines) and mobility devices (such as chariots, trucks, and the trappings of horses and elephants).

Striking Weapons

During the ancient and medieval period, nearly all striking weapons and most missile weapons were **hand weapons** (that is, weapons employed and carried by a single individual). The most primitive of the hand weapons were bludgeoning weapons. The **club,** simply a heavy stick that is thicker at one end than the other, was the forerunner of weapons like the mace and the morning star. The **mace** consisted of a handle of either metal or wood and a metallic head that was generally spiked. It became the favorite weapon of clergy in the Middle Ages because of the canonical law that forbade priests to shed blood. The mace, which was used to bludgeon and bruise, later became a symbol of royal authority—especially in England. Today it is a part of many English political ceremonies and an important symbol of authority in the United States House of Representatives and the Canadian House of Commons. Another of the bludgeoning weapons of the Middle Ages was the **morning star.** A variation of the mace, its usually spiked metal head was attached to a handle by a short length of chain. It was an effective weapon against the fully armored warriors of the Middle Ages.

A distinction between cutting and thrusting weapons is difficult to make, as is the seemingly easy distinction between some cutting and thrusting weapons and missile weapons. When prehistoric man picked up the first pointed stick in the heat of a brawl, it is unclear whether he intended to cut his opponent, stab his opponent, hurl the stick at his opponent, or bludgeon his opponent about the head and shoulders. The intention is generally reflected in the design of the weapon, but when intentions are not known or usage differs from intention, it is not possible to classify some early weapons accurately. Similarly, when the doctrine that applies to the prescribed use of some weapons is not fully understood, the classification of the weapon is difficult. At other times, a single weapon can be used in a variety of ways that make its definitive classification impossible. Such problems exist with the spear, one of the most significant weapons of early Western civilization.

The spear is one of the most basic of weapons. Although it is often used as a thrusting weapon, many spears are designed primarily as missile weapons. A **spear** is defined as a weapon consisting of a long shaft and a sharp, pointed head that has no auxiliary blades or points. It was the principal weapon of the Greek *hoplite.* A slightly longer spear, the *sarissa,* was the principal weapon of the Macedonian foot soldier of the fourth century B.C. The Numidian cavalrymen, who were mercenaries in Hannibal's Carthaginian army of the third

century B.C., carried the **lance,** which is a spear carried by mounted warriors and used primarily as a thrusting weapon. Lances and **lancers,** soldiers who carry the lance, remained an important part of military fighting organizations until the present century. The Roman legionnaire was armed with the *hasta,* a thrusting spear used in the phalangeal legion and in the third wave of the later manipular legion. The first two waves of the manipular legion were armed with the *pilum,* a spear intended primarily for throwing. It was five to five and a half feet in length, and consisted of a wooden shaft, a soft metal tip that bent to preclude its being easily extracted and reused after it struck a hard target, and an iron rod one-third its length that connected the arrow-shaped tip to the wooden shaft. It could be thrown accurately at ranges up to 25 yards, and could pierce a double shield thickness or stop a cavalry charge when thrown in volley. Because of its effectiveness, the *pilum* remained a principal weapon of the Romans until the last days of the western Empire.

The *pilum* was ultimately replaced by the **javelin,** a shorter and lighter spear that could be thrown much farther than the *pilum* and hence could compete with the bow, which was widely used by the warriors of western Asia. Some Byzantine light infantry continued to show a preference for the javelin over the bow as late as the sixth and seventh centuries A.D. Frankish warriors of the eight and ninth centuries, too, used the javelin as a weapon of war.

A variation of the spear that was developed during the Roman period and rose to prominence during the sixteenth and seventeenth centuries was called the **pike.** Like the spear, it consisted of a long wooden shaft and a pointed steel head. It was much longer, however, than most spears, and, unlike most spears, was used primarily in a defensive manner. Roman pikes of the fourth century A.D. were 12 to 14 feet in length and were used to protect infantry from cavalry. By the fifteenth and sixteenth centuries, most pikes were 16 to 22 feet long; their role, however, had changed little.

The weapon that has transcended the ages probably more than any other is the sword. A **sword** is a hand weapon consisting of a long, sharp, pointed blade; a sharp edge on one or both sides of the blade; and a **hilt,** or handle. The sword probably evolved from the **dagger,** a short, narrow-bladed, sharply-tipped hand weapon that was used chiefly for close fighting, but could be thrown. Flint daggers were used by prehistoric man. More sophisticated versions were carried by the Byzantine heavy infantry of the sixth century, Frankish warriors of the eighth century, and sea pirates of the nineteenth century.

The sword has been carried by warriors of nearly every age and branch of service, but in modern times it has been the favorite of the cavalry. It became a symbol of rank in the eighteenth and nineteenth centuries, and is still used for ceremonial purposes. In ancient times, the sword was generally subordinated to the spear, but for close fighting the Romans preferred the fine Spanish-made, flat-bladed sword of bronze and iron called the *gladius.* Its double-edged iron blade came to a gradual point, and its total length was between 22 and 24 inches. Spanish mercenaries in the Carthaginian Army used a cutting and thrusting sword as one of their principal weapons, and after the third century A.D., auxiliaries of the Roman Army used the *spatha,* a sword longer than the *gladius* and single edged. Swords of the early Middle Ages were made of soft iron; added protection for the hands was gained by the full development of the crossbar, or **quillons.** Byzantine cavalrymen of the sixth century carried a large, two-handed sword known as a **broadsword.** The **Viking sword** and its successor, the medieval **Knight's sword,** usually had a straight, double-edged, pointed blade, with a shallow hollow down the center that provided lightness and rigidity. The **saber,** a sword that is primarily a cutting weapon, has a curved blade sharpened on the convex edge only. Originating in the East, sabers were not widely used in the West until after the introduction of gunpowder. Rapiers, scimitars, and other types of swords were developed during later ages.

Striking weapons that are used predominantly for cutting include the ax, battleax, taper ax, broadax, and *francisca.* Although they rarely dominated the battlefield, they significantly influenced both the outcome of many battles and the development of subsequent weapons.

The **ax** (sometimes spelled **axe**), or **battleax,** was widely used in the sixth century by both Frankish and Byzantine warriors. It consisted of a wooden or metal handle and a heavy but often hollow head with a large cutting edge. The Frankish ax was often referred to as a *francisca.* These weapons, dating from Roman times, were capable of shearing helmet or shield, and were often used as missile weapons. In the tenth and eleventh centuries, Anglo-Saxon warriors used a **broadax,** which had a long handle and a heavy head, for close fighting and a **taper ax,** which had a short handle and light head, for throwing. The taper ax was also used as a means of land measurement; one unit was defined by King Canute as the distance that a taper ax could be thrown.[2]

Missile Weapons

Although some cutting, thrusting, and bludgeoning weapons are used as missiles, missile weapons include only those weapons that are exclusively intended to be thrown or that hurl a projectile. Small, round stones were probably the first missile weapons. A considerable advancement over the thrown stone was the sling. The **sling** consisted of a piece of leather with a small hole in the center and two pieces of cord,

each about one yard long. A smooth stone, placed in the leather and stabilized by the small hole, was swung rapidly from the ends of cord, thereby attaining considerable speed. When one of the cords was properly released, the stone traveled with great force toward its target. In this manner, the First Book of Samuel reports, David slew Goliath.

Small javelins, intended exclusively for throwing, were called **darts.** These small, pointed weapons were the principal weapon of the Numidian cavalry of the third century Carthaginian Army. Darts can also be shot from blowguns, air guns, and gunpowder weapons.

The missile weapon that has played a dominant role in warfare over the longest period of time is the **bow.** Like darts and slings, spears and swords, the bow dates from prehistoric times. It consists of a narrow piece of wood, horn, bone, sinew, or a laminated combination of these materials that is bent so that its ends can be connected by a **bowstring** made of cord, sinew, or gut. The side of the bow closest to the bowman, or **archer,** is the **belly;** the far side is the **back.** In or near the middle is the grip, or **handle.** The ends, or **tips,** have notches for the string that are called **nocks.** Bows, which were used by ancient warriors, became the favorite weapon of western Asian fighters. The Byzantines relied heavily upon the **composite bow,** a bow made of various laminated materials. The missile fired from the bow is called the **arrow,** and consists of a straight wooden shaft, a sharp head of stone or metal called an **arrowhead,** feathers to improve the quality of its flight, and a notch, or **nock,** to guide the arrow to and from the bowstring.

In the eleventh century, the **crossbow,** a bow set crosswise on a stock, was developed. Early models could be armed by hand or with a simple pulley. The arms of the bow were later made of metal, and eventually reached such a size that elaborate arming devices were required. Variations of the crossbow were developed, and by the Middle Ages, crossbows were effective weapons that were relatively simple to operate and lethal at a range of several hundred yards. As a result of the latter characteristic, the Church forbade their being used by Christians against other Christians. Moreover, crossbows posed a threat to cavalry, and worse, to the knight. That a highly trained soldier like the knight could be defeated by a virtually untrained soldier with a crossbow was considered unjust and unchivalrous—at least by knights and their retainers.

The missile fired by the crossbow was called a **bolt.** It resembled a short arrow, generally had an iron tip, and was of square cross section. Because of the square section, it was also called a *carreau* or **quarrel;** both words are derivatives of *carre,* the French word for square. Most bolts used wooden or leather **flights** rather than feathers for stabilization.

Each of the missile weapons just mentioned is a hand

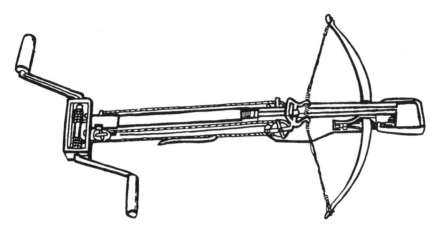

Crossbow and Arming Device

weapon—as were nearly all of the striking weapons of the ancient and medieval period. Yet, some **crew-served weapons** (weapons that require more than one individual to operate them properly) existed as early as 400 B.C. These crew-served weapons were predominantly missile weapons, and were used almost exclusively in siege operations. Collectively, they were called **siege engines, engines of war, siegecraft,** or simply **machines.** The one crew-served weapon that did not hurl a projectile was the Roman **ram.** Made from a large tree trunk that sometimes was tipped with an iron head, the ram was suspended from a covered frame by ropes that allowed it to be swung back and forth in order to batter down walls of fortifications. It was powered by as many as hundreds of men, some of whom worked under a cover, called a *testudo* or **tortoise,** to prevent their being injured by burning oil, molten lead, arrows, rocks, and other objects that the besieged would cast upon them.

The missile-throwing engines were used effectively as early as the fourth century B.C. by Alexander the Great. Unfortunately, the nomenclature of the different engines was confused by writers on the subject, and a great deal of confusion remains to this day. The **catapult** was apparently the most general term applied to engines of war; according to some sources, it included all engines of the ancients, regardless of the missile or the means of propulsion. Other reliable sources refer to the catapult as the machine that hurled stones or other objects against fortified positions. The catapult consisted of a long arm that terminated in a hollow bowl, which held the missile, and twisted cords, which provided torsion-power to hurl the missile.[3] Most of the ancient catapults had a sling of rope and leather attached to the arm. The sling increased the range of the catapult by at least a third, with the result that an eight-pound stone could be hurled nearly 500 yards. The large Roman catapult, developed in the fourth

Onager

century A.D., was called an *onager,* having been named after the wild ass because of its kick. The *ballista,* (sometimes considered a type of catapult, sometimes distinguished from the catapult, and sometimes considered a generic term that includes all catapults) was probably the arrow- or javelin-firing engine that functioned like a large crossbow. A smaller model, the *carro-ballista,* was pulled by mules and assigned to each century of the Roman field army.

The secret of creating the torsion required for an effective siege engine was lost to a great extent in the centuries after the Romans copied the Greek engines. As a result, the trebuchet was introduced for throwing stones. The **trebuchet** consisted of a great weight (a basket or box of stones or sand) connected to the short arm of a lever that was constructed from a sizeable tree trunk. When the weight was released, the long arm swung upwards with great velocity, and the missile, contained in a sling and weighing from 200 to 300 pounds, was hurled as far as 300 yards. Used primarily to breech walls, the trebuchet was also used to throw dead and diseased horses or humans into a besieged fortress in order to cause pestilence and hasten surrender. In addition to the occasional putrid animal that was hurled from the trebuchet, one fortress was reported to have been assaulted with bags of gold; resistance from within halted abruptly.[4] **Incendiary projectiles,** or fireballs, were also used by siege engines, but round stones and clay balls containing many smaller stones, which scattered on impact, were more commonly used.

Trebuchet

Mobility

Technology provided little to improve the mobility of armies during the period under discussion. Beasts of burden included horses, mules, elephants, camels, and oxen. The **saddle** enhanced the comfort and stability of the mounted warrior after about the fourth century A.D., and by the eighth century A.D., the introduction of the **stirrup** to Western Europe increased the power, stability, and effectiveness of the mounted soldier.

On the battlefield of the ancients, the foot soldier prevailed, although soldiers mounted on horses and elephants figured prominantly in many battles. The **chariot,** a two-wheeled cart often furnished with sharpened, jagged points that protruded from the axles, was only occasionally significant on the battlefield. After the fifth century A.D., the mounted horseman, whether a knight or a barbarian, became an increasingly formidable opponent; by the eleventh and twelfth centuries, most pitched battles were decided by the mounted arm.

Naval Objects

The ancient and medieval periods were a part of the age of oars. Ships were small when compared with ships of later periods, and were powered by oars and relatively small sails. Few provisions could be carried, for the hold was occupied by oarsmen—often slaves, chained to their oars and to each other. The **galley,** a long, low, single-decked ship, was the prominent warship in use, and two types of galleys prevailed. The *trireme* had three banks of oars on each side and could carry a boarding party of 18 to 30 soldiers. The *quinquereme* had five banks of oars on each side; it composed the major

Ballista

portion of the Roman fleet. Sea battles were typified by ships attempting to ram and sink their opponents and by the attempts of boarding parties to overwhelm enemy ships. This latter task was facilitated by the *corvus,* a boarding bridge or gangway that contributed to Rome's success in her naval battles against the Carthaginians during the First Punic War.

Defensive Items

Defensive items afford protection to either individuals or groups. When all the protective items of an individual, including his weapons, are considered, they are collectively referred to as a **panoply.** From ancient times, **shields** were a common part of the panoply. Early shields, made of leather or bronze, were carried on the left arm, but because early firearms required the use of both arms, shields were discarded with the advent of gunpowder weapons. Helmets, too, were widely used in the ancient and medieval periods, and after a period of disuse from the seventeenth through the nineteenth centuries, helmets returned as an important part of the soldier's panoply. The head protection of knights, which was referred to as the **heaume** or the **helm,** consisted of hats that covered the entire head. When smaller iron hats were used, they were called **helmets,** or little hats. Protection of the chest was provided by the **cuirass,** a corselet of metal or molded leather that was used from the Greek period until the nineteenth century. The Roman cuirass was made of **mail**—small, flexible plates or links of metal or hardened leather—and was called a **lorica.** By the time of Charlemagne, the lorica had become a sleeveless, close-fitting jacket, or **jerkin,** of mail. The English referred to thin mail body armor coats as **byrnies.** This term was also widely used on the Continent during Charlemagne's reign. The **hauberk** was a longer mail coat that evolved from the lorica. Short-sleeved, it extended from the knees and had a split to the crotch that enabled the wearer to mount a horse. Most hauberks had a hood with an opening for the face, and they often were worn over a **gambeson,** a quilted garment that softened the blows to the mail. The gambeson was the sole protection worn by foot soldiers of the Middle Ages. Leg protection was afforded to the ancients by **greaves,** which were made of metal or leather and extended from the foot or ankle to the top of the knee. They had fallen into disuse by the time of the cohortal legion (100 B.C.) **Gauntlets,** protective coverings for the hands, were worn by the Byzantine cavalry as early as the sixth century A.D., and in the Middle Ages, gauntlets of leather, covered with metal, were a part of the knight's panoply. Protection for a knight's horse was also provided; in fact, so much mail was used that the horse seemed trapped, and his accoutrements were hence called **trappings.** They consisted in part of **frontlets,** pieces

that protected the front of the horse, and **poitrels** or **poitrails,** collars that protected the neck and breast of the horse.

Group protective devices are often associated with military architecture and military engineering. The **mole,** which was used by Alexander the Great at the siege of Tyre (333 B.C.–332 B.C.), was less a protective device than an engineering item. Constructed of stone, the **mole** was a barrier built in the water to protect ships, harbors, and beaches from the force of waves. It provided Alexander with a causeway to the fortress at Tyre, which was built on an island just off the coast. Protection could be provided for the troops on the mole by a **cat,** a movable building with a steep roof that served the same purpose as the *testudo* of the Romans. One of the earliest constructed group protective devices was the **palisade,** a system of defense that used long, pointed stakes in a fence-like arrangement. Each Roman soldier on campaign carried two stakes to form palisades as part of the nightly defense. The Romans were also great builders of stone walls for protective purposes, and few medieval towns could long survive without extensive protective walls. In the ninth century, the Franks began to build fortified places, or **castles,** to provide protection against Viking intruders. The earliest castles consisted of little more than a ditch, or **moat,** and palisades atop the **rampart,** the broad embankment inside the ditch that often was built of earth taken from the ditch.*The crusaders (eleventh–thirteenth centuries), however, brought far more elaborate military architectural ideas from the Near East. Double and triple defenses around the core of the fortress; **brattices,** or wooden galleries, extending out from the ramparts; and **flanking towers** set at intervals along the straight portion, or **curtain,** of the walls were introduced to Western thinking. After the Crusades, a castle was far more than a **keep**—the strongpoint that was located within an enclosure of plain walls.

Operations

Modern terminology and doctrine are far more closely related to the descriptions and accounts of past **operations** (those military activities that occur when combat is underway or being rehearsed) than they are to past concepts of professionalism, organizations, or technology. Perhaps operational concepts have changed very little in the course of recorded history, or perhaps the parameters of operations are so imprecise that the language of one era can easily be made to describe operations in another era. Regardless of the reason,

---

*See page 54 for an illustration of the profile of a later defense system that employed rampart, ditch, and moat.

the understanding of modern doctrinal concepts is facilitated when current terms are used to describe the past. For example, the term **forced march,** which by current definition is a march requiring "the expenditure of more than the normal effort in speed, exertion, hours marched, or a combination of these,"[5] means the same thing whether applied to a modern infantry unit or the foot companions of Alexander's army.

Some general terms that today are omitted from doctrinal sources, even though they are still in use, were precisely defined in the past. For example, the terms "camp" and "bivouac" are still widely used, but are not included in the current *Dictionary of United States Army Terms* (AR 310-25). Earlier dictionaries, however, offered the following definitions, which are useful and appropriate today: a **camp** is "a group of tents, huts, or other shelter set up temporarily for troops"; a **bivouac** is "a pre-selected piece of terrain, generally in rear areas but out of direct contact with the enemy where a command rests and prepares for further movement."[6] A camp is more permanent than a bivouac. An **encampment** is a temporary camp in the field. It generally involves more troops than a bivouac, and is established for a longer period of time. Where temporary buildings, rather than tents, are utilized to house troops, the location is called a **cantonment.** Buildings that house troops but are more permanent than those found in cantonments are called **barracks.** Where barracks exist, troops are generally **stationed,** or reside permanently, and the location is called a **post** or, in United States usage, a **fort** (such as Fort Bragg, Fort Benning, Fort Knox, and Fort Hood). In a more general sense, a **fort** refers to a strong, or fortified, place that is protected by walls and ditches. **A fortress** is also a fortified place, but is generally larger than a fort and often includes a town within its fortified perimeter.

Some concepts important to the understanding of the military profession have remote usefulness in modern times but were essential to the soldier of the past. Such concepts fill gaps in the evolution of doctrine, and hence provide foundations for a more complete understanding of the present. Terms like "campaigning season," "winter quarters," and "castrametation" are included in this category. Campaigns were generally conducted only during those periods when weather and the trafficability of terrain were conducive to the conduct of major military operations. The **campaigning season** in northern Europe thus referred to that period of time from the end of spring rains in late April or May until the hard freezes of winter in late November or December. However, geography, politics, social factors, economic factors, technology, and many other considerations also influenced the campaigning season. Modern technology, for example, has produced weapons, communications, transportation, and protective clothing that lessen the effects of extreme heat and cold. Hence, the term campaigning season has lost much of its former usefulness.

When the campaigning season ended, commanders placed their troops in **winter quarters,** a camp or cantonment in which troops trained and passed the days between campaigns. George Washington and the Continental Army were in winter quarters at Valley Forge during the American Revolution. Today, **winterization,** the preparation of materiel for operation at extremely low temperatures, has taken the place of winter quarters.

A final example of a military term that is rarely used in modern times is **castrametation,** the making or laying out of the military camp. Since ancient times, camps have been vulnerable to surprise attack, and the proper preparation of the camp has been essential to the survival of the command. When frontlines became clearly defined, as in World War I, camps moved to the rear, and castrametation left the military lexicon. The undefined lines of contact experienced in Vietnam and the technological developments that allowed the attack of camps from above revived the importance of castrametation, but not the word.

Selected Bibliography

General military history works that cover the period from the ancients to the gunpowder revolution include Lynn Montrose's *War Through the Ages;* Richard A. Preston and Sydney F. Wise's *Men in Arms: A History of Warfare and its Interrelationships with Western Society;* Oliver L. Spaulding, Jr., Hoffman Nickerson, and John Womack Wright's *Warfare: A Study of Military Methods from the Earliest Times;* and Hans Delbrück's *History of the Art of War.* Works that treat military topics in more specific periods include Elmer C. May, Gerald P. Stadler, and John Votaw's *Ancient and Medieval Warfare;* Frank Adcock's *The Greek and Macedonian Art of War;* Graham Webster's *The Roman Imperial Army of the First and Second Centuries;* A.V.B. Norman's *The Medieval Soldier;* John Beeler's *Warfare in Feudal Europe 730–1200;* and the pioneer work upon which the others rely heavily, Charles Oman's *A History of the Art of War in the Middle Ages.* An abridged version of Oman's work is available under the title *The Art of War in the Middle Ages: A.D. 378–1515.*

The evolution of military weapons from clubs to nearly present-day arms is detailed in Edwin Tunis' *Weapons: A Pictorial History.* More scholarly, more current, and beginning earlier than its title suggests is Bernard and Fawn Brodie's *From Crossbow to H-Bomb.* Jac Weller's *Weapons and Tactics* treats the relationship between technology and tactics from early Greek warfare to the beginning of the American involvement in Vietnam.

Notes

[1] A.V.B. Norman, *The Medieval Soldier* (New York, 1971), p. 146.

[2] Edwin Tunis, *Weapons: A Pictorial History* (Cleveland, 1954), p. 36.

[3] This definition is consistent with such authorities as Sir Charles Oman, *A History of the Art of War* (New York, 1898), p. 545; Graham Webster, *The Roman Imperial Army* (London, 1969); pp. 231–236; and Sir Ralph Payne-Galloway, *A Summary of the History, Construction and Effects in Warfare of the Projectile Throwing Engines of the Ancients* (London, 1907), pp. 3–41.

[4] Payne-Galloway, *A Summary of the History of the Ancients,* p. 30.

[5] United States Army, FM 21-18, *Foot Marches,* January 1971, p. 4.

[6] United States Army, Special Regulation 320-5-1, *Dictionary of United States Army Terms,* November 1953.

Gunpowder, Renaissance, and Reason

4

The first flash of gunpowder in Western Europe shed little light on the era known as the Dark Ages, but gunpowder, along with numerous other discoveries, did eventually usher in a new age. People's ideas about their environment changed. Their concepts of religion and art changed. Their political, social, and economic institutions changed, and their military institutions changed. But change was slow.

Major Themes of the Age

Although there are claims that gunpowder was used earlier, Roger Bacon is generally credited with having invented it in the middle of the thirteenth century. During the Hundred Years' War, gunpowder was used with little effect. By the middle of the fifteenth century, however, it was able to destroy the walls of Constantinople.

At the end of the fifteenth century, a New World—at least a world new to European society—was discovered. Paralleling the age of discovery and exploration of the New World was the period now known as the **Renaissance,** during which civilization was reborn. New art, like that produced by Leonardo da Vinci and Michelangelo, was acknowledged, and new wealth accumulated in centers of commerce. Early in the sixteenth century, the **Protestant Reformation,** a movement that destroyed the unity of the medieval Church, encouraged different ideas about learning and humanity. Skepticism brought men like Copernicus, Galileo, and Newton to revolutionary understandings of the universe in the sixteenth and seventeenth centuries. In the eighteenth century, the term **Enlightenment** described the way learned men thought about their contribution toward the progress of man through time. The Enlightenment was an age in which thinking men were dominated by the influence of science, logic, and philosophy. In the midst of this expansion of knowledge, the influence of kings and princes expanded, too. Large land areas—and, with the land, its people and wealth—were organized, administered, and defended under the direction of dynastic sovereigns like the Bourbons in France, the Hohenzollerns in Prussia, and the Romanovs in Russia. Their power increased through organization and unity brought about by agreements that were often the result of war. But change was slow.

No precise line separates the Middle Ages from the modern world. Thus, the long period of transition, beginning as early as 1250 with Bacon's gunpowder and ending late in the eighteenth century, is sometimes called the "dawn"—the dawn of a new era, or the dawn of modernity. The gunpowder revolution, the renaissance of knowledge, and the Age of Reason are all myths as far as the general society was concerned, but they are useful myths that further the understanding of the past and the meaning of the present.

Within the military world, changes were spurred by the external factors discussed above. Gunpowder helped to end the dominant role played by the mounted warrior in the warfare of the Middle Ages. Other technological developments (like the longbow and crossbow) and changes in tactics (such as the return to the phalanx-like formation) also contributed to the demise of the fully armored, mounted warrior. As artillery improved, castles could no longer provide protection against besieging forces, and the art of fortification became far more complex. The accumulation of wealth and the consequent growth of banking in the centers of commerce enabled the feudal arrangement of exchanging military service for land use to be supplemented by a system of exchanging military service for money or gold. Thus, in the fifteenth and sixteenth centuries, mercenary armies were reborn, along with the classical forms of art and architecture. As sovereigns attempted to extend their temporal powers, they found that well led and well equipped armies were an essential prerequisite. Because larger armies were needed to defend extended frontiers, recruiting practices necessarily changed. By the eighteenth century, the poorly trained and poorly equipped rabble of in-

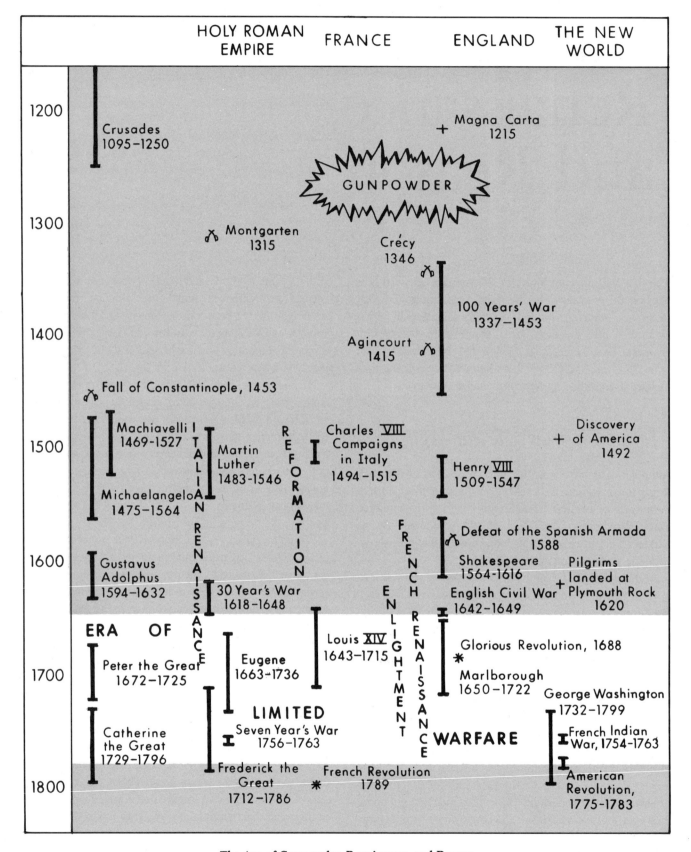

The Age of Gunpowder, Renaissance, and Reason

fantry that characterized the forces of the Middle Ages had evolved into highly disciplined formations. Although recruited from the rabble of society, these men were superbly trained at the expense of the state. Absolute kings ruled absolutely, and their armies reflected the order that they attempted to bring to their societies. Wars were common; the most powerful sovereigns extended their control to boundaries that remain familiar today, as in the case of Spain and France, or to boundaries that subsequently receded, as in the case of the Austrian and Ottoman Empires. The armies of Europe thus forged what was to become the modern state system of both Europe and America.

Participants in the Profession of Arms

The date that Western civilization began its emergence from the Dark Ages cannot be precisely fixed. Nor can the specific acts or causes of the emergence be stated with certainty. Some things, however, are at least representative of the far-reaching changes that occurred in Europe during the thirteenth and fourteenth centuries. First, the crusaders brought new ideas and new materials from the Near East. New ideas about military fortifications, for example, influenced warfare. New products from the Near East also influenced warfare, but in a far more indirect way. The new materials, such as spices and silk, were in such demand that trade was encouraged, supported, and perpetuated by merchants in Western Europe. Cities were established along the trade routes, and the wealth concomitant with trade accumulated. That wealth was used to encourage artists and writers, and was also used to pay military men to provide protection for the cities and trade routes. These soldiers were motivated primarily by their desire for money. Hence they were mercenaries, and the most famous of the mercenaries that flourished in the fourteenth century were the Italian *condottieri.*

Since Italy was at the crossroads of trade between Western Europe and the Near East, it became not only the center of trade and of Renaissance art and thought, but also the birthplace of many modern military terms and ideas. The condottieri mercenary bands prevalent in Italy, however, were not peculiar to that collection of city-states. Similar organizations were established by the Swiss, Germans, and French. The German units were called *Landsknechte,* and the French adopted the name *gens d'armes,* or men of arms. In England, agricultural prosperity was largely responsible for the creation of the **yeomen,** a class of small farmers who filled the ranks of the English armies and made the longbow famous. Organizations of *condottieri, gen d'armes, Landsknechte,*

and yeomen formed the basis for subsequent permanent fighting organizations, the regulars of their respective areas. The English yeomen became the only regulars who did not have a mercenary background.

Where political organizations were weak, **irregulars,** or armed individuals who were not a part of a regular force, comprised the principal fighting organizations of the region. Members of such irregular forces were called *Jäger* in many German regions, and the famous irregulars of the Alps region known as the Austrian Tyrol were called *Pandours.* Early **riflemen** were often irregulars because the high cost of the rifle, the skill required to use early rifles, and the slow rate of fire made the rifle an inappropriate weapon for regular forces.

Within the three broad categories of regulars, irregulars, and *Landsknechte* or *condottieri,* soldiers were also categorized by the equipment they used. One of the early titles of this sort from the period between the Dark Ages and the late eighteenth century was the **arbalester**—the soldier who used the **arbalest,** a heavy crossbow that was replaced by the longbow in England and by early firearms in other areas. **Arquebusiers** were those armed with the arquebus, and **musketeers** were those armed with the musket.* The **grenadier** was initially the soldier armed with grenades; by the late eighteenth century, however, grenades were no longer used, but because highly selected and well-trained individuals had thrown the early grenades, the term grenadier was applied to selected elite troops armed with muskets.

Mounted soldiers were distinguished in this period as hussars, dragoons, and *cuirassiers.* The term **hussar** was orginally used to describe fifteenth century Hungarian cavalry who raided trade routes. By the eighteenth century, the term was generally applied to light cavalry. **Dragoons** were mounted infantrymen of the sixteenth century who were armed with a short musket called a **dragon** or **dragoon.** The term is generally applied to soldiers who dismount to fight, but has also been used to describe heavily equipped cavalry soldiers. The *cuirassier* was the mounted soldier who wore a heavy metal breastplate, or *cuirass.* The term is also used to describe heavy cavalrymen in general.

In the era aptly called the dawn of modern warfare, fighting armies became increasingly more formal, and commanders increasingly sought to maintain close control over the movement and actions of their troops. Subordinate commanders were designated to simplify the requirements for control during battle, and before battle was undertaken, the senior commander often called his immediate subordinates

---

*Arquebus and musket are defined in the Technology section of this chapter.

together to discuss the situation and to decide on a course of action. Such a gathering of a commander and his subordinates was called a **council of war.** Until the early twentieth century, when wire and radio communications allowed commanders to exchange ideas over great distances, councils of war were frequent and essential components of military campaigns. Ironically, commanders who held councils of war have been widely criticized for their lack of conviction or courage. Most of the information exchanged at a council of war is today relayed routinely to commanders in the form of telegraphic and radio reports.

Councils of war primarily involved commissioned officers. A **commissioned officer** was one who held a **commission,** that is, a document signed by the principal authority in a military or political organization that gave the person named the rank, authority, and responsibility to act in behalf of the principal authority. In England, commissions are granted by the King or Queen; in the United States, they are granted by the President. Military rank has not always been granted by commission. The term **captain,** for example, was originally used to denote the leader who was either nominated or elected to serve as the head of a relatively small military organization. The word comes from the Latin word *caput,* which means head. Another of the early terms used to indicate rank was **colonel,** which comes from the Italian word *colonello,* meaning column. The colonel commanded a small column. The Spanish word for colonel is *coronel,* and may explain the English pronunciation of the word. The term **lieutenant** means one who takes the place of another (*lieu* meaning "place"; *tenant* meaning "holding"). A **general** was one who was generally in command, that is, he commanded troops of all arms rather than troops of a single arm. A **lieutenant general** was one who took the place of a general in the absence of the general, and a **lieutenant colonel** was one who took the place of a colonel in the absence of the colonel. Hence, a lieutenant general is subordinate to a general, and a lieutenant colonel is subordinate to a colonel. Any member of a military force who holds a commission is an **officer.**

The term **private** was applied to the private man, that is, one who was not obligated to another because of a commission. A **corporal** was a degree above a private and commanded a small body, or corpus. When a private man served for a considerable period of time, he was designated a **sergeant,** a term derived from the Latin word *servire,* meaning to serve. Because a sergeant, by virtue of his service rather than a commission, has gained a position of responsibility and a measure of authority, he can also be referred to as **non commissioned officer. Non commissioned** officers are commonly called **NCOs.**

Titles that applied to positions, rather than ranks, also came into use during the early modern period. One of these

titles was the **chief of staff,** a senior commissioned officer whose principal task was to supervise the officers assigned to a commander's staff. The **staff** is an organization consisting of those officers whose principal task is to assist the commander in areas of specific responsibility, such as personnel matters or supply matters. The word may be derived from the idea that a general relies on his military staff just as he may lean on a wooden staff or cane. One early member of the military staff was the **quartermaster,** the officer primarily concerned with supplies and services. The term originally applied to the individual who was master of a district, or quarter, and who had to furnish lodgings to the troops assigned to the district. In the German armies of later periods, the **quartermaster general** performed the functions of the American chief of staff. In the nineteenth and twentieth centuries, the **first quartermaster general** was the principal assistant to the **Chief of the German General Staff,** who by World War I was the senior uniformed military individual in Germany. The *maréchal des logis* in the French Army performed the same tasks as the quartermaster, but originally he was a non commissioned officer whose principal duty was to insure that horses were properly cared for. Today, the English equivalent of *maréchal des logis* is sergeant. The term *maréchal,* however, is translated as **marshal,** the highest ranking officer in many modern armies and the one who was second-in-command to the King in armies of the seventeenth and eighteenth centuries. Ironically, *maréchal* originally meant horse servant. The term *maréchal de camp* referred to the marshal who commanded a campaign. He was immediately superior to the colonels who commanded the columns or forces on campaign. The modern equivalent of the *maréchal de camp* is the brigadier general in the United States Army. The German equivalent of *maréchal de camp,* however, is *Feld-Marschal,* but in Germany, the *Feld-Marschal* is the highest ranking officer in the Army. In England, too, the **field marshal** outranks all other officers. The **inspector general** was another of the staff officers commonly encountered in the seventeenth century; his function was to insure that orders from headquarters were being complied with in the field. The name of one famous and presumably very effective inspector general in the seventeenth century French Army, Jean Martinet, has been immortalized in the English language—a **martinet** is a harsh and strict military disciplinarian.

Discipline was also enforced by **courts-martial,** which were and are tribunals of commissioned officers appointed to hear cases against and to designate punishments for violators of martial law. The present systems of military law in England and the United States were derived from the first military act passed after the accession of William the Conqueror to the throne of England in 1066.

Principal Organizations

The new titles of rank associated with the dawn of modern warfare appeared in the same period during which soldiers were organized into new and more formal groupings. The trend toward more formal organization and the concomitant trend to move from the mass formations that characterized the ancient and medieval armies to more linear formations were evident from the fourteenth through the eighteenth centuries. Informal organizations like the **battle,** generally a group of several hundred men, mounted or on foot, persisted until the early sixteenth century. The informal organization termed the **detachment,** which can refer to any number of troops of any arm or combination of arms, as long as they are a subgroup of a larger organization, was also widely used, and remains in use today. Task forces and wings are also units whose size and composition vary considerably according to the situation. A **task force** is a temporary grouping of units under a single commander whose purpose is to carry out a specific mission. The term **tactical tailoring** is used to describe the breaking up or combining of specific permanent units into smaller units, larger units, or units with a greater variety of weapons, for the purpose of accomplishing a short-term task for which a standard organization would be inappropriate. In current United States Army parlance, a task force is formed when at least one company of one arm (such as a company of tanks) is attached to a battalion of another arm (such as a mechanized infantry battalion).

The trend toward more formal military organizations was signaled by the appearance of the **Swiss phalanx** early in the fourteenth century. The formation consisted of from 600 to 2,000 men who were initially armed with the halberd, and later armed with approximately equal numbers of pikes and halberds. Late in the fifteenth century, the Spanish leader Gonzalo de Cordoba experimented with formations. Cordoba grouped infantry elements into battles of 400 or 500 men that could also include cavalry and guns. By the early sixteenth century, the experimental system had become standardized in the Spanish Army, the grouping of several battles being designated a *tercio,* or **Spanish square.** The *tercio,* ranging in size from 1,500 to 3,000 men, was generally made up of pikemen and arquebusiers in equal proportions.

The large tactical formations of the fifteenth and sixteenth centuries, like the *tercio,* were often broken down into administrative units known as companies. The company was

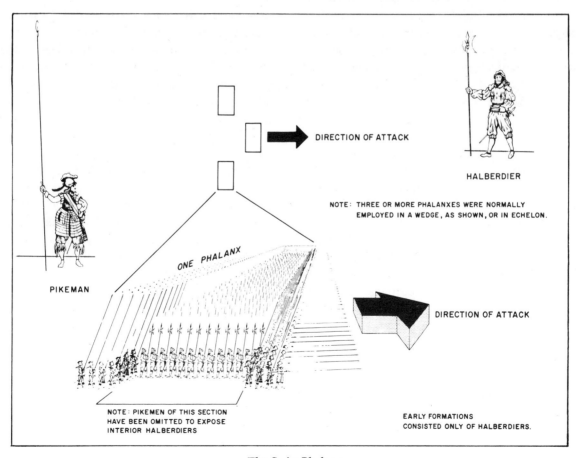

DIRECTION OF ATTACK

HALBERDIER

NOTE: THREE OR MORE PHALANXES WERE NORMALLY EMPLOYED IN A WEDGE, AS SHOWN, OR IN ECHELON.

ONE PHALANX

PIKEMAN

DIRECTION OF ATTACK

NOTE: PIKEMEN OF THIS SECTION HAVE BEEN OMITTED TO EXPOSE INTERIOR HALBERDIERS

EARLY FORMATIONS CONSISTED ONLY OF HALBERDIERS.

The Swiss Phalanx

composed of those men, about 100 to 200 strong, whose feeding and quartering was the responsibility of a single individual. In many armies, this individual was only administratively responsible for his troops; hence, he was not involved in their preparation for or conduct of battle. In combat, the company became directly responsive to commanders of larger units. In Spain, companies were formed into *tercios* or *escuadrons,* which varied in size from 600 to 3,000 men. Even though no modern cavalry unit is known by the term company—the term **troop** currently applies to the cavalry unit of company size—many earlier armies referred to their cavalry forces that numbered from 60 to 100 men as companies. The term troop was not officially used in the United States Army until 1883. The artillery unit of company size is called a **battery,** a term that comes from the verb "to batter." In addition to referring to the company-sized artillery organization, the term battery also is applied to the place where artillery is positioned. The guns of a battery, in the first sense, are rarely separated from one another, and are generally under the control of a single commander. Machineguns, cannons, howitzers, mortars, and searchlights have been commonly organized into batteries, and four to six like items of these major pieces of equipment generally make up one battery. A unit that has a number of one major item (such as an 8-inch gun) and a number of another major item (such as a 175-mm gun) is called a **composite battery.** The subordinate unit to the company, troop, or battery is the **platoon.** The term is derived from the French word *peloton,* which means simply a group of men. The term originally referred to the small groups who fired their weapons in unison in early organizations.

In addition to the trends toward greater formality and greater linearity in military organizations in the early modern period, there started a trend toward the establishment of organizations made up of smaller organizations. *Tercios* were the subordinate organizations of the Spanish field army of the fifteenth, sixteenth, and seventeenth centuries, and each *tercio* was made up of a number of companies. By the late eighteenth century, battalions, brigades, regiments, and divisions had all been interposed in the command structure between companies and armies of most modern states. The **battalion,** which Maurice of Nassau patterned after the Roman cohort, consisted of 500 to 700 men, and was made up of several companies. In the Prussian Army of the eighteenth century, it was made up of five musketeer companies and one grenadier company. Two or more battalions formed a **brigade** in Gustavus Adolphus' Swedish Army, but in later Prussian armies, two regiments formed a brigade. Battalion and regiment were synonymous terms in some armies. In other armies, two or more battalions made up the regiment, and the regiments formed a brigade. The **division** was composed of two or more brigades or regiments. The term division did not appear until the late eighteenth century, although there was a less formal and more general use of the word much earlier to denote a division, or simply a part, of any larger organization.

Many of the organizations that were developed during the early modern period used the **combined arms concept**—that is, their subordinate organizations consisted of cavalry, artillery, and infantry forces. The term **combined arms** refers to any force made up of either two or all three of these principal combat arms.

The Technology of the Age

The discovery of gunpowder did not revolutionize the implements and methods of war. It did, however, mark the beginning of a trend toward the greater significance of missile weapons and the lesser significance of striking weapons. Technological change occurred slowly.

Striking Weapons

From the discovery of gunpowder until the late eighteenth century, the striking weapons of earlier periods continued to be used and refined, and a few new striking weapons were developed. The lance and the pike, both of which dated from Roman times, remained important against mounted warriors. But a new and more sophisticated family of weapons, known as **pole arms,** brought a new dimension to the unmounted warrior's fight against the cavalryman. The simplest of the pole arms, called a **bill,** was a slightly hooked blade mounted on a pole. It was used to prune limbs from trees, and its military uses were strikingly similar. The pole arm that combined a sharp spear-like point with a head intended to be used primarily as a bludgeon was called the **goedendag.** In the fifteenth and sixteenth centuries, an eight-foot pole arm known as the *halberd,* or *halbert,* was used with great success by Swiss infantry. The head of the *halberd* had a spear-like point for thrusting, a curved blade for cutting, and barbs or hooks that could be used to pull a rider from his mount or to pull the reins from a rider's control. The **spontoon,** or **espontoon,** was a half pike; it was carried in the eighteenth century as a symbol of rank.

Swords and sabers continued to be used widely, but like the *halberd* and the bill, they became symbols of rank and authority as their practical value decreased. The characteristic thrusting sword of the sixteenth and seventeenth centuries was called the **rapier.** It had a long, narrow, double-edged blade and a short grip. The pike, even though its form

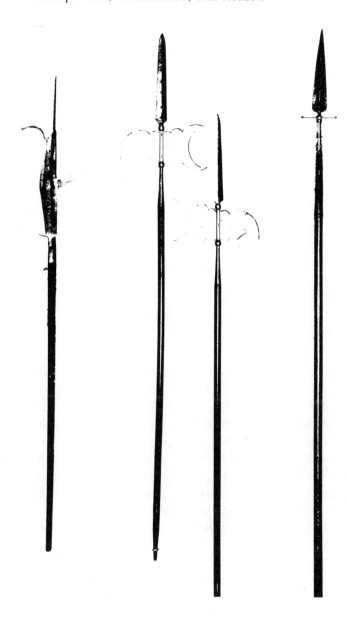

Pole Arms: Bill, Ceremonial Halberds, and Spontoon

later, steel. For close fighting in wooded areas, the tomahawk was superior to the sword, and American colonial troops often carried the tomahawk in place of the sword.

Missile Weapons

Gunpowder slowly changed the nature of warfare, eventually making the principal weapons of the pre-gunpowder period obsolete. The change, however, was evolutionary rather than revolutionary, and established missile weapons like the bow, crossbow, and trebuchet remained important for centuries after the discovery of gunpowder. The most popular of the crossbows, and the one most widely used by continental soldiers until the sixteenth century, when it was superseded by firearms, was the arbalest. In England, the arbalest was replaced in the fourteenth century by the **longbow,** a six-foot bow made of yew or ash. The longbow became an effective weapon in the hands of well trained Welsh mercenaries and English foot soldiers. It could shoot farther—to ranges of 300 to 400 yards—and more rapidly than the best crossbow.

Gunpowder, however, was the propelling force of the future, even though its early military uses were of little significance. One of the first weapons to use gunpowder on the battlefield was the **hand cannon,** a small iron tube, 12 to 15 inches long and 2 to 4 inches in diameter, that was mounted on a pole. A stone placed in the tube and atop finely ground **gunpowder**—a mixture of sulphur, charcoal, and saltpeter—was propelled from the tube when the gunpowder was ignited by placing a glowing **match**—a cord that had been soaked in a solution of saltpeter and allowed to dry—into the powder through the **touchhole**—a hole in the base of the cannon. The hand cannon was the antecedent of both individual firearms, like the musket and rifle, and crew-served firearms, like mortars and artillery.

The first of the individual firearms to succeed the hand cannon was called the **arquebus,** a tube closed at one end except for the touchhole and mounted on a wooden **stock,** a device that allowed the weapon to be held horizontally against the shoulder or chest. The arquebus was the first **shoulder weapon,** or **shoulder arm,** that is, a hand weapon fired from a shoulder-supported position. When a glowing match was placed in the touchhole, the arquebus fired a small stone or iron ball in the general direction in which the tube was pointed. Since it was difficult to aim the **piece,** a term used to refer to any firearm, and to place the match to the touchhole simultaneously, devices called **locks** were developed to assist in igniting the **charge,** which was the measured amount of powder necessary to fire the weapon. The first lock to be developed, the **matchlock,** consisted of a lever extending through the stock and pivoted within the stock. Because of its

changed markedly, was the one weapon that did not lose its practical value as gunpowder weapons supplanted the striking weapons of earlier periods. In the late seventeenth century, in Bayonne, France, a device known as the **bayonet,** which had a short handle and a steel blade with a tapered point, was developed to fit into the muzzle of a musket, thereby transforming the musket into a short pike. These early bayonets were called **plug bayonets** because they plugged the muzzle of the firearm to which they were affixed.

An effective and primitive striking weapon that was also used as a missile weapon was encountered by Europeans when they arrived in the New World. American Indians were adept in the making and use of the **tomahawk,** a hatchet-like device with a short wooden handle and a head of stone or,

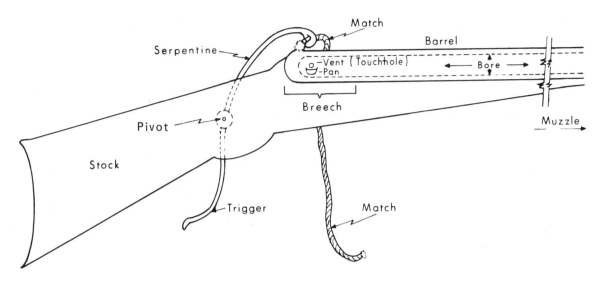

The Matchlock

S-shape, the lever was called a **serpentine.** The end of the serpentine extending below the stock served as the **trigger,** the part of the firing mechanism that caused the ignition of the charge. The end of the serpentine above the stock held the burning match. When the trigger was pulled toward the stock, the upper part of the serpentine lowered the match into the touchhole, or, in later models, into a **pan,** a small cup that held priming powder. When ignited by the match, the priming powder sent a flash through a hole, called a **vent,** to ignite the main charge. The main charge and the ball were loaded from the muzzle into the **breech,** the rearmost part of the barrel.

Early in the sixteenth century, a lock called the **wheellock** was developed. It ignited the charge in the pan, called the **primer,** by sending into the pan a shower of sparks that was created by the striking of a spring-wound steel wheel against a piece of iron pyrite or flint. Because the wheellock was expensive, it was rarely available to foot soldiers and was never mass-produced. However, it became the favorite of cavalrymen, not only because they could afford the piece, but also because it was far easier to handle on horseback than the heavy matchlock arquebuses, which had been used by earlier cavalrymen with the aid of a fork attached to the saddle.

Most wheellocks were used on **pistols**—a term applied to firearms that were designed to be held, aimed, and fired with one hand. The next major lock was developed early in the seventeenth century and was called the **flintlock.** It served as the principal method of small arms ignition until the middle of the nineteenth century. In the flintlock, a piece of flint, held in a **cock,** was released by pulling the trigger. The flint struck a coverplate, called the **frizzen,** over the pan, uncovered the pan, and created sparks that ignited the primer.

The early handguns were often known by the name given to the method of ignition (such as the matchlock, wheellock, or flintlock), but more general terms also applied. The most general term was the **gonne, gunne,** or by current spelling standards, **gun,** the name applied to any weapon that has a tube or barrel from which a projectile is thrown by a force that is created by exploding gunpowder. Hand cannon, wheellocks, flintlocks, pistols, and many later and larger firearms can all be generally referred to as guns. **Musket** is also a general term that refers to the smoothbore, muzzle-loading shoulder arms of the sixteenth through nineteenth centuries. The term **small arm** applies to any firearm capable of being fired while held in one or both hands.

The **matchlock musket** was the first firearm to be issued in

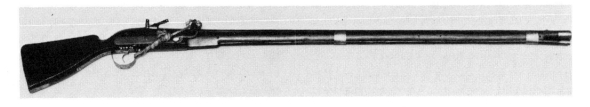

Sixteenth and Seventeenth Century Matchlock Musket—Caliber .80

Circa 1750 "Brown Bess" Flintlock Musket and Bayonet—Caliber .75

large numbers. It was inaccurate and extremely slow to load. The match was difficult to keep ignited, especially in wet weather, and at night it revealed the position of its users. Since it was very inexpensive to produce, however, it was used for almost two centuries.

The flintlock musket popularly called the **"Brown Bess"** was more formally called the British long land pattern musket. Well trained British troops could load and fire six rounds a minute from this weapon. During the early years of the American Revolution, it was used by American troops, but after 1778, French muskets replaced the "Brown Bess" in the American service. With only minor changes, the British used the "Brown Bess" from about 1730 to 1840.

Nearly all of the individual firearms in use before the American Civil War had a smooth bore and were loaded from the muzzle with charge, bullet, and wadding. The **wad-**ding, which was made of either linen or paper, served to seal the powder in the **chamber,** the rearmost interior portion of the barrel where the powder and ball were held until the weapon was discharged. A **ramrod** was used to pack the powder and to force the bullet and wadding against the charge. The first **iron ramrods** were used by the Prussian Army during the eighteenth century.

One of the popular missile weapons of the seventeenth and early eighteenth centuries that was abandoned until the middle of the nineteenth century was the **hand grenade,** a small explosive missile thrown by hand. Due to the difficulty of throwing the grenade, foot soldiers of large stature and considerable strength were selected to serve as grenadiers.

In addition to the family of individual firearms, or handguns, that evolved from the primitive hand cannon of the fourteenth century, there evolved a family of crew-served

Vasi or *Pot de Fer*

Bombard Mounted on a Swivel With Elevating Capability

firearms. This second family has on occasion been referred to as **artillery,** a term that originally was applied to any contrivance for throwing missiles. Since the sixteenth century, the term has been applied only to large guns that are served by a crew of five or six soldiers. The term **cannon,** which was in wide usage from the sixteenth through the nineteenth century, was applied initially to any tubed weapon, and later to only the largest of tubed weapons. (The French word *canna* translates into reed, pipe, or tube.) Today, the term cannon is generally used to refer to the smoothbore, crew-served pieces of the nineteenth century and earlier, and to very large guns of the twentieth century.

The earliest of the crew-served missile-firing weapons was called the *vasi* or *pot de fer.* It had a large base, in which a

Fifteenth Century Breech-Loader

powder charge was held, and a small neck, from which an arrow or quarrel was fired when the powder in the base was ignited through a touchhole. The *pot de fer* was succeeded by the **bombard,** which had a short, thick barrel that was of greater diameter at the muzzle than at the breech.

In the fifteenth century, many attempts were made to produce a breech-loading weapon capable of firing large projectiles. As long as the charge was small, such weapons fired satisfactorily, but the range and projectile weight were quite limited. Even though **breechblocks,** the portion of the breech removed to allow breech loading, were tightly wedged in place, gas still escaped, and the breech was not strong enough to withstand the strain that resulted when larger charges or heavier projectiles were used. These problems were not solved until the mid-nineteenth century, when significant advances were made in metallurgy. Until the nineteenth century, most cannon were made of bronze because it was less brittle than iron and easier to mold. In about 1520, gunpowder was improved when coarse grains supplanted the earlier fine-grained forms. The coarse grains allowed fire to travel more quickly through the air spaces in the gunpowder, creating a more powerful detonation.

In accordance with the whim of either the cannon maker or his employer, who was often the king, sixteenth century cannon were made in all shapes and sizes. Each of the variety of sizes had its distinctive name. This variety is but suggested by the following chart, which shows only English guns. The French and Spanish had different types, names, and

specifications. Moreover, the specifications for a given **caliber** —the diameter of the **bore,** or interior of the barrel—often varied from one **foundry,** or place of manufacture, to the next.

One of the most common means of describing a gunpowder weapon is by stating the size of its bore, and the term caliber, which comes from the Latin *qua libra* meaning "what pound," generally applies to this important measure. The caliber of many early firearms, especially the crew-served arms, was determined by the weight of the solid iron ball that could fit comfortably into the bore of the weapon. The weapon was named according to the weight of the ball in pounds; hence, a **3-pounder** was a weapon into the bore of which a 3-pound solid iron ball would fit. For small arms, the term **gauge** was used as a measure of caliber. The gauge was determined by finding the iron ball that properly fit the bore of the weapon, and then determining the number of balls that it took to weigh one pound. Hence, a 12-gauge weapon had a bore size into which a 1/12-pound iron ball would fit. This cumbersome method is still used today to designate the size of **shotguns,** which are smoothbore shoulder weapons that fire shot pellets or slugs and are used primarily for sporting purposes. A more direct method of determining the caliber of a weapon is simply to measure the diameter of the bore. This measurement is expressed in inches, where imperial measures are common, or in millimeters, where the metric system is used. When caliber is concerned, fractions of inches are expressed as decimals; hence, a gun with a ¾-inch bore is re-

Name	Caliber (inches)	Length Feet	Length Inches	Weight of Gun (pounds)	Weight of Shot (pounds)	Powder Charge (pounds)
Rabinet	1.0	—		300	0.3	0.18
Serpentine	1.5	—		400	.5	.3
Falconet	2.0	3	9	500	1.0	.4
Falcon	2.5	6	0	680	2.0	1.2
Minion	3.5	6	6	1,050	5.2	3
Saker	3.65	6	11	1,400	6	4
Culverin Bastard	4.56	8	6	3,000	11	5.7
Demiculverin	4.0	—		3,400	8	6
Basilisk	5.0	—		4,000	14	9
Culverin	5.2	10	11	4,840	18	12
Pedrero	6.0	—		3,800	26	14
Demicannon	6.4	11	0	4,000	32	18
Bastard Cannon	7.0	—		4,500	42	20
Cannon Serpentine	7.0	—		5,500	42	25
Cannon	8.0	—		6,000	60	27
Cannon Royal	8.54	8	6	8,000	74	30

Principal English Guns of the Sixteenth Century

SOURCE: Albert Manucy, *Artillery Through the Ages: A Short Illustrated History of Cannon, Emphasizing Types Used in America* (Washington, D.C.: National Park Service, 1949), p. 35.

Leather Gun, 1630

ferred to as a .75-caliber weapon. The term caliber is also used to designate the length of a gun in terms of the diameter of its bore. An 8″/55, or 8 inch × 55, caliber gun has a bore diameter of 8 inches and a length of 55 × 8 inches, or 440 inches.

Because of the great variety of crew-served, missile-firing weapons, names were often designated for groups of such weapons. For example, the English sixteenth century pieces could be grouped as cannon, which had calibers of six to eight inches; **culverins,** which had calibers of four or five inches (earlier, the name culverin was applied to a hand cannon mounted on a stock); **periers,** which had still smaller calibers (the perier was the ancestor of the howitzer); and **mortars,** a German invention characterized by a short barrel and a high angle of fire.

In the seventeenth century, guns were assigned to smaller units; concomitantly, they became more mobile. In the Swedish Army, a **leather gun** that fired a 3-pound iron ball was developed and assigned to each regiment. The leather gun had a copper tube that was strengthened with parallel iron bars and wrapped with several layers of cord. These cords, in turn, were covered by an outer shell of tough

leather. After firing a maximum of 10 or 12 rounds, the tube had to be unscrewed for cooling. This shortcoming led to the development of lighter, all-metal guns.

By the eighteenth century, the terms gun, howitzer, and mortar were used to distinguish the various types of crew-served artillery weapons. Guns were the largest pieces in size, weight, and caliber and had the greatest muzzle velocity. Hence, they had the lowest **trajectory,** a term used to describe the curved path a projectile follows upon being fixed from an artillery piece. **Howitzers** were next in size, weight, caliber, and muzzle velocity; their maximum angle of elevation was about 65 degrees. Because of the size differences, the howitzer was more mobile than the gun and was better suited for field use. Guns were more generally used for siege purposes and for the defense of fortified places. The term **mortar** has consistently been applied to those pieces characterized by an extremely high angle of fire.

Ammunition

Ammunition refers to any of the various projectiles that can be discharged from a firearm and to the powder, primer, and

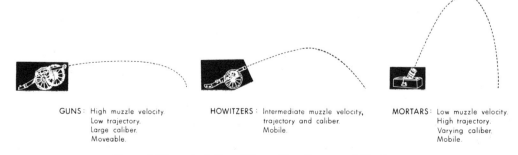

GUNS: High muzzle velocity. Low trajectory. Large caliber. Moveable.

HOWITZERS: Intermediate muzzle velocity, trajectory and caliber. Mobile.

MORTARS: Low muzzle velocity. High trajectory. Varying caliber. Mobile.

General Characteristics of Guns, Howitzers, and Mortars

other materials used to discharge the projectiles. Three common types of ammunition were used in early crew-served gunpowder weapons—shot, canister, and grapeshot. **Shot** refers to a solid projectile that has no exploding charge inside. Stone and iron shot were both commonly used. **Hot shot,** which refers to iron shot that is heated to incandescence prior to loading, was used for setting fire to buildings or wooden ships. **Canister** refers to an antipersonnel projectile made up of many smaller projectiles that are enclosed in a casing of burlap or metal and that leave the casing just beyond the muzzle. Another term for canister is **case shot.** When the smaller projectiles in an antipersonnel projectile are held together by a combination of rods, wires, or plates, **grape** or **grapeshot** results. Even though grapeshot is generally distinguished from canister and case shot in military dictionaries, the terms are often used indiscriminately. **Barshot** consists of two iron balls connected by an iron bar or of one ball with a bar extending from it. **Chain shot** consists of two iron balls connected by a short length of chain. Barshot and chain shot increased the effective damage area of the projectile.

Mobility

There were few technological innovations in the area of mobility during the period from the late Middle Ages until the late eighteenth century. Horsepower, manpower, wind, and water were the chief means of moving armies and their supplies. Some cannon, and especially the howitzers of the late sixteenth to eighteenth centuries, were placed on carriages with large wheels to facilitate their cross-country movement. By increasing their mobility, these weapons could accompany troops on campaign and on the march, and operate beside them on the battlefield; such mobile, crew-served weapons were referred to as **field guns** or **field artillery.** When every member of a field artillery unit was mounted on a horse, the term **horse artillery** was applied. When light artillery and its impedimenta were organized for long marches, they formed an **artillery train. Siege trains** were made up of the numerous oxen, horses, and wagons necessary to haul heavy artillery, called **siege artillery** or **siege guns,** and related impedimenta from place to place. **Baggage trains** carried more general items: ammunition, parts, tents, the personal effects of officers, and other supplies.

Artillery pieces were supported on their carriage by pivots, called **trunnions,** which formed the horizontal axis about which the piece rotated when it was elevated. When the piece was moved over long distances, its carriage was often connected by means of **trails** (the part of the carriage that rested on the ground and stabilized the piece when it was fired) to a **limber,** which consisted of two wheels, an axle, and generally an ammunition chest as well.

Naval Objects

The age of oars gradually gave way to the age of sail. The most notable changes of this period of naval evolution were twofold. First, fighting vessels became more mobile and were able to stay at sea longer; second, weapons that used gunpowder affected naval tactics to the same degree as they did land warfare.

The late medieval galleys had both oars and sails. They were rowed into battle and, unlike earlier galleys, which had an underwater ram, the late medieval galley, like those used at the Battle of Lepanto in 1571, had a **spur** at the **bow,** or front of the ship. The spur could shear oars from an opposing ship and could also badly damage an enemy's hull. The spur could also be used as a means of boarding the ship. Early in the age of sail, cannon were placed on ships, and the practice of boarding another's ship to fight an "infantry" battle became outdated. Naval battles became artillery battles, and the spur was replaced by the **bowsprit,** the boom on the front of a ship to which sails could be attached. Rowed vessels, or **long ships,** were favored by the Mediterranean powers because of the relative calmness of the Mediterranean. The Atlantic powers favored **round ships,** which as early as the thirteenth century relied on the exclusive use of sail power. The earliest round ships had a single mast, and when sailors added masts and sails to help maneuver their vessels, they found that the speed of the ship increased as well. The popular Portugese three-masted vessel that had a low freeboard and mounted cannon just above the waterline was called the **caravel.** The caravel and most of the merchant ships of the sixteenth century were **castled,** that is, the **fore** (forward) and especially the **aft** (rear) sections of the ship were built higher than the main deck to allow archers or arquebusiers to fire down on other ships and on any soldiers who boarded the main deck of the vessel. The **carrack,** a longer, more maneuverable, and higher castled ship, succeeded the caravel. Evolving from the carrack and consisting of three or four decks, the **galleon** became the major sailing warship of the sixteenth century. In the Mediterranean, a ship smaller than the galleon, and different from the oared galley because of a deck over the rowers, was called the **galleass.** It was the last of the oared warships.

The dominant ship in the English Navy of the sixteenth and seventeenth centuries was a galleon that was often referred to as a **man of war.** The term, however, also refers to any warship of a recognized navy. **Ship of the line** is another imprecise naval term that refers to warships that are among the largest ships in a powerful navy. A man of war was a ship of the line. **Frigates** and **corvettes,** which were smaller armed sailing ships, were not ships of the line.

Defensive Items

Individual items of protection generally decreased in importance as gunpowder weapons increased in importance. The need for greater individual movement, the expense of providing armor to larger numbers of soldiers, and a belief that God would protect the righteous were reasons for the decline in the use of body armor. In the fourteenth and fifteenth centuries, knights continued to use elaborate suits of full armor, but the foot soldier was often protected by no more than a gambeson, the quilted coat that had earlier been worn under an iron vest. In the seventeenth century, full armor was worn primarily during sieges and parades; however, **three-quarter armor,** which extended to the knees, was still worn by heavy cavalry. Spanish and Portugese troops wore **half armor,** which consisted of a full breastplate and backplate. In the eighteenth century, the *cuirass* was the only significant piece of armor in wide use.

Group protective items increased considerably in importance and sophistication as gunpowder weapons developed. Because the walls of castles could not withstand the impact of projectiles fired from large guns, new architectural designs were created to provide greater protection to towns and other places of military significance. One of the early developments was the **bastion,** which was a construction of earth and stone that projected from the main enclosure of a fortification. From the bastion, fire could be directed along the curtain. The area behind the most forward construction of a fortification, where a standing man could not be seen by a besieger, was called the **covered way.** More generally, the term referred to any area near the enemy where troops could move and not be seen. Other parts of the fortification included the **glacis,**

which was the slope outside the covered way along which the principal defensive weapons could fire; the **parapet,** which was the raised earth or stone portion that protected the guns: the **loopholes,** or **embrasures,** which were the holes through which the guns were fired; the **scarp,** which was the inner wall of the ditch; and the **counterscarp,** which was the outer wall of the ditch.

Any temporary group protective work constructed by an army in the field is called a **field work.** Elaborate field works were called **field fortifications,** but today the term is used in lieu of field works. A construction primarily of earth is an **earthwork.** When the field work was sufficiently substantial to afford protection to a man in a standing position, and was built wholly or partly above the ground, it was called a **breastwork.** Field works that were open at the **gorge,** or rear, were called redans or lunettes. The **redan,** shaped like the letter V, pointed toward the enemy. The **lunette** was a redan with flanking walls extending from the arms of the V. An irregularly shaped field work that provided all-round protection was called a **redoubt.**

By the late seventeenth century, fortifications had become so significant and formidable that protection for attackers was an important consideration. Again, digging provided a solution. By digging extensive **trenches,** or ditches, called **siege lines,** the besieger could provide protection for attacking troops and weapons. The zigzag trenches that generally led toward the fortification were called **approaches** or **formal siege approaches.** The trenches that were perpendicular to the approaches were called **parallels.** Usually, three parallels were dug. The first parallel was that closest to the fortification, while the third was that farthest away. Artillery and mortars occupied the third and second parallels respectively, while the final assault by foot soldiers began from the first parallel. Sieges, however, often ended before the final assault. The lines directed toward the fortification were sometimes referred to as **lines of contravallation.** Occasionally, an army also built a line, called a **line of circumvallation,** that faced out-

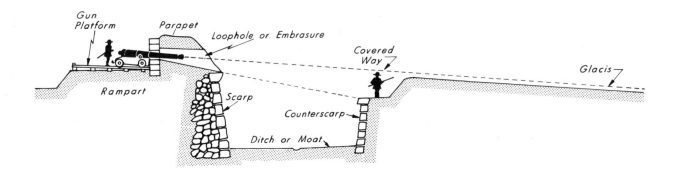

Profile of an Eighteenth Century Permanent Fortification

ward and protected the besiegers from a **relief force,** that is, a unit seeking to come to the aid of the besieged. If the relief force had to travel a great distance, it was called a **relief expedition.** The study of sieges was of great importance in the seventeenth and eighteenth centuries, and was known as **poliorcetics.**

Operations

As the introduction of gunpowder slowly altered the conduct of war, it also altered the operational terms and concepts used to describe the conduct of war. "Fire and movement" and "fire and maneuver," for example, are terms that have been used in various ages to explain the method of attack in which one element of a unit moves while being supported by the fire of another element. **Fire and movement** is sometimes distinguished from **fire and maneuver.** The former term applies to the procedure employed by an individual soldier. First, he selects a position a short distance away that preferably offers **cover,** defined as protection from small arms projectiles, and hopefully offers **concealment,** defined as protection from observation. Next, he **rushes,** or runs the short distance in a crouch. Finally, he prepares to either fire or advance to another position. The term fire and maneuver applies to small units that move a portion of the unit, possibly using fire and movement techniques, while the remainder of the unit provides **fire support,** which refers to any assistance from missile weapons that is intended to facilitate the activities of combat elements.* In fire and maneuver, the maneuvering element becomes the fire support element, and the roles continue to alternate until a designated objective is taken or the technique is abandoned.

When the troops of a unit fire their weapons in unison, the fire is referred to as **volley fire.** When troops fire simultaneously or sustain a rapid firing of weapons, a **fusilade** results. When all guns—especially naval guns—are fired at the same target simultaneously, the fire is referred to as **salvo fire.** A prearranged barrier of fire, other than small arms fire, that is intended to impede the enemy's advance is called a **barrage.** If the impacting missiles in a barrage remain in the same relative position but move incrementally toward a suspected target, the barrage is known as a **creeping barrage.** Any simultaneous firing by a number of weapons at a common target is called a **concentration.** When the trajectory of

missiles from gunpowder weapons is perpendicular to the line of enemy troops or perpendicular to the long axis of any other target, the fire is called **direct fire.** If the fire is parallel to the long axis of the target, the fire is called **enfilade fire,** or **raking fire.** The latter term is often used when a ship or a column of ships is the target. **Plunging fire** occurs when a weapon is fired from a position considerably higher than the target. **Oblique fire** occurs when a piece is fired at an angle to the long axis of the target. When a missile fired from a horizontally-held weapon never rises above the top of a target before striking the target, the target is said to be in **pointblank range.** When two or more lines of gunfire intersect at a target, the target is in a **crossfire.** Intersecting lines of gunfire are also referred to as **interlocking fire.** The **field of fire** is the area that a weapon or group of weapons can cover effectively from a given position.

Field Artillery

According to modern United States Army doctrine, field artillery units are assigned tactical missions that define the unit's responsibilities. Among these missions are direct support and general support. Historically, these terms generally defined the relationship between supporting arms and supported units. **Direct support** denotes that the supporting unit responds to the requests of a specific commander of a regimental or lower-echelon force. **General support** denotes that the supporting arms are controlled by the overall commander of a large independent force.

Cavalry

Although gunpowder was a significant element in the decline of the cavalry, many other factors also were involved. For example, the discipline of infantry increased; when standing armies were created, foot soldiers had time for drill, and thus learned to form squares when attacked by cavalry. The **square** was a solid, or nearly solid, formation of infantrymen, all of whom faced toward the outside of the square and prepared to break the cavalry charge by strength of numbers, volley of muskets, or point of bayonet. **Pottes,** another of the defenses employed against mounted soldiers, also enjoyed some success. Pottes were simply camouflaged holes, dug to a depth of two or three feet, which could easily cause a break in a horse's leg.

Because of the awkwardness and size of early gunpowder weapons, few could be readily used by the mounted warrior. Pistols, however, could be used with ease. Accordingly, armies developed a technique known as the **caracole,** in which

---

*This definition of fire support allows bows and arrows, slingshots, and other **supporting arms** (weapons that provide assistance to combat elements) that predate gunpowder weapons to provide fire support. The term fire support undoubtedly was first applied to gunpowder weapons, but is occasionally, and probably erroneously, used in reference to weapons that do not utilize gunpowder.

the cavalryman rode to the enemy, fired one pistol, wheeled his horse to fire the pistol held in the opposite hand, and returned to a safe distance to reload. Disciplined infantry, pole arms, pottes, muskets, and bayonets, however, were all effective countermeasures to the short-range cavalry firearms.

Infantry

As gunpowder weapons became more effective in the hands of foot soldiers, the fighting formations and tactics of battle changed. The general trend from the Dark Ages to the present century has been toward more linear formations, which allow greater freedom and a more effective use of manpower, firepower, and technology. **Linear tactics** referred especially to those arrangements in which successive lines of soldiers were employed to meet an enemy in battle. By the eighteenth century, a formation of three lines of soldiers was used. Like many earlier linear formations, it relied on the **countermarch,** a technique in which soldiers, after firing from the first rank, moved to the rear to reload and advance in succession until they were again in the first rank. Together, the three lines of soldiers made up a **line of battle.** An army would generally form two lines of battle; the second, about 300 paces behind the first, was used as a reserve. The means of passing from one position to another—either within a line of battle, from one line to another, or, more usually, from one formation to another—was referred to as an **evolution.** The **wheeling movement,** in which a line or a portion of a line of battle pivoted on a common point, was one of the standard evolutions of the seventeenth and eighteenth centuries. When lines of battle were successively committed to an assault in the same area, the lines were referred to as **waves.**

Logistics

The introduction of gunpowder and standing armies greatly increased logistical requirements. The term logistics is derived from the ancient Greek word *logistikos,* meaning "skilled in calculating." Although there was a military administrative official with the title "logista" in the ancient Roman and Byzantine Empires, the term logistics did not enter military terminology until the eighteenth century. The *maréchal des logis* was the staff officer charged with such duties as arranging for marches, organizing encampments, and quartering troops. As warfare grew increasingly formal and elaborate, the duties of the *maréchal des logis* were increased to include the responsibility for security and reconnaissance measures; the stocking of **depots** (places where general supplies are stored) and **magazines** (places where arms and ammunition are stored); the hiring of transport and movement of supply convoys; the requisitioning of food, forage, fuel, horses, and labor; the custody of prisoners; the maintenance of camp and march discipline; the paying of troops; and the regulation of the accompanying trains of provisions, ammunition, cattle, and **camp followers** (those individuals, such as **sutlers**—contractors or salesmen—and women and children, who followed the troops when they were on campaign). The *maréchaux des logis,* like their Prussian counterparts, the *Quartermeister* or the *Generalquartermeister,* became the equivalent of chiefs of staff when additional staff officers were assigned to assist them in specific duties.

In order to reduce the logistical requirements, or the **logistical tail** (distinguished from the fighting tooth in recent years), of units in the field, many armies encouraged **foraging,** which refers not only to animals' acquiring food in the area where they happen to be, but also to the procuring of more general supplies by the same means.

Seapower and Sails

The coming of the age of sail brought new opportunities for commerce and exploration. Concomitantly, **seapower**—which encompasses **naval power** (that is, seagoing military power consisting of ships, armaments, and troops), commercial and economic interests and capabilities, and imperialistic interests in the ocean—became a concern of governments. The age of sail also introduced new terms and concepts to the naval lexicon. The wind took on greater significance, and terms like **windward,** which refers to the side against which or to the direction from which the wind is blowing, and **leeward,** which refers to the side protected from the wind or the direction toward which the wind is blowing, became more significant in the navigation of ships. The term **on the weather gage** (or gauge) applied to a ship that was windward relative to another ship or any enemy ship. **On the leeward gage** (or lee gauge) applied to a ship that was leeward relative to another ship or an enemy ship. When a ship attempted to sail into the wind, its sails were said to be **close hauled.** The sailing direction of a ship was often expressed according to the method of points, a reference to the 32 points of a compass. Each point is 11¼ degrees, and the various directions, or points, are termed north, northeast, north-northeast, east-northeast, etc. "Six points into the wind" described a vessel whose direction of travel formed an angle of 67½ degrees with the direction from which the wind was blowing. Among the many other terms important to naval commanders in the age of sail were line ahead, broadsides, and blockade. **Line ahead** refers to a column of ships. **Broadside** refers to the side of the ship above the waterline, and **broadsides** refers to the array of guns on one side of the ship or to the simultaneous discharge of those guns. **Blockades** are operations, generally naval, undertaken by vessels or entire navies to preclude the entry of enemy vessels, contraband, or commercial vessels into certain ports or areas. A ship that attempts to deliver goods in spite of a blockade is called a **blockade runner.** Blockading ships seek to **interdict**—that is, to destroy, cut, or damage with firepower—enemy lines of reinforcement, supply, or communication.

Selected Bibliography

One of the most widely used general military history works that deals in part with the era from the use of gunpowder through the eighteenth century is Theodore Ropp's *War in the Modern World,* the first three chapters of which treat, respectively, land warfare from the fifteenth through the eighteenth centuries, naval warfare during the same period, and colonial America through the American Revolution. A second work that deals with the entire period is *The Dawn of Modern Warfare.* The first chapter, written by Gerald P. Stadler, discusses military events, thought, technology, and organizational changes from the early fourteenth century until the early seventeenth century. The second chapter, by Jerome A. O'Connell, is descriptively entitled "Early Developments in Naval Warfare." Chapters three through six, written by Dave Richard Palmer, discuss the period from Gustavus Adolphus through the age of Louis XIV. The final two chapters, written by Albert Sidney Britt III, deal with Prussia and Frederick the Great. The American military experience from its colonial beginnings until the eve of the American Civil War is the topic of Dave Richard Palmer and James W. Stryker's *Early American Wars and Military Institutions.* Other popular works that deal with the American military experience from its origins include Walter Millis' *Arms and Men: A Study in American Military History,* Maurice Matloff's *American Military History,* and Russell Weigley's *The American Way of War.* Naval affairs are the focus of Carlo M. Cipolla's *Guns, Sails and Empires,* Garrett Mattingly's *The Armada,* and Alfred Thayer Mahan's classic, *The Influence of Sea Power Upon History, 1660–1783.*

Two works contain selections from the early modern period. Thomas R. Phillip's *Roots of Strategy* includes extracts from the military writings of Sun Tzu (500 B.C.), Vegetius (A.D. 390), Marshal Maurice de Saxe (1732), Frederick the Great (1747), and Napoleon. A collection of essays edited by Edward Mead Earle is entitled *Makers of Modern Strategy: Military Thought From Machiavelli to Hitler;* its first section deals with Machiavelli, Vauban, Frederick the Great, Guibert, and Bülow. A revised edition, published nearly 45 years later, is edited by Peter Paret.

Works that treat more specific topics within the period under discussion include C.T. Allmand's *Society at War: The Experience of England and France During the Hundred Years War;* Michael Mallett's *Mercenaries and Their Masters,* which discusses mercenary warfare in Italy in the thirteenth, fourteenth, and fifteenth centuries; Charles Oman's *A History of the Art of War in the Sixteenth Century,* a sequel to his earlier classic, *A History of the Art of War in the Middle Ages;* Geoffrey Parker's *The Army of Flanders and the Spanish Road,* which deals with the Spanish *tercio* and the logistical system required to sustain it; C.V. Wedgewood's standard, *The Thirty Years War;* Michael Roberts' two definitive volumes entitled *Gustavus Adolphus: A History of Sweden, 1611–1632;* Leonard Krieger's *Kings and Philosophers, 1689–1789;* Winston Churchill's *Marlborough;* Gerhardt Ritter's *Frederick the Great;* R. Don Higginbotham's *The War of American Independence: Military Attitudes, Policies and Practice, 1763–1789;* and John Shy's *A People Numerous and Armed.* A comprehensive reference to terms, people, and places of the American Revolution is found in Mark Boatner's *Encyclopedia of the American Revolution.*

Technological developments discussed in this chapter can be examined in depth in such works as William W. Greener's *The Gun and Its Development;* Albert Manucy's *Artillery Through the Ages: A Short Illustrated History of Cannon, Emphasizing Types Used in America;* and Sebastian Le Prestre de Vauban's republished *Manual of Siegecraft and Fortification,* which was originally published in the eighteenth century. Information on participants in the profession of arms and organizations is found in Mark M. Boatner's *Military Customs and Traditions;* Elbridge Colby's *Army Talk: A Familiar Dictionary of Soldier Speech;* and Hubert Foster's *Organization: How Armies Are Formed for War.*

Field manuals are the essential sources of modern United States Army doctrine. In addition to the work FM 100-5, *Operations,* primary source information on infantry and artillery employment can be found in FM 7-20, *The Infantry Battalion,* and FM 6-20, *Fire Support in Combined Arms Operations.*

The Age of Napoleon

The term "limited warfare" has been used to describe the military campaigns in Europe that took place from the end of the Thirty Years' War in 1648 until the wars of the French Revolution, which began in 1792. Some writers have referred to the warfare of the period after 1792 as the first of the modern wars, while others have referred to it as the return to total war. No term, however, better describes the last decades of the eighteenth century through the early decades of the nineteenth century than the "Age of Napoleon."

Major Themes of the Age

Even in France, Napoleon was not widely known until 1795, when he restored order in Paris on behalf of the **Directory** (the executive rulers of France from 1795 until 1799), or 1796, when his successes at the head of the Army of Italy were publicized in Paris. But, from the time of his birth in 1769 until long after his death in 1821, Napoleon's life was analogous to and later synonymous with the major events—and especially the military events—of Europe and the world.

The late eighteenth century was an age in which science and the quest to expand man's knowledge of himself and his environment dominated the lives of learned men, and science and learning were beginning to affect the lives of ordinary men. The *philosophes,* intellectuals of the eighteenth century, challenged existing beliefs and existing authority and worked to extend their influence beyond their narrow realm. They met in **salons,** where they shared their ideas concerning the fundamental equality of all men and the right of all individuals to participate in the political systems that affected their lives. Their ideas emanated from the salons in the form of books, tracts, and the spoken word. The dynastic government of France, and to a lesser extent those of other Euro-

pean states, tried to maintain the traditional rule over the people in order to demonstrate the efficacy of long-established practices, but the economic requirements of the lavishly living nobility and the limited resources of the common people brought internal strife to France and wars of reaction from the conservatively dominated powers of Europe. Although Napoleon's exposure to the scientific and philosophical movements of the intellectuals was limited, he attended formal schools where he demonstrated his own intellectual talents: an aptitude for mathematics, the language of science, and geography, the language of eighteenth century power diplomacy. In the early days of the Revolution, he was politically inactive, but his military exploits—his influence at the siege of Toulon and his dispersal of the crowd in Paris with a "whiff of grapeshot"—soon brought him to the attention of high-ranking politicians. His life was analogous to the age in which he lived. Just as the Government brought new energy in the form of mass armies to the pursuit of its interests, Napoleon brought new energy to the troops he commanded and the tasks he undertook.

Between 1799 and 1804, the analogies between individual and state were fused, for Napoleon the commander became also Napoleon the instrument of state. New energy and new ideas created new institutions of war and peace until the wave of reaction overwhelmed the forces of change.

Yet neither the name of Napoleon nor the institutions of war that he influenced have ever been effectively suppressed or forgotten. He is remembered by some as a liberator and by others as a despot, by some as a creator and by others as a destroyer. He is an enigma, and hence an exciting object of study; and in spite of one's judgment concerning Napoleon the man, his influence on the language of the military profession and on the profession itself cannot be denied. His name will not be forgotten, nor will the age that rightly claims his name.

Scale Varies		
1770	Birth of Napoleon Bonaparte: August 15, 1769.	
	Death of Louis XV; Accession of Louis XVI; May 10, 1774	
1780		American Revolution 1775–1783
	† Death of Frederick the Great: August 17, 1786.	
1790	‡ Estates General met at Versailles: May 5, 1789. † Bastille taken: July 14, 1789.	
	† Louis XVI guillotined: January 21, 1793. † Levée en masse: August 23, 1793. † Whiff of grapeshot: October 5, 1795. ✗ Arcola: November 15-17, 1796. ✗ Rivoli: January 14, 1797. ✗ Naval Battle of the Nile: August 1, 1798. † Coup d'état of Brumaire: November 9-10, 1799.	Reign of Terror 1792–1793 Italian Campaigns 1796–1797 Egyptian and Syrian Campaigns 1797–1798
1800	✗ Marengo: June 14, 1800. † Concordat with Pope Pius VII: July 15, 1801. † Napoleon voted Consul for Life: August 1, 1802.	The Consulate 1799–1804
	† Napoleon crowned Emperor: December 2, 1804. ✗ Capitulation of Austrians at Ulm: October 17, 1805. ✗ British naval victory at Trafalgar: October 21, 1805. ✗ Austrians and Russians defeated at Austerlitz: December 2, 1805 ✗ Prussians defeated at Jena-Auerstadt: October 14, 1806. ✗ Eylau: February 8, 1807.	The First French Empire 1804–1804
1810	✗ Friedland: June 14, 1807. ✗ Austrians defeated at Wagram: July 5-6, 1809. † Imperial marriage: April 1, 1810.	Operations in Spain 1808–1813
	† Invasion of Russia: June 21, 1812. ✗ Borodino: September 7, 1812. ✗ Leipzig: October 16–19, 1813. ✗ Abdication: April 11, 1814. † Return to France: March 1, 1815. ✗ Waterloo: June 18, 1815.	War of 1812 1812–1814 Governor of Elba 1814–1815
		Exile on St. Helena 1815–1821
1820	† Death of Napoleon: May 5, 1821.	

Some Key Events During the Napoleonic Era

Participants in the Profession of Arms

During the **French Revolution** (the events of 1789–1792 that immediately contributed to the overthrow of the Bourbon monarchy in France and to its replacement by a republican government), citizens participated in affairs of violence to a greater extent than they had in previous eras. When the more conservative European powers declared war against the Revolutionary French Government, the participation of citizens in affairs of violence was formalized in order to provide France with the forces necessary to halt the coalition against the Revolution. *Levée en masse*, which translates into "mass raising" (of the population), was instituted. The *levée en masse* obligated every able-bodied male citizen, within specified age limits and with some exceptions, to serve in the defense of his country. *Levée en masse* hence connotes a **universal military obligation.** By imposing an obligation for military service on all citizens, a new sense of belonging and community became a component of French society. This sense of community transformed the state, which previously had been characterized by the rule of one class over other classes, into a **nation,** a large political and territorial organization that has common bonds and ideals between those governing and those governed. **National armies** are those armies that serve the nation. Often large, they use **conscription,** the compulsory enrollment of men for military duty, as the principal means of raising troops. The term that refers to the nation that possesses a system of conscription is **nation in arms.** Another term used since the Napoleonic era for the selection of individuals for compulsory military service is **draft.** The term **class** is applied, especially in France, to all those individuals who are eligible for the draft in the same year. An alternative system to the draft is afforded by **volunteers,** those individuals who agree in the absence of obligation to serve in military forces. A volunteer who is primarily motivated by opportunity for material gain is a **mercenary.** Conversely, a **patriot** can be either a conscript or a volunteer, and serves primarily to protect and promote the ideals and beliefs of the political group to which he belongs. Any soldier without a commission who has just become a part of the military service, whether as a volunteer or a conscript, is called a **recruit.**

All military service requires an understanding and acceptance of the concept of **duty,** which refers to the actions and attitudes that are expected and required by legal and moral force because of the holding of a position of responsibility and trust. In most modern societies, the soldier's first duty is to the nation he serves. Duty is closely related to **obligation,** which is a condition or feeling of being bound legally or ethically. An obligation, however, generally applies to an immediate constraint, whereas duty suggests a more general constraint based on profound feelings and beliefs. In the Napoleonic era, some men fought because of obligations to individuals or to contracts, but, increasingly, soldiers fought because of duty to the nation.

When the administrative requirements that governments had to fulfill in order to raise and maintain large standing armies increased beyond the capabilities of the ruler's advisers and a small staff, a department of government, known in different countries as the **War Ministry,** the **War Office,** or the **Department of War,** was established. Headed by the **Minister of War,** these organizations became early bastions of government bureaucracy. The Austrian ministry was called the *Hofkriegsrat* and had authority in political, economic, and judicial spheres. In the United States today, the successor of early war ministries is the **Department of Defense.**

Titles of Rank and Position

With the return of mass armies, many new terms were applied to soldiers, many terms established by tradition were retained, and some new terms came into being. Because of increased participation in the service by society, the Napoleonic example, or, perhaps, the expanding belief in the efficacy of education, greater attention was paid to the study of the military past. In particular, military studies focused on the most distinguished group of commanders of all time, who collectively came to be known as the **Great Captains.** According to many writers, Napoleon was the last of the Great Captains. In the years since Napoleon's rise, **supreme commander,** which admittedly is a term of less distinction, has been applied to the commander designated to lead all the military forces of a **combined command,** a group of forces that come from two or more nations. The highest rank in the French Army, marshal, was used in the Middle Ages, and was first used to denote the highest rank in the Army during the sixteenth century. Abolished in 1792, the rank was restored by Napoleon in 1804 and given to the commanders of corps, the largest components of the Napoleonic army. The corps system, also, was established with the founding of the French Empire in 1804. Prior to Napoleon's coronation as Emperor, the highest ranking officers were referred to as **generals of brigade** and, one step higher, as **generals of division.** These terms applied to rank rather than assignment. Those corps commanders who were not marshals were called **generals in chief,** and by 1815, the term **lieutenant general** was often applied to these individuals. The term "general in chief" also applies to the highest ranking general in an area or organization that has several general officers.

Napoleon's personal headquarters was called the *Maison.* It was made up of a number of **aides** or **aides-de-camp,** who,

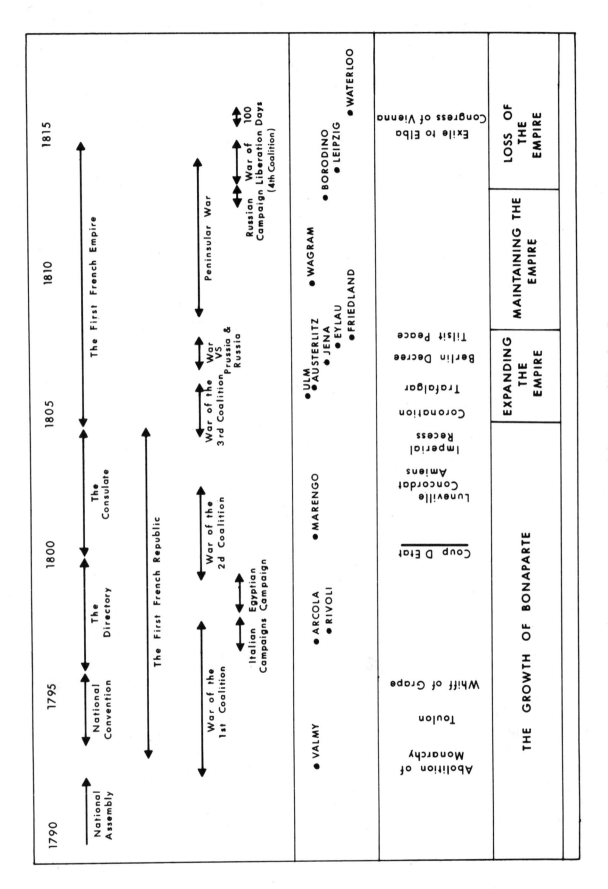

Wars of the Republic and Empire and the Career of Napoleon Bonaparte

as a rule, were general officers. These aides delivered, interpreted, and sometimes supervised the execution of orders. In addition, they were sometimes empowered to negotiate with the enemy. The **general headquarters** handled personnel, intelligence, and security matters. Another headquarters was farther to the rear; there, the **intendant** was responsible for routine logistic and administrative support. Aides and **liaison officers,** whose function was to insure mutual understanding and unity of purpose and action between units employed in a common area or assigned to a common mission, were the essential communication links in the Napoleonic armies. In the Prussian Army, communications were facilitated late in the Napoleonic period by the creation of the **general staff.** Its partial purpose was to provide doctrinal unity for students, who became the **general staff officers** assigned to advise field commanders.

Infantry

During the era of Napoleon, infantry forces were often referred to as **soldiers of the line** or **line troops,** and were further classified as **line infantry** or **light infantry.** Both types were similarly organized and equipped, but the light infantry had a tradition of dash and mobility that encouraged their use as skirmishers and in security missions. Elite troops in the line infantry were known as grenadiers, while the elite of the light infantry were known as *carabiniers.* The remaining line infantry troops were *fusiliers,* and most light infantrymen were referred to as *chasseurs à pied,* which translates literally into "hunters on foot." In 1804, a new company of elite infantry, called *voltigeurs,* was established. The *voltigeur* was too small to be selected as a grenadier, but was supposedly able to keep pace with cavalry at the trot and was habitually used as a skirmisher. A more general term applied to a French skirmisher was *tirailleur.*

Elite troops of other nations were given specific titles, too. Russians called their elite *streltsi;* elite raiders from the Ukraine were *cossacks;* and skilled Spanish irregulars were called *guerrileros.* The term **shock troops** applies to any elite force that is intended to damage heavily any enemy force that it encounters. Shock troops are especially well suited for offensive operations and often rely on **shock action,** the method of attack used by mobile units in which the suddenness, violence, and massed weight of the first impact have a major effect on the outcome of the attack. In the Napoleonic era, cavalry forces were often associated with shock action.

Cavalry

Many terms were used to describe the cavalry forces of the Napoleonic era, and distinctions were often as slight as a difference in uniform or plumage. *Cuirassiers, carabiniers à cheval, gendarmerie,* and *gendarmerie d'élite* were heavy cavalry. The *cuirassiers* wore steel helmets and the heavy breastplate called the *cuirass.* The *carabiniers à cheval* were armed with the carbine and began to use armor in 1810. The *gendarmerie* were the descendants of the *gens d'arms,* or men-at-arms, of the fourteenth century. The *gendarmerie* were still mounted in the Napoleonic army, but rather than being the focal point of battle as the *gen d'armes* had been in the Middle Ages, their principal task was to maintain security in rear areas. Individuals belonging to modern organizations formed in accordance with a military pattern and capable of serving as an auxiliary or diversionary force with civic police duties are termed *gendarmes* or **paramilitaries.** The *gendarmerie d'élite* was an elite heavy cavalry force concerned with security and other military police duties near Napoleon's headquarters. **Dragoons** in the Napoleonic army generally fought mounted, even though they were also trained to fight on foot. *Chasseurs à cheval,* literally "hunters on horse," hussars, and lancers made up the light cavalry, or *chevaux légers.* The hussars were more elaborately dressed than the *chasseurs à cheval,* but in organization and equipment, these two types of light cavalry did not differ. The **lancers** were armed with the lance, which gave them a special advantage in bad weather when firearms were too wet to fire. Against other cavalry armed with the saber, however, they were at a distinct disadvantage in close fighting.

Artillery

Artillery was classified as foot artillery, horse artillery, coast artillery, and artillery for garrison duty in interior fortresses. Contract civilians were employed in some artillery units, but the concept of nation and the unreliability of these contractors led to their replacement by about 1800 with military troops. The artillery included **pontoon battalions,** which gave the army an inherent river-crossing capability. The **munitions,** or weapons and ammunition, for the artillerists were prepared by individuals known as **artificiers.**

Principal Organizations

The rapid conversion of masses of recruits into effective fighting forces required skilled leadership and new organizational arrangements. Between 1792 and 1803, demi-brigades of infantry existed in the French Army. Each **demi-brigade** consisted of one battalion of troops from the old Regular Army and two battalions of recruits raised in response to the *levée en masse.* This combination advantageously merged the steadiness and experience of Regulars with the enthusiasm of the *levée en masse,* but too often each of the battalions of different origin acquired the bad traits of the other. In 1803, Napoleon ordered that the term ''regiment'' be used in lieu of ''demi-brigade.'' After 1804, the smallest force of all arms in the French Army was the **corps,** which was generally commanded by a marshal and consisted of two to four infantry divisions, a brigade or division of light cavalry, and a company or two of artillery, engineers, and service units. None of Napoleon's corps or divisions had a standard organization. The size and composition of each depended upon the capabilities of the commanders, the strategic situation, and the available troops. **Engineers,** or soldiers who specialized in construction, demolition, and the building of roads, bridges, and fortifications, were organized into battalions. **Topographical engineers** were charged with surveying and mapmaking. **Service units** included medical personnel, supply-train troops, communications personnel, military police, and naval units that were used during campaigns on rivers and lakes for security and transportation purposes. The division consisted of two or more brigades, each of which was composed of one or more regiments and one or two companies of artillery. The infantry regiment was made up of four battalions of six companies each, a training battalion of four companies, a medical detachment, a band, and a supply train; in 1809 and 1812, the regiment also contained a cannon company. Each regiment carried a **regimental eagle,** a representation of an eagle that was placed on a standard with distinctive numbers or insignia, and thus served as a means of identification. In addition, it provided the unit with a physical and moral point about which to rally. The **battalion of the train,** or simply the **train,** was organized to bring the system of purchasing and contracting under military control and thereby end the abuses of the civilian supply system, which often enriched the contractors while starving the Army. The term **logistic trains,** or **wagon trains,** refers to the units, and especially their vehicles, whose function was to provide supplies and services to an army. When the vehicles of the trains moved along a common route under the control of a single authority, they formed a **convoy.**

Smaller nonstandardized units of the Napoleoic army have been referred to as combat teams and long-range reconnaissance patrols, but these terms postdate the Napoleonic era. A **combat team** is a temporary organization of small combat units from different arms. In modern usage, a team is generally made up of either a platoon of tanks attached to a company of infantry or a platoon of infantry attached to a company of tanks. The term **long-range reconnaissance patrol (LRRP)** can be applied to any small detachment of infantry or cavalry that performs a reconnaissance mission, generally of at least several days duration, in an area remote from its parent unit. Formal LRRP organizations existed in the United States Army during the Vietnam War. The term **tactical command post** is another modern term that has been applied to the Napoleonic period. It refers to the small group of staff officers, and often aides and security forces, that accompany a commander to forward areas during the conduct of operations. **Command post** or, in French, the *poste de command* or *P.C.,* refers to the location in which a commander and his staff perform their routine military duties, especially in the field.

The term **expeditionary force** refers to those armed forces of a nation that are organized to accomplish a specific mission in a foreign country. Generally, an expeditionary force travels to its destination by sea.

The Technology of the Age

Because nearly all the weapons and other materials of war employed during the Age of Napoleon had been widely used during earlier periods, the technological innovations of the times were not significant. However, in the late eighteenth century, the **Industrial Revolution** (the changeover from hand power to machine power in manufacturing) was underway in France and Great Britain. The resultant mass production of weapons was essential to the provision of arms for the new mass armies.

Striking Weapons

Few striking weapons of significance remained in the Napoleonic era, for gunpowder weapons prevailed on the battlefield. Swords and sabers, however, were carried by cavalrymen, elite infantrymen (*grenadiers* and *voltigeurs*), artillerymen, train troops, and the Imperial Guard. The heavy cavalry generally preferred a long, straight, thrusting sword, while the light cavalry preferred a heavy, curved saber. Lances were used by some cavalrymen, especially after 1809, but their effective use required extensive training. Bayonets, most of which were about 15 inches long, were the most effective of the striking weapons still in use.

Missile Weapons

The most common small arm used by the French infantry was the muzzle-loading, smoothbore, shoulder-supported, black powder, flintlock musket. The most popular model was called the Charleville pattern, or *Charleville,* because most of

French Musket—Model 1763—Caliber .70 (Charleville)

the muskets of its design were made at the Charleville Arsenal in France. About 400,000 of these muskets were exported to the United States during the American Revolution. They became such favorites that when America opened its first arsenal in 1795, an almost exact copy of the *Charleville,* known as the **Model 1795 Musket,** was produced.

The caliber of French muskets varied from .69 to .71. The muskets had a maximum range of over 1,000 meters, an effective range of about 200 meters against formed troops, and an effective range of less than 100 meters against individuals. The **dragoon musket,** or **musketoon,** was shorter and lighter than the infantry musket and was issued to dragoons,

engineers, foot artillery, and *voltigeurs.* The **carbine** was slightly longer than the musketoon, but shorter than the musket. It was used by most mounted and some dismounted troops, and was effective only at short range. Pistols, carried by some cavalrymen, were accurate to only about 10 yards, but were heavy enough to make good clubs. During the **Hundred Days,** the period between Napoleon's return to Paris after his exile on Elba until his departure from Paris after his defeat at Waterloo, many citizens who volunteered to fight once again for Napoleonic France armed themselves with **fowling pieces,** a term referring to any weapon used to shoot fowl and small game.

Eighteenth Century Field Gun

During the late eighteenth century, artillery remained smoothbore and muzzle-loading. However, it profited from the work of General Jean Gribeauval, who became France's first Inspector General of Artillery in 1776. Gribeauval reorganized French artillery into field artillery, siege artillery, garrison artillery, and coast artillery. He used lighter materials in the construction of guns, provided interchangeable parts wherever possible, and improved the **elevating screw,** a large screw that was placed beneath the breech and had small handles to facilitate the turning of the screw. Earlier gunners had changed gun tube elevation with a **wedge block,** a triangular piece of wood that was inserted beneath the breech and hammered toward the muzzle to depress the elevation of the gun tube. Guns were traversed (pivoted) by shifting their trails. A **hand spike** was a bar of wood shod with iron that was inserted into a ring at the base of the trails to facilitate traversing.

A great variety of artillery was employed by the French during the Napoleonic era, the most common being the 4-, 6-, 8- and 12-pounder guns and the 6-inch howitzer. The British used 3-, 6-, 9-, 12-, 18-, and 24-pounders. The range of these weapons was generally less than 1,000 meters with shot, and about half that distance when canister was fired. Trained crews could fire no more than two rounds per minute, for a variety of tools, equipment, and procedures had to be used each time the gun was fired. Powder was placed in some guns with a **ladle,** a cylindrical scoop on a long handle. Most guns in use during the Napoleonic era, however, used powder that was packaged in wool bags, called **cartridges,** and both bag and powder were inserted into the gun as a single unit. The more finely ground primer powder was brought to the gun in a **passing box** and placed in the vent, or touchhole. The **rammer,** a wooden plug on a long handle with marks to show when each part of the load was properly seated, was used to push the powder, wad, and shot down the barrel. The match, which was held in a **linstock**—a pole with a clamp or a slot at one end—was then touched to the powder through the touchhole. After firing, the bore was swabbed: while one member of the crew inserted the **sponge,** a long-handled wooden plug covered with sheepskin or cloth and doused in water, into the bore, the gunner pressed his thumb, protected by a leather **thumbstall,** on the vent to smother any sparks remaining in the tube. The **wormer,** a double iron screw on a long handle, was used when bits of old wad and unburned portions of the cartridge had to be removed from the bore. After the bore was cleared, the **pick,** a short wire with a handle or loop, was used to clear the vent. The pick was also inserted into the vent to poke a hole in the cartridge when bagged powder was loaded.

The **rocket** (a missile that is propelled by hot gases ejected rearward by a burning charge), was first used as a weapon in Europe during the late medieval period and was reintroduced to the British Army by Sir William Congreve during the first decade of the nineteenth century. The Congreve rockets' "red glare" over Fort McHenry in Baltimore Harbor during the War of 1812 was immortalized by Francis Scott Key in "The Star-Spangled Banner." Rockets were again abandoned as a military weapon in the late nineteenth century and again returned in World War II. Today, their successor, the missile, forms an important part of every major nation's arsenal.

Ammunition

Artillery ammunition included shot, hot shot, canister, and grapeshot, all of which had evolved during earlier periods. New developments included the **explosive shell,** which was shot filled with an explosive charge, and **spherical case-shot,** which was a hollow, round shot filled with explosive and musket balls that scattered when the explosive charge was ignited by an adjustable fuse. Spherical case-shot was also called **shrapnel** after its British inventor, Lieutenant Henry Shrapnel. Today the term shrapnel is applied to any munition fragment. After 1790, **semifixed ammunition** (a round shot fixed to a wooden shoe, or **sabot,** that served as the wadding) and **fixed ammunition** (shot, sabot, and powder bag in one unit) were both available.

Mobility

Although some technological innovations were newly available to armies in the Napoleonic Age, the standard items employed relied on the proven methods of the past. Manpower moved armies, and horsepower moved supplies. Limbers and **caissons,** the two-wheeled, horse-drawn ammunition wagons that were attached to the limber, provided the field artillery with mobility, while **hospital wagons** provided mobility to the fledgling medical service. **Pontoons,** or **pontons,** as the term is used in the United States Army, were floats across which the deck of a bridge could be laid. Pontoons were also used as rafts to transport troops or equipment across rivers. **Balloons,** the lighter-than-air craft that gave man access to the air—the third dimension of his environment—had great potential for reconnaissance. However, after using them in 1799 to impress the Egyptians with the military prowess of the French, Napoleon disbanded the balloon companies, which had been a part of the Revolutionary Army since 1794. Although steam engines had been developed in the late seventeenth century, they found no significant military or commercial application until the mid-nineteenth century.

Eighteenth Century Field Gun Attached to Limber

Naval Objects

A variety of names and terms were applied to the many naval vessels, from small, oared **flatboats** used to ferry infantrymen and supplies across rivers, to **capital ships,** a term applied to the warships of the heaviest armament and largest size during any given era. The smallest armed vessels were generally called **gunboats.** Those gunboats capable of navigating shallow waters and used for close-in coastal defense were called **shallow draught (draft) gunboats.** Some terms applied to intermediate-sized ships. **Corvettes,** sometimes called **sloops of war** in the United States, were smaller than **frigates,** which were in turn smaller than **cruisers. Brigs,** short for **brigantines,** were two-masted square-rigged sailing vessels. The largest and most heavily armed—and, in later periods, also the most heavily armored—of the warships were the **battleships. A flagship** was any ship from which the commander of a larger naval unit exercised command. The flagship was generally the largest ship in the command.

During the eighteenth and nineteenth centuries, privately owned ships that were employed by a sovereign in his fight against enemy commerce or warships were called **privateers.** The individuals who manned privateers were called **pirates,** and those pirates from the Barbary Coast of Africa who were authorized by their leaders to prey on the commercial vessels and to harass the shores of Christian nations were called **corsairs.** The practice of attacking merchant ships was known as **commerce raiding.**

Communications

Even though Napoleon dismissed the balloon companies that might have been a boon to communications, he authorized the construction of a mechanical telegraph system that could relay messages from Paris to Strasbourg. The system consisted of a series of stations, each in sight of the next (in clear weather). Each station had a pole of moderate height to which movable arms were attached. Messages were relayed by assigning meaning to the relative position of the arms.

Operations

The late eighteenth and early nineteenth century military lexicon contained a variety of terms that described the activities in which the forces of the era participated. Some activities involved only the highest ranking officers. Other activities were the domain of the increasing number of specialists who served in the complex military organizations of the Napoleonic Age.

High Command

At the higher levels of command, terms like strategic points, line of operations, lateral communications, and concentric advance were particularly meaningful.* **Strategic points** referred to every location in the theater of war that was of great importance to an army conducting operations. **Key terrain** and **critical terrain** are the more current terms applying to

---

*These terms and many other nineteenth century military terms were given precise definition by the Swiss soldier and writer Antoine-Henri Jomini. See especially his *Art of War.* Also see Thomas Wilhelm, *A Military Dictionary and Gazeteer* (Philadelphia, 1881) for further explanations of nineteenth century doctrinal concepts.

strategical points. The **line of operations** refers to the routes used by an army to reach its **objective point,** which is generally a strategic point that is initially held by the enemy. When the defeat of the enemy army became a more decisive element than the holding of key terrain, the term **objective** supplanted the term "objective point." Often, and especially in nineteenth century works, the terms are used interchangeably. An army moving in a given direction with all its parts united, or with its parts within supporting distance, is said to use a **single line of operations** or a **simple line of operations.** When obstacles or distance place the parts of an army out of supporting distance, the army is said to employ **double** or **multiple lines of operations.** All lines of operations are also lines of communication, which are *all* the routes that connect an army with its base. Where combat forces are concerned, however, the term "lines of operations" is generally applied. When supplies, reinforcements, services, and communications are being considered, the terms "lines of communication," **supply line,** and **lines of supply** are more commonly applied. **Lateral communications** refers to those routes that are parallel to the line of engagement or exterior to the enemy's position. A **concentric advance** implies that separated parts of a force will employ multiple lines of operations and seek to concentrate on the battlefield. When two forces employing a concentric advance are stopped before they can concentrate on the battlefield, a **two-front war** or an **engagement on two fronts** results, and the possibility of defeat in detail exists. Similarly, a **war on multiple fronts** is a war in which the parts of an army or coalition are unable, through design or accident, to concentrate prior to making contact with the enemy. When a force is fighting on several fronts, it may try to halt the opposing force's multiple advances by establishing **blocking positions,** which are positions organized to deny the enemy access to a given area or to prevent further advance of the enemy. When blocking positions or other defensive actions are undertaken by nearly all the forces of a command, the command is said to have adopted the **strategic defensive.** When hostilities between belligerents cease through mutual agreement, a cease-fire, truce, or armistice occurs. A **cease-fire** is usually local and often lasts for only a few hours or a few days. A **truce** is similar to a cease-fire but is usually of longer duration; in both cases, the temporary cessation of hostilities occurs for a specific reason—to allow commanders to discuss terms of surrender, to allow units to bury dead, or to allow troops to celebrate religious holidays. During a truce, it is highly unlikely that equipment would be **mothballed**—that is, treated in order to protect it from deterioration during a long period of disuse. An **armistice** is a general cessation of hostilities for a considerable length of time and applies to all belligerents in a given war. Armistices are usually concluded when a **treaty** (a written

agreement between two or more political authorities) is signed by the belligerent sovereigns. Treaties sometimes establish **demilitarized zones (DMZs),** areas in which military personnel and military activities are prohibited.

Tactics

A variety of defensive terms became popular during the Napoleonic era. The **reverse slope defense,** a technique employed by the British under Wellington, describes a defense in which forces await an enemy advance on the side of a hill that is farthest from the enemy. The military authorities of states that border the sea became familiar with **coastal defense,** which refers to all the weapons and techniques employed to protect the seacoast, and especially harbors, from enemy actions. When the variety of defensive forms failed to stop an advancing enemy, **breakthroughs** occurred. If the broken force was subsequently surrounded or encircled, its goal became a **breakout** (from encirclement) and a **linkup** (with friendly forces).

At the lower echelons of command in the Napoleonic army, commanders were often concerned with such matters as open-order tactics, **close-order tactics,** attacks from march column, standing operating procedures, and the missions that are today referred to as covering, protecting, and screening missions. The term **open-order tactics** refers to the formations in which soldiers either were irregularly spaced, as when skirmishing, or were formed in ranks that were about three yards apart. This contrasts with **close-order tactics,** in which the ranks were separated by about half a pace. The different formations that French units used also were referred to as thin orders *(ordre mince),* deep orders *(ordre profond),* and mixed orders *(ordre mixte).* The **thin order** was a linear formation of the type used by most eighteenth century armies. In the **deep order,** a column of troops, rather than lines of troops, engaged the enemy, and the advancing column relied more on shock than firepower for success on the battlefield. The **mixed order** referred to the simultaneous employment of columns, lines, and skirmishers, and could be adapted to various enemy formations and terrain. (See diagram, page 69.)

When a unit does not have time to **deploy** (move into a position, known as an **attack position** or **assault position,** where it awaits engagement), it must **attack from march column.** Units should have time to deploy, however, for it is a **standing operating procedure,** (a prescribed activity of a unit) to have forces assigned to determine the location of enemy forces and to give the main body time to deploy. These units are referred to as the advance guard, and in the modern United States Army, they are assigned the mission of cover-

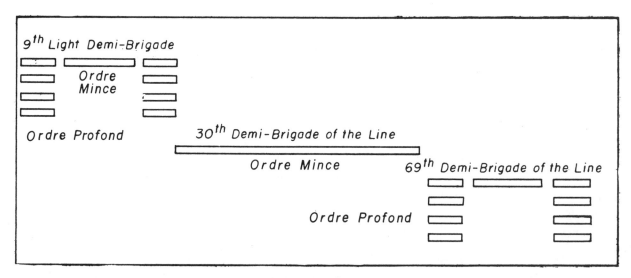

The *Ordre Mixte*

ing, protecting, or screening on behalf of the main body. The term **cover** implies that the advance or covering force will be able to engage an enemy force successfully and preclude the enemy's advance for a specified time. **Protect** implies that a security force will be able to engage an enemy force successfully, and will be able to impede his advance for an unspecified period unless the enemy is reinforced. **Screen** implies that a security unit will preclude the enemy from gaining precise information about friendly dispositions. Covering, protecting, and screening missions are routinely assigned to modern United States Army reconnaissance forces. When such a mission, or any other explicit instruction is given to a unit, it often includes **implied missions,** which are tasks that, although necessary, are unstated in verbal or written orders. **Mission-type orders** are very broad in scope, and leave the commander receiving them a great deal of flexibility; normally, they contain a number of implied missions. A commander considers all missions when he makes an **estimate of the situation,** which is the reasoning process that is used to consider the circumstances that affect the military situation and assist in choosing the best course of action. The term **estimate,** which refers to the computation of expenses to be incurred by all or parts of an army during a specified period of time, should not be confused with the term "estimate of the situation."

Revolutionary Warfare

The activities discussed above generally occur between the forces of legitimate governments. When a force employs political action and violence to threaten the existence or extent of power of a legitimate government, however, revolutionary warfare results. The activities of revolutionary warfare include violent means (such as terror, murder, arson, riots, guerrilla war, and conventional military operations) and political means (such as propaganda, persuasion, political competition, psychological war, mass organizations, and infiltration). Carefully blended, these activities interact to produce the power needed by weak or conquered peoples to seize political control from hostile occupying powers, or by rebellious minorities to seize control from their rulers. In the past, such activities have taken the name **partisan war, small war, insurgency,** and **people's war.** Those who support such activities are invariably guilty of **treason,** which is the offense of attempting by overt acts to overthrow the government to which the offender legitimately owes allegiance.

During the late eighteenth century, revolutionary warfare was prevalent in France and in the many parts of Europe that were occupied by Napoleonic armies, notably Spain and the Tyrol. Revolutionary warfare also prevailed in the 1790s in the area of western France known as the Vendée. Many of the same factors that existed in those regions have contributed to making the second half of the twentieth century especially susceptible to revolutionary warfare. Today, revolutionary warfare is often referred to as an insurgency or, simply, as **low-intensity conflict.** These forms of warfare are most prevalent in modern **developing nations**—nations that are progressing beyond a traditional society, and thus are experiencing the turbulent process of economic, social, military, political, and psychological change.

As a part of low-intensity conflicts, military forces often conduct the following activities: intelligence operations, psychological operations, tactical operations, and advisory assistance. **Intelligence** refers to the product that results from the collection, evaluation, analysis, integration, and interpretation of all available information that is significant to military planning and operations. **Intelligence operations** are those activities that involve the creation of intelligence. Many of the functions performed by modern **intelligence services,** which are those agencies involved in intelligence operations, were formerly accomplished by the *coup d'oeil,* the art of distinguishing by a rapid glance (wink of the eye) the weak points of an enemy's position, and of discerning the advantages and disadvantages offered by an area for a camp, a battle, or any other military activity. Intelligence also involves **reconnaissance,** the examination or observation of an area, territory, or airspace, either visually or with mechanical, electrical, or photographic devices, to secure information about the terrain, the enemy, or the results of friendly activities. Reconnaissance is often performed from **observation posts,** which are positions that afford good visibility and have good communications potential. **Pickets**—soldiers designated to perform reconnaissance duties—can be positioned in fixed observation posts or mounted. Originally, pickets were mounted on horses; in later periods, horses were replaced with vehicles or aircraft. Those mounted on horseback were known as **cavalry pickets.**

Psychological operations are other activities that military forces might be called upon to perform during low-intensity conflicts. **Psychological operations (PSYOPs),** or **psychological warfare,** consist of the broad range of political, military, economic, and ideological actions conducted to influence the opinions, emotions, attitudes, and behavior of foreign target audiences.* PSYOP themes and messages use words familiar to the target audience and aim to be clear, easily understood, and repeated frequently. Though particularly useful in low-intensity conflict, PSYOPs are important dimensions of war at all levels. Napoleon, for example, used the press effectively in his conduct of psychological operations.

In low-intensity conflicts, tactical operations are often offensive in nature and are characterized by a high degree of mobility. The object is to find, fix, destroy, or capture insurgents. Such operations include raids, reconnaissance in force, and deliberate and hasty attacks. The military forces in low-intensity conflicts are largely concerned with **developing the situation**—that is, with learning more about enemy dispositions and intentions. Once an enemy force is encountered, conventional forces can **pile on,** or bring overwhelming numbers and munitions to bear against it.

Advisory assistance is another activity in which military forces have become increasingly involved in the twentieth century. **Advisers** manage resources, assist in the education and training of troops, program and monitor foreign military sales and equipment that is furnished as gratis aid, and coordinate the activities of those forces providing assistance with those of the host country.

*See United States Army, FM 33–1, *Psychological Operations,* August 1979.

Selected Bibliography

The best single volume that focuses on the military events of the Napoleonic period is David Chandler's *The Campaigns of Napoleon.* Another single volume that focuses on the military aspects of that age is Albert Sidney Britt III's *The Wars of Napoleon. A Military History and Atlas of the Napoleonic Wars,* prepared by Vincent J. Esposito and John Robert Elting, is even more specific in its treatment of military campaigns and battles, and contains a good annotated bibliography of battles and memoirs. Yorck von Wartenburg's *Napoleon As a General* is a classic work on the period, although readers should be aware that it is dated and has a Prussian bias.

Among the more popular works on the life of Napoleon are Felix Markham's *Napoleon,* Octave Aubry's *Napoleon,* Vincent Cronin's *Napoleon Bonaparte: An Intimate Biography,* and Owen Connelly's *Epoch of Napoleon.* Napoleon's early military training and his career through the Battle of Lodi form the focus of Spenser Wilkinson's *The Rise of General Bonaparte.* French tactics of the period are covered well in Robert S. Quimby's *Background of the Napoleonic Wars,* while Prussian tactics, as

well as the role of Prussia in the Napoleonic era, are presented in detail in Peter Paret's *Yorck and the Era of Prussian Reform.* An excellent general history of the period is Geoffrey Bruun's *Europe and the French Imperium, 1749–1814.* Naval affairs of the period are covered in Piers Mackesy's *The War in the Mediterranean 1803–1810,* Alfred Thayer Mahan's *The Influence of Sea Power Upon the French Revolution and Empire,* and Oliver Warner's *Trafalgar.* The exploits of Napoleon's 26 marshals are the topic of R.F. Delderfield's *Napoleon's Marshals.*

Two recent works that afford an excellent opportunity to view the Napoleonic experience in the context of warfare through the ages are Michael Howard's *War in European History,* which provides a brief but excellent examination of war's connection with and effect upon technological, social, and economic change during the last 1,000 years, and John Keegan's *The Face of Battle,* which focuses on the personal aspects and effects of battle by examining memoirs and other first-person reports from the Battles of Agincourt (The Hundred Years' War), Waterloo, and the Somme (World War I).

From Napoleon Through the American Civil War

6

The 50 years that followed Napoleon's defeat at Waterloo in June 1815 cannot be understood without an appreciation of the impact of Napoleon and Napoleonic warfare on Western civilization. Themes that were either capitalized upon by Napoleon or encouraged in response to his policies dominated governments, economics, societies, attitudes, and military institutions for at least the next half century. During that half century, several wars occurred in Europe and America. While the level of violence in the European wars did not approach that reached during the Napoleonic era, the American Civil War did reach the Napoleonic level of violence and destruction. For this reason, and because of tactical and technological innovations, the focus of this chapter is the Civil War.

Major Themes of the Age

Nationalism and Conservatism

Perhaps the greatest force unleashed in the Napoleonic era was **nationalism,** the impetus derived from the attitudes, feelings, or beliefs of people of common culture, language, background, and interests that causes them to exalt their heritage and to willingly sacrifice their lives to preserve that heritage for their progeny. Napoleon did not use armies of professionals and mercenaries who represented kings and fought on behalf of rulers and their subjects. Rather, his soldiers were conscripted from and represented the masses, who were imbued with revolutionary, patriotic, and nationalistic ardor that far exceeded that known in earlier modern societies. In the decades after Napoleon's fall, those nations in which the people were united by strong nationalistic bonds tended to grow in power and influence, while those nations whose subjects were made up of many nationalistic groups tended to be hampered by internal strife and external threat. Even in the United States, two divisive nationalistic forces, which might be described as "Northernism" and "Southernism," contributed to the internal strife known as the American Civil War.

Although nationalism was a force of change, the victors of the final Napoleonic campaign were governments that were old in comparison with the defeated, upstart French Empire. Hence, these old governments and ideas dominated the **Congress of Vienna,** the gathering of the diplomats of the European powers in Vienna, Austria in 1814 and 1815 that decided the future of post-Napoleonic Europe. **Conservatism,** which is the force representing the attitudes, feelings, and beliefs of those who advocate the preservation of the established order and who tend to view proposals for change critically and usually with distrust, prevailed in Vienna. Nationalism, liberalism, democracy, and industrialization, however, were too pervasive to allow conservatism to remain dominant for long.

Democracy and Liberalism

Democracy, which is the belief in rule by the people, was a common theme of the eighteenth century *philosophes.* It found expression in the American Revolution, the French Revolution, the charters of the fledgling government of the United States, and certain political parties in each of the major Western European powers. Democrats believed in rule by an entire people (in practice, by all white, male adults through their representatives), and abhorred the existence of social and economic classes and the granting of special privileges to any favored individual. The height of the democratic movement was reached in the United States during the administration of Andrew Jackson (1829—1837).

73

1815	⚔	Congress of Vienna Waterloo: June 18, 1815		
1820				
	†	Death of Napoleon: May 5, 1821		
1825	†	Uprising in Russia: December 1825		
1830	†	Revolution in France: July 1830		Presidency of
	†	Reform Bill in England: 1832		Andrew Jackson
1835	†	Zollverein in Germany: 1834		1829–1837
1840				
1845				Mexican War 1846–1848
1850				Revolutions in Europe 1848–1849
	†	Death of Wellington: September 14, 1852		
1855	⚔	Balaclava: October 24, 1854		Crimean War 1853–1856
				Napoleon III, Emperor of France, 1852–1870
1860	⚔	Fort Sumter: April 12, 1861		American Civil War 1861–1865
	⚔	Gettysburg: July 1-3, 1863		Prussian-Danish War
1865	⚔	Surrender at Appomattox: April 9, 1865		1864

The Half Century After the Defeat of Napoleon

Akin to the democratic movement, but in many respects distinct from it, was liberalism, a term that comes from the Latin word *liber,* meaning "free." **Liberalism** encompasses the attitudes, feelings, and beliefs of those who seek freedom from both foreign domination and the despotic rule of their own government. Liberals objected to governmental restrictions on trade and sought constitutional guarantees from governments to insure the rights of individuals. Largely a movement of the middle class, liberalism had considerable influence in France, Great Britain, and the United States.

Industrialization and Growth

The magnitude and influence of the Industrial Revolution increased dramatically during the early nineteenth century. Machine power and interchangeable parts were essential to the arming, equipping, and maintaining of the mass armies that were formed by and in response to Napoleon. The movement continued after the defeat of Napoleon, and domestic applications of machine labor were readily found and widely adopted. Industry encouraged the growth of both railroads and cities. It also encouraged the search for and exploitation of raw materials, and the exchange of raw materials for finished goods. Especially in the United States and Africa, industrialization meant the rolling back of the **frontier,** that imaginary line which separates the civilized world from the uncivilized, the known world from the unknown. The resultant growth of civilization was attributed by many Americans to **Manifest Destiny,** the mythical force, often regarded as inevitable, that leads a people, race, or nation to expand its territory to boundaries that appear to be natural.

The 50 years after Napoleon's fall were an extension of and a reaction against the forces that had marked the Napoleonic Age. Although military institutions were internally dominated by conservatives, liberals, working in conjunction with conservatives, were able to drastically reduce the size of armies from their wartime levels. Perhaps the sole enduring military benefits of the half century that followed Napoleon are the writings of the Swiss soldier Antoine-Henri Jomini and his Prussian counterpart, Carl von Clausewitz. Each possessed a profound understanding of history and firsthand experience gained in the Napoleonic wars. Jomini's focus was operational; hence, his writings have become dated. Clausewitz' focus was philosophical; hence, his writings retain their vitality.

Armies improved technologically, but the victorious armies of the age, when called upon to perform their special functions of restoring order and winning wars, prevailed either because of overwhelming power (as in the internal revolutions in Europe or the Indian Wars in the United States) or in spite of blunders that were exceeded only by the blunders of their opponents (as in the Crimean War or the American Civil War). The bravery, heroism, deprivation, and sacrifice characteristic of individual soldiers in all ages remained a part of the history of war, but in terms of leadership, professionalism, and doctrine, conservatism affected the military art in a negative manner.

Participants in the Profession of Arms

As in previous wars of considerable magnitude, many terms were used to describe the tasks and positions of individuals who participated in the American Civil War. As in other wars, too, some terms retained the meanings assigned to them in former times, some found new meaning, and some came into use to suit the peculiar needs of the day. Some terms might be called "veterans" after the **veteran** soldiers of considerable experience on campaign and in battle.

Throughout the war, **enlistment,** or the voluntary commitment to military service, continued to be a popular means of obtaining soldiers. However, conscription was used after 1862 by the North, and later by the South. The draft proved to be an inequitable system of calling citizens to arms; however, because **substitutes** (individuals who served in the place of someone who was called) and **commutations** (payments to the government in lieu of service) were alternatives to service for men drafted from families of means and influence. The soldiers who fought for the South were known as **rebs, rebels,** or **Johnny Rebs** by northerners, who generally felt that a southern soldier was rebelling against his rightful government. Those individuals who lived in a border state, but who nevertheless supported the southern cause, were known as **sympathizers.** Conceivably, some northern sympathizers lived in the South, but the term "sympathizer" generally connotes a southern sympathizer in a border state. Contraband is another term that was widely used in a specific sense during the Civil War. In general, **contraband** refers to any goods or merchandise that is declared illegal by a legitimate power under certain conditions, often war. During the Civil War, however, a contraband was a Negro slave who had escaped to or was brought within Union-controlled territory.

As in the past, the armies of the Civil War were often plagued by camp followers: those civilians such as wives, mistresses, prostitutes, and children, who traveled behind the army on campaign and made camp in the vicinity of the army's camp. Salesmen, called sutlers, were generally among the camp followers; often, sutlers were authorized to establish shops, or general stores, within army encampments. Those

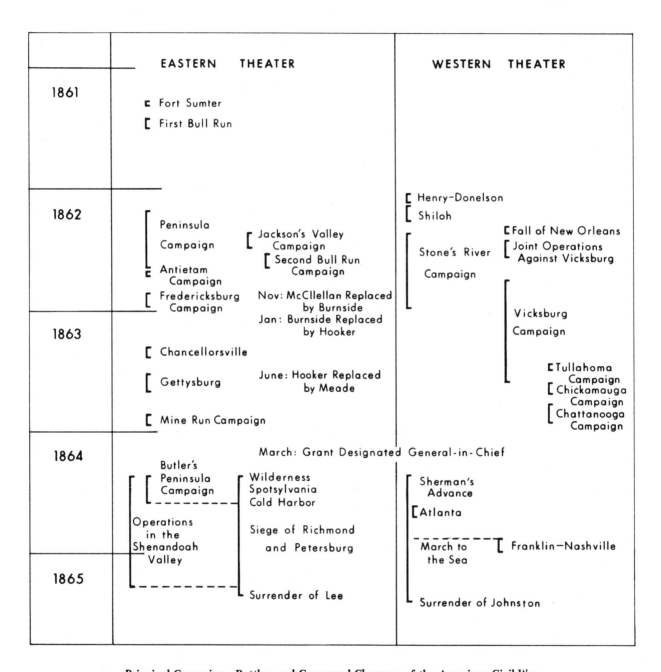

Principal Campaigns, Battles, and Command Changes of the American Civil War

civilians who were driven from their homes for ideological reasons or because their homes were destroyed or threatened by war were called **refugees.** The areas in which refugees found temporary shelter and food were called **refugee camps.**

Titles of Rank and Position

A great deal of confusion exists regarding the use of terms that refer to nineteenth century military rank. Part of the confusion results from imprecise usage or incomplete understanding on the part of authors. In their defense, however, titles were often misused in official parlance, changed from time to time, or retained as one name in official usage and as another name in popular usage. Hence, some difficulty is inherent in any study of the Civil War **chain of command,** a term that refers to the succession of commanding officers, from a superior to a subordinate, through which command is exercised. The term **command channels** refers to the same succession of command.

A **deputy commander** assists the commander and commands in the absence of the commander, but is not a member of the chain of command. A deputy commander is often referred to as the **second-in-command** and, especially in the British and Australian Armies, the second-in-command often is referred to by the acronym **2-I-C.** Another term used to refer to the second-in-command, especially with reference to ships, posts, and units from company to brigade size, is the **executive officer,** who is also sometimes called the **executive** or, more commonly, the **exec** or **XO.**

At the head of the chain of command of all United States forces is the **Commander-in-Chief,** who by constitutional authority is the President of the United States. The title, however, is often applied to the uniformed officer of highest rank in the army. In the American and British Navies, the title is also applied to the senior officer in command of a fleet. In 1821, the officer of highest rank in the American Army was designated the **Commanding General of the Army.** This position was held initially by Jacob Brown and subsequently by Winfield Scott and Henry Halleck. In 1864, Congress created the higher position of **Commanding General of the Armies of the United States.** Ulysses S. Grant was the first assigned to this position, making him the senior commander in the Army. A new advisory position, the **Chief of Staff of the Army,** was created to better describe the office that Halleck was filling in the War Department. The position of Commanding General of the Armies of the United States is no longer used; today, the Chief of Staff of the United States Army is the senior uniformed officer in the chain of command.

Field commanders are those members of the chain of command who direct large troop units engaged in field duty. Sometimes, however, the term refers to any large unit commander whose place of duty is removed from the highest headquarters of his service. All field commanders are general officers. During the Civil War, considerable confusion existed in the ranks of generals because of the system of **brevetting,** the awarding of rank as an honorary title for gallant or meritorious action in time of war. In theory, brevet rank had none of the authority, precedence, or pay of real rank, but regulations were so vague that controversies arose from the time the system was promulgated by an Article of War of 1806 until the article was repealed in 1869. Of the 1,978 generals appointed either during or immediately after the Civil War, about 1,700 were of brevet rank. Until the appointment of Grant as Commanding General of the Armies of the United States and his designation as a lieutenant general (whose insignia consisted of three stars), the only authorized generals were **brigadier generals** (one star) and **major generals** (two stars). The term **brigadier** sometimes refers to a brigadier general; in the British Army, however, it refers to the officer who commands a brigade.

The authorized ranks and order of precedence of Union officers and non commissioned officers during the Civil War were officially as follows:

1st	Lieutenant General	11th	Sergeant Major
2nd	Major General	12th	Quartermaster
3rd	Brigadier General		Sergeant
4th	Colonel		of a Regiment
5th	Lieutenant Colonel	13th	Ordnance Sergeant
6th	Major		and
7th	Captain		Hospital Steward
8th	First Lieutenant	14th	First Sergeant
9th	Second Lieutenant	15th	Sergeant
10th	Cadet	16th	Corporal

Officer Rank and Title in the Union Army[1]

In the Confederate Army, general officers were referred to as either brigadier generals or **full generals,** a term that today refers to the highest rank attainable in peacetime (four stars). Although this rank is formally known simply as "general," all general officers are addressed as "general."

The officers below general officers were known as **field officers** or **field grade officers,** and consisted of the ranks of colonel, lieutenant colonel, and major. In the sixteenth century, the **major** was the officer who commanded a battalion and was therefore known as a sergeant major ("big sergeant"). In the modern United States Army, battalions are generally commanded by lieutenant colonels, and most majors serve as staff officers rather than **line officers** (that is, members of the chain of command or **line of command).** **Company grade officers** are those who hold the rank of captain or lieutenant. The term **company officers** can also refer to all captains and lieutenants, or it can refer only to those company grade officers who are assigned to duty in a company organization. The word **subalterns,** which is applied to all lieutenants, is rarely used in today's American Army.

During the Civil War, the rank structure of the Navy differed entirely from that of the Army. Those officers comparable to general officers were called **flag officers;** they composed the ranks of **admiral, rear admiral,** and **commodore.** Ranking beneath the commodore in descending order were the **captain, commander, lieutenant commander, lieutenant,** and **ensign.**

Many duty positions in the Civil War armies were also assigned specific titles. The senior engineer officer within each army, for example, was called the **chief engineer,** and the senior artillery officer, who organized and assigned duties to all the artillery units of each army and advised his commander on the employment of artillery, was called the **chief of artillery.**

Those staff members who a general in command chose to have report directly to him, rather than through his chief of staff, became part of the commander's **personal staff.** Those members of the staff who worked under the direction of the chief of staff constituted the **coordinating staff.** Today, the coordinating staff in a division and higher headquarters is called a general staff. The **provost marshal,** the officer who supervised the activities of military police and advised his commander on military police matters, was a member of his commander's personal or coordinating staff. Similarly, the **surgeon,** the senior medical officer who was in charge of a commander's medical organization and advised his commander on medical matters, was also a member of his commander's personal or coordinating staff. The **commissary officer,** who dealt almost exclusively with the supply of food, was usually a member of the coordinating staff. In today's Army, the provost marshal and surgeon are a part of a commanding general's **special staff,** which is made up of those staff officers, usually technical specialists, who are not a part of the general coordinating staff or personal staff.

In the modern Army, commanders below the rank of general have an executive officer rather than a chief of staff, and a special staff rather than a coordinating or personal staff. Such staffs, found at battalion and brigade level, are generally made up of: an **adjutant,** who is responsible for all official correspondence except combat orders, for personnel and other records, for the distribution of orders, and for other administrative duties; an **intelligence officer,** who handles all intelligence services for his commander; an **operations officer,** who prepares combat orders, training plans,

and organizational changes; and a **supply officer,** who procures, stores, and issues supplies needed by his unit. The officer who performs the tasks of the adjutant at the headquarters of a general is called the **adjutant general.**

The titles of positions filled by enlisted soldiers in the Civil War were little different from titles used in earlier periods. Infantrymen, artillerymen, and engineers continued to perform in traditional roles. **Teamsters** (soldiers who drove a team of horses or any other vehicle), **pontoniers** (engineers who handled the pontons during river crossing operations), and **pioneers** (engineers especially trained for road building, temporary bridging, and demolitions) were all specialists who had been well known to armies of previous eras.

Professional Institutions

The establishment and growth of professional institutions occurred slowly. Most of the institutions associated with military professionalism in the United States today were either fledgling institutions, like the United States Military Academy founded at West Point, New York in 1802, or simply did not exist prior to the Civil War. The survival of the Regular Army itself was in jeopardy, and those who sought to reform the military institutions of the nation received little support from Congress, and often less support from the executive branch of the Government. The need for officers educated in military skills, for example, was recognized by only a few, and **schools of practice,** which would teach such subjects as gunnery and road construction to young commissioned officers, were proposed. One such school was established at Fort Monroe, Virginia, in 1826, but it was short-lived. In response to the need for **junior officers,** a term that includes all officers below the grade of major, the South relied on **military academies,** schools intended to provide preparatory training essential to commissioning. However, few officers who served in the Civil War had received significant formal military schooling. Many relied on **tactical manuals,** that is, handbooks prepared by the War Department and intended to provide the most current and reliable guidance available for the accomplishment of tasks that an officer would be most likely to encounter. The enlisted soldier learned by rote and by experience. The meager body of knowledge that he was able to understand and practice was called the **school of the individual soldier.**

Morale

In terms of responsibilities and expectations, the participants in the profession of arms during the American Civil War differed little from earlier generations of soldiers. **Morale,** defined as the soldier's attitude toward military life and everything associated with it, remained a necessary and essential component of combat power. That morale was not as good as commanders desired was indicated by the persistence and frequency of **desertion** (deliberate absence from a place of duty without intention to return), **flogging** (physical abuse administered for breaches of discipline), **straggling** (failing to keep up with a unit on the move), **fraternization** (the unauthorized exchange of anything from greetings to supplies with the enemy), **relief** (removal from command or other positions of authority), and **courts of inquiry** (tribunals called to investigate possibilities of dereliction of duty). Along with morale, *casualties*—those soldiers whose duty cannot be performed because of injury (either **hostile,** that is, caused by the enemy, or **nonhostile**), death, or capture—detracted both from a unit's combat power and from its combat strength. **Combat power** refers to a unit's moral *and* material measures of strength. **Combat strength** refers only to the quantifiable measures of a unit's strength, such as numbers of men and supporting weapons.

Those men who were seriously **wounded** (injured by a weapon of war), rarely returned to the ranks. A soldier who was injured as a direct result of hostile enemy activity was said to be **wounded in action,** or **WIA.** A soldier who was killed as a direct result of hostile enemy activity was said to be **killed in action,** or **KIA.** Those who were captured were called **prisoners of war,** or **POWs.** Prisoners of war were either **ransomed** (set free after payment of a fee determined by the captors), **paroled** (set free with the promise not to fight again), **exchanged** (returned to their own lines upon the release of an enemy captive), or **incarcerated** (placed in physical confinement). In today's terminology, places in which POWs are detained are called **POW cages, POW camps,** or **concentration camps.**

To offset the deleterious effects of the hostile environment that accompanies war, activities to increase morale were encouraged and supported by the chain of command. For example, during lulls in the fighting, units occasionally conducted **reviews,** or ceremonies designed to honor some official (who in nearly all circumstances was present as the **reviewing officer,** that is, the officer before whom the troops marched and rendered salutes) or some individual or individuals from the unit conducting the review. In addition, **furloughs, leaves of absence,** or in today's parlance, simply **leaves**—terms that refer to authorized absences from one's place of duty—were granted to enhance morale. Morale was also a factor in allowing troops that had been on extended field duty to move into a **garrison,** that is, a town or camp that was occupied by troops, for a considerable period of time. In garrison, the troops could **stack arms,** a slang term

that meant "relax" but officially referred to the procedure of stacking shoulder weapons in a group, upright, with the butts on the ground.

Morale was also enhanced by *esprit de corps,* which translates from the French into "spirit of the group" and refers to the pride of individuals in their unit, service, and nation. During the Civil War, **distinctive insignia,** cloth patches or metallic emblems that uniquely identified a soldier as a member of a particular unit, were introduced to the United States Army. The peculiar dress and embellishments adopted by units of earlier armies had served similar purposes—they allowed quick identification of an individual and his unit and contributed to the soldier's pride in and identification with his unit. Flags, banners, standards, colors, jacks, pennants, pendants, pennons, and guidons were also adopted as symbols to enhance esprit and contribute to morale. A **flag** is a general term that refers to a piece of fabric of distinctive design that is used as a symbol of a unit, organization, state, or nation, or that serves as a signaling device. A flag is usually displayed hanging free or from a staff, or **halyard,** to which it is attached at one edge. A **flag of truce** is an all-white flag that signals an intention to convey to the enemy some nonhostile communication, often concerning surrender. The term **banner** is a synonym for flag, and is frequently used in situations involving emotional ties and appeals. A **standard** is often an elongated flag associated with an individual, a party, or a cause. The term **colors** (nearly always plural) generally refers to the national flag or flags, one of which is the flag that a military unit is authorized to display. Colors are displayed in garrison, and during the Civil War were carried in the field as well. They were both a physical and moral rallying point on the battlefield, and disgrace accompanied the unit whose colors were lost without the greatest of sacrifice. To **strike the colors,** that is, to be forced to remove the colors from a garrison or fort to preclude their capture, was a symbolic action that indicated a cessation of resistance from within. A **jack** is the flag indicating nationality that is flown from the bowsprit of a ship. **Pennants** and **pendants**—the latter spelling is favored in Great Britain—are triangular flags used to signal and identify units and modern athletic teams. A **pennon** was generally a very narrow flag that was often hung from a lance. A **guidon** is generally a swallow-tailed flag carried to identify a company or a cavalry troop. Flags in their various forms have been associated with military activities since at least the early Middle Ages.

Principal Organizations

The military organizations found in the Union and Confederate Armies during the American Civil War were similar in name and size to most earlier organizations, especially at corps and lower echelons. At higher echelons, like the War Departments, organizations were frequently renamed, new ones were created, and old ones were moved, combined, and occasionally deactivated.

Adding to the organizational confusion, the land forces of the Civil War armies were divided into **territorial (or geographic) organizations** and **operational organizations.** Moreover, in a few cases, an organization within one of these groups had the same name as an organization in the other group. The territorial organizations included, from smallest to largest, military posts, territorial districts, territorial departments, and territorial divisions. The **department** (such as the Department of the Ohio, the Department of the Cumberland, and the Department of the Tennessee) was the most common territorial subdivision, and generally gave its name to the army operating within its boundaries. Boundaries of the departments were changed in response to the military situation. A total of 44 military departments existed at one time or another during the war, but when Grant assumed command of the armies in March 1864, there were only 19 departments. The **territorial division,** or **military division,** was the largest of the territorial organizations. The boundaries and existence of the territorial divisions, like those of the departments, changed from time to time. The commander of the military division commanded all the troops within his division. For example, Grant was placed in command of the Military Division of the Mississippi, organized in October 1863, and Sherman was given command of the Middle Military Division, which was organized in August 1864.[2]

The largest operational organization was referred to as an army, which on the Union side generally took the name of its department. The department, in turn, was generally named after the area's major river—for example, the Ohio, the Potomac, or the Tennessee. The Confederates usually named their armies after the states or regions in which they were active—for example, the Army of Northern Virginia or the Army of Tennessee. Sixteen armies existed at one time or another during the war on the Union side, while 23 existed on the Confederate side. Corps were the next smaller operational organization. However, corps organizations did not exist on the Union side until after the first Battle of Bull Run (June 1861), and did not become a part of the Confederate system until November 1862. When Union General Burnside reorganized the Army of the Potomac in the fall of 1862, he combined his corps into **grand divisions,** but the grand divisions were eliminated when General Hooker succeeded Burnside in January 1863. Corps were made up of two or more divisions, and divisions were composed of two or more brigades. Brigades were made up of about five regiments.

Northern brigades were designated by numbers, while Confederate brigades were known by the names of their commanders or former commanders.

Volunteer infantry regiments were generally composed of 10 numbered companies. The regular infantry regiment consisted of two or more battalions, with eight companies to a battalion. Each cavalry regiment or artillery regiment usually comprised 12 companies. Cavalry companies within the regiment were lettered in alphabetical order, with the letter "J" omitted. Some tales suggest that "J" was omitted because of a "J Unit" whose members either disgraced it beyond redemption or were all killed in a single battle, but the unromantic fact is that "J" is not used because it is too easily confused with "I" in writing.[3] The cavalry regiments of the period were also referred to as **mounted regiments,** and those armed with the rifle were known as **mounted riflemen regiments.** Some regiments, known as **composite regiments,** were made up of troops from more than one arm. The regiment was the organization with which the soldier most closely identified. Volunteer and militia regiments were named after their state of origin and given a numerical designation within the state—for example, the 20th Maine or 9th Massachusetts. Many regiments were known by colorful nicknames, such as the "Teacher's Regiment," the "Persimmon Regiment," and the "Iowa Temperance Regiment," whose members allegedly did not touch, taste, or handle any alcoholic beverages—except at "such times as they were under the overruling power of military necessity."[4]

Like the territorial and operational organizations of the Civil War, the War Department was characterized by a lack of standardization and frequent change. For example, in the Union War Department, the highest echelons of the technical services were organized into bureaus, departments (not to be confused with territorial departments), and corps (not to be confused with operational corps). **Bureaus** were specialized administrative units of the War Department. Some, like the Military Information Bureau and the Bureau of Military Justice, used the word "bureau" in their official titles. Other bureaus, or technical services administrative units, were referred to as **departments;** these included the Subsistence Department, the Quartermaster Department, and the Medical Department. Other technical service administrative units were referred to as **corps;** these included the Corps of Engineers, the Military Telegraph Corps, and the Military Balloon Corps. The **bureau chiefs** were the heads of the various technical service administrative units. Working in concert with the general staff, they determined policy and provided the means by which the Army was organized and sustained.

The Technology of the Age

The Industrial Revolution, the benefits of commerce, and man's continuing quest to improve the tools of his age all contributed to the significant changes that occurred in the manufacture and design of the implements of war during the 50 years between Waterloo and the close of the American Civil War. The developments that had the greatest impact on the conduct of war were railroads and rifling. Railroads provided a reliable means of moving large numbers of troops and supplies over great distances in a short and predictable period of time. **Rifling,** which refers to the helical grooves cut in the bore of a gun that impart spin to the projectile when the piece is fired, increased the range and accuracy, and hence the lethality, of gunpowder weapons. **Arsenals,** or places where arms and other military equipment are manufactured, repaired, stored, or issued, focused on rifled weapons, and railroads hurried the products of the arsenals to **armories,** where weapons are sometimes manufactured but are more generally stored for subsequent issue. An armory often stores weapons and has a large drill room in which training with weapons can be conducted. A **standard** is an officially accepted item or an item by which other like items are measured. The standard weapon of the Civil War became the **rifle musket** or **rifled musket,** a muzzle-loading, shoulder-fired gunpowder weapon with rifled bore.

Striking Weapons

Few changes occurred in the striking weapons employed by the Civil War armies. Cavalrymen and officers continued to carry the saber, and infantry shoulder arms continued to mount the bayonet. Lances of about nine feet in length were used early in the war by the 6th Pennsylvania Cavalry, but were discontinued in May 1863 because of the missions and terrain in which the cavalry was employed.[5] The sword and bayonet continued to be worn as **sidearms,** weapons worn at the side or in the belt. Pistols and the new revolvers, however, were so improved over earlier models that the practical value of the striking weapons decreased markedly.

Missile Weapons

The most effective missile weapons of the Civil War were **rifles,** a term that applied to both rifled shoulder arms and rifled artillery. Effective rifles had been used by hunters and some soldiers since about 1750, but they were muzzle-loading, slow to load, and expensive to produce. Some models required a mallet to pound the bullet into the chamber. **Bullet** is the name given to the projectile used with small arms. Derived from the French *boulette,* it means "little ball." Breech-loading bypassed the difficulties inherent in muzzle-loading rifles. However, the United States **Hall rifle,** the first breech-loading shoulder arm issued in large quantities to any army, was unpopular because of the amount of gas that leaked at the breech and, subsequently, into the firer's face when the rifle was aimed and fired.

The problem of gas leakage was largely overcome by the introduction of the **percussion lock,** an ignition system in which a **cap,** which was about the size of half a pea, was loaded with a detonating mixture and fitted over a nipple. A hollow area within the nipple led to the chamber. When a blow of the **hammer** exploded the detonating mixture, the explosion ignited the main charge. The detonating mixture, which was a crystalline compound made by the reaction of mercury, alcohol, and nitric acid, was called **mercury fulminate** or **fulminate of mercury.** The best percussion lock of the early nineteenth century was invented by Joshua Shaw, a Philadelphia artist. The percussion cap reduced the amount of flash and gas leakage at the breech.

The percussion idea was used by the German armorer Johann von Dreyse in developing an effective percussion-lock, breech-loading rifle. His rifle, called the **Dreyse needle gun, Prussian needle gun,** or simply **needle gun,** used a **paper cartridge** (a bullet and powder charge wrapped in paper that served as wadding). A long needle, when struck by the hammer, pierced the paper, passed through the powder, and finally struck the fulminate, which was contained in a wood or pasteboard sabot and packaged between the powder and bullet.

The major disadvantage of the needle gun was that the needle was very weak and had to be replaced after only a few

Hall Rifle—Model 1819—Caliber .52

Dreyse Needle Gun—Model 1854—Caliber .60 (Carbine Shown)

hundred shots. Nevertheless, the needle gun was the Prussian standard until after the Franco-Prussian War (1870–1871). During that war, the French used the *chassepot*. This was much like the Prussian rifle except that the fulminate was at the rear of the cartridge, where it logically seemed to belong, and its needle, or **firing pin,** was therefore shorter and more durable. Prussia's decisive victory over France serves as a reminder that the technological superiority of even the most important weapon in a conflict does not assure victory. The

chassepot was a marked improvement over the needle gun, but Prussian organization, leadership, and spirit prevailed.

The United States Army had earlier adopted a percussion-system weapon similar to the French and Prussian models as its standard, but with a muzzle-loading rifle rather than a breech-loading rifle. The key to the efficient use of a muzzle-loading rifle was the development of an undersized bullet, called a **Minié ball.** Named after its French developer, Claude Minié, who did much of his work in the employ of the

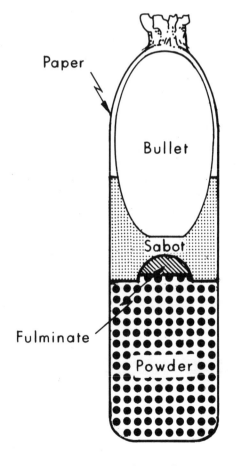

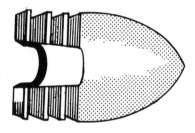

BEFORE FIRING

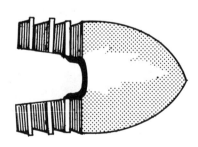

AFTER FIRING

A Sectional View of a Needle Gun Cartridge   **A Sectional View of a Minié Ball**

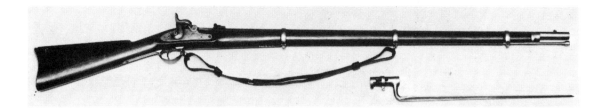

United States Special–Model 1861–Rifle-Musket–Caliber .58

British, the Minié ball had a hollow base into which an iron thimble was placed. When the weapon was fired, the thimble was driven into the bullet and the lead walls of the **cylindrical-conoidal** (rather than round) **bullet** were expanded into the rifling. While this system greatly increased accuracy and firepower, muzzle-loaders of any type were still nearly impossible to load on horseback or from a prone position.

The **U.S. Special** utilized both the percussion lock and the Minié system. Still a muzzle-loader, it was the most advanced weapon in wide use by Civil War infantrymen.

The first American percussion, breech-loading rifle was a six-shot **revolver,** a small arm utilizing several chambers that rotate around an axis and are discharged by a common lock. Manufactured by the Colt Company, the Colt revolver pistol—the **six-shooter** that, according to legend, tamed the Wild West—was successful. However, the revolver principle did not work as well in rifles, in which the chambers that leaked excessive amounts of gas were near the face and the left arm was held in front of the point from which the hot gases and residue were spewed.

One of the few successful breech-loaders used during the Civil War was the **Sharps** or **Sharps carbine.** The Sharps breech block, or **bolt**—a term synonymous with breech block but more commonly used than breech block when small arms are concerned—was lowered by pulling down a lever that also served as a trigger guard. Lowering the block exposed the chamber into which a paper or linen cartridge was placed. This weapon initially used the percussion cap ignition system, but the 1863 model used the **Lawrence patent primer,** which

fed a primer to the nipple on a tape of primers that was advanced each time that the hammer was manually cocked. The Sharps was a **single-shot weapon,** that is, only one cartridge could be placed in the weapon, or **loaded,** prior to each firing. **Repeating rifles** could hold several cartridges and could fire those cartridges successively without reloading between shots. While Colt revolver was the first repeating rifle, the first popular model was the Spencer. The **Spencer carbine** and **Spencer rifle** (the former for cavalry and the latter for infantry) utilized the same firing mechanism, which fired the first American metallic, self-contained cartridge. Seven copper **rim fire cartridges** (cartridges that held the fulminate of mercury in the rim rather than the center of the cartridge) were loaded through a trap in the butt and were fed into the chamber by the action of the trigger guard. The hammer was cocked manually. Christopher Spencer patented his weapon in 1860 and tried to sell it to the Union Ordnance Department, but it was rejected on the grounds that it would use too much ammunition. Not easily discouraged, Spencer obtained approval to demonstrate the weapon on the White House lawn for President Lincoln. By 1864, Spencers were in full production. The **basic load** (the amount of ammunition authorized to be carried by the individual armed with the weapon) was 28 rounds, **rounds** being the term used to define the projectile and those components necessary to propel it.

The **Henry rifle** was another of the popular Civil War repreating rifles. Although it was never officially adopted, many soldiers purchased Henry rifles with their own money. The rifle held 16 cartridges below the barrel and could be

Sharps Carbine–New Model–1863–Caliber .52

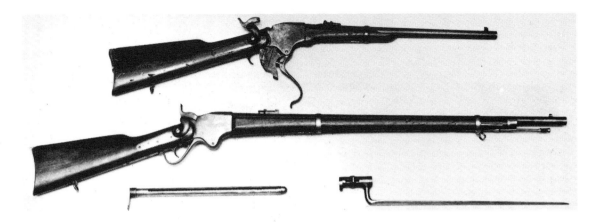

Spencer Carbine and Rifle—Caliber .50-.56

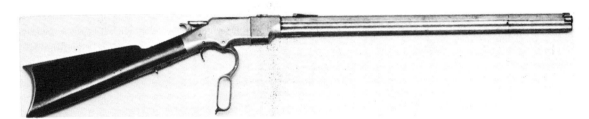

Henry Rifle—Caliber .44

Napoleon Gun Howitzer—United States Model 1857

fired more rapidly than the Spencer. Confederate soldiers referred to it as the "gun that the Yanks could load on Sunday and fire all week." The Henry was similar in design to that of modern **Winchesters** because the Winchester Arms Company bought the Henry patent. Winchester also bought the Spencer Arms Company and discontinued the Spencer design.

Like the small arms of the period, the artillery of the American Civil War era was influenced by the introduction of rifling, new ignition systems, and breech-loading weapons. The most popular artillery piece on both sides, however, was the **12-pounder Napoleon,** or the **Napoleon Gun Howitzer,** which was a smoothbore, muzzle-loading piece developed at the direction of Napoleon III, Emperor of France from 1851 until his capture during the Franco-Prussian War.

Most **rifle artillery,** which were generally called **rifle guns** during the Civil War, were loaded from the breech, but the popular **Parrott guns,** which varied in size from 3 to 10 inches, were rifled muzzle-loaders. The Parrott guns were named after their designer, Robert P. Parrott, who was superintendent of the West Point Iron and Cannon Foundry at Cold Spring, New York from 1836 to 1867. Parrott's technique of reinforcing the breech, where the greatest pressure developed inside the barrel, was to heat shrink a heavy wrought-iron band around the breech. These bands gave Parrott guns a distinctive appearance. Because they were

rifled and muzzle-loading, they used an undersized projectile, somewhat similar in design to that of the Minié ball, with copper bands around the base that expanded when the gun was fired. A third popular artillery piece on the Union side was the **3-inch Ordnance Gun,** which was an 1863 Ordnance Department modification of the 10-pound Parrott.

The principal gun used by the Navy during the Civil War was the **Dahlgren gun,** which was named after its inventor, John A. Dahlgren. The gun was much thicker at the breech than most guns of the period, and because of the resemblance of the taper to the soda-water bottles of the period, it was nicknamed the **soda-water bottle** or **soda-bottle gun.** The Dahlgren was a smoothbore muzzle-loader.

Like the North, the South experimented with breech-loading rifled guns, but toward the end of the war, both sides relied increasingly on the simpler smoothbore muzzle-loaders. Among the breech-loading rifled guns used by the South were the Blakely gun and the Brooke gun. The **Blakely gun,** or **Blakely rifle,** was imported from England in a variety of calibers. Some were used in the field (3.1–inch and 8–inch), while larger calibers were used for coastal defense. The **Brooke gun** was invented by John Brooke, the Confederate Chief of Ordnance and Hydrography. It resembled the North's Parrott and was manufactured in a variety of sizes.

In the early nineteenth century, an improved method of firing cannon used a **friction primer** instead of the glowing

Parrott Rifle—United States Model 1861

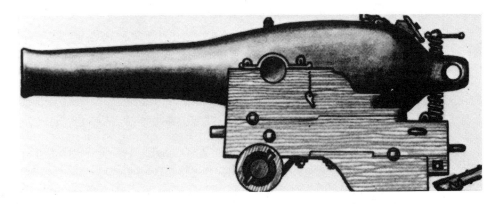

9-Inch Dahlgren Gun on a Marsilly Carriage

match and touchhole method of the Napoleonic era. The friction primer consisted of a tube of primer that was placed in the touchhole, and a roughened twisted wire that was inserted perpendicularly to the long axis of the primer tube. The end of the tube, where the wire passed, contained a sulphurous substance, like that found on the tips of modern household matches. When the wire was pulled from the tube by a yank on the **lanyard** (a short rope attached to the wire) the friction created by the movement of the twisted wire caused the sulphurous substance to ignite. The primer in the tube ignited in turn, carrying the flame into the main powder charge.

A **percussion primer** for cannon was developed as well. In this system, a plunger, hammer, or **striker,** was released when the lanyard was pulled. The striker, in turn, hit a percussion cap, which fired the main charge.

Ammunition and Other Implements of Destruction

The standard types of ammunition in use during the Civil War were shot (or **solid shot**), grape, canister (or case shot), shell (also called exploding shell), chain shot, and bar shot. The term **shell,** or **exploding shell,** referred to a projectile that contained a bursting charge of powder. **Shell guns** were rather ineffective because of poor fuzes* and because the hollow projectile did not fragment into small, uniform pieces. The term "shell," which is generally used to denote any type of projectile, may also be applied to the metallic cartridge cases that came into use during the Civil War. A shell or cartridge case that was loaded with powder but not with a bullet or projectile was referred to as a **blank.** A shell or projectile that was slowly delivered to its target (relative to other projectiles) —for example, one that was released from an aerial platform or a mortar—was called a **bomb.**

*Both fuses and fuzes are used to set off explosives. A **fuse** is a cord that contains high or low explosives and is used primarily in blasting and demolition work. A **fuze** is a device that initiates an explosion in an item of ammunition.

When a shower of projectiles is directed against a well defined target, a **shelling, bombardment,** or **cannonade** results. Those rounds or projectiles that do not reach the identified target, especially because of malfunction or human error, are called **shorts.** When the target is a group of artillery pieces that has already begun its own bombardment, **counter-battery fire** results. When a group of artillery pieces or several small arms fire to harass the enemy or to obtain more information about the enemy in a particular area, the technique is called **reconnaissance by fire.**

Mobility

Few changes occurred in the methods commanders used to gain mobility on the mid-nineteenth century battlefield. However, movement off the battlefield and between areas of hostility—often referred to as **strategic mobility**—was greatly influenced by the steam engine, which was put to use as a means of propulsion on both riverways and railways. Earlier forms of strategical movement, such as foot and horseback, persisted, and the movement of supplies, especially in rugged areas, continued to rely in part on **pack animals,** such as horses and mules. For a short time, even camels were used by the United States Army in the semi-arid and arid territories of the Southwest.

As the nation increased in size and population, the road network along which the lifeblood of commerce flowed improved markedly. **Turnpikes** were the best roads of the day. Where roads crossed low marshy areas, they usually were **corduroyed,** that is, built of logs laid transversely to the direction of travel. Bridges were few. Small rivers were generally crossed at **fords,** or shallow places where horses could wade across and wagons would not be swept away by the current. Larger rivers were generally crossed by ferry.

The first railroad in the United States was built in 1828, and by the time of the Civil War, the railroad linked key centers of commerce and provided a relatively high-speed route across mountains, valleys, and streams. **Railroad bridges,** which carried the iron rails over deep ravines and rivers, became vulnerable targets for raiding cavalry, and even the rails themselves were sometimes destroyed to impede the enemy's logistical efforts. **Rolling stock,** that is, the locomotives, passenger cars, freight cars, and any other wheeled vehicles used on the railroad, also provided vulnerable targets for sabotage, as did the **rail junctions** (any place where two rail lines met).

Naval Objects

During the Civil War, the opposing navies introduced a number of innovations, including steampower, ironclad vessels, rifled ordnance, rams, and torpedoes. The introduction of steampower marked the beginning of the end of the age of sail, which dated from the sixteenth century. The sailing navy survived at sea for a short time after the Civil War. During the war, however, ships like **schooners,** the two-masted sailing ships that often were armed with mortars and guns in support of land forces, were subordinate to steampowered vessels. Most steamboats had a rotating wheel made up of a number of flat paddles that pushed the ship through the water; hence, these early steamers were called **paddle wheels,** or **paddle-wheelers.** When the wheels were on the side of the vessel, the vessel was called a **side-wheeler.** When the wheel was mounted in the rear, it was a **stern-wheeler.** The invention of the **propelling screw, propeller,** or **screw** allowed steampower to be more effectively used on larger vessels, thereby enabling steamships to become associated with ocean travel as well as river travel.

In order to protect the naval vessels from enemy weapons, **ironclad boats,** or **ironclads** (vessels that had iron plating affixed over wooden construction) were developed. Cheaper means of producing iron made it competitive in cost with wood. In addition, iron allowed for greater variation in design. During the Civil War, the small, ironclad river boats that mounted small guns were called **armored gunboats.** The lightest of the ironclads—nicknamed the **t inclad**—drew less than two feet of water, could allegedly travel anywhere that the ground was a little damp, and had less than one inch of armor on its **bulwarks,** the side of the ship above the deck. In addition to ironclads and tinclads, the South employed some **cotton clads,** which used bales of cotton for protection. **Rams** were the ironclad vessels that had a massive iron bow designed to damage other ships, especially woodenhulled ships, by ramming.

While many sophisticated innovations appeared on the rivers during the Civil War, simpler devices were also widely employed. The simple **flatboat,** a shallow draft vessel that could be rowed or poled, was a sure and tested means of crossing streams and rivers. Even **rafts,** which were simply a collection of logs or lumber fastened together for transportation by floating, were often used. **Fire rafts,** which were rafts on which a fire had been started, were used as a weapon against wooden-hulled ships, ponton bridges, and **booms,** which were timbers lashed together and extended across a river to stop or impede the movement of vessels on the river. During the Civil War, **torpedoes,** a term that applies to any device intended to blow up an enemy ship, were metal cases containing explosives that detonated when struck by a vessel. Torpedoes were also employed to harass and thus impede enemy naval movement. Unlike modern torpedoes, which are aimed and propelled through the water toward their targets, Civil War torpedoes were like modern **mines,** which passively await the arrival of a target.

Large merchant ships, often called **transports,** were generally unarmed and tended to convert to steampower more slowly than naval ships. Because of the difficulty of maneuvering large ships in harbors and rivers, **lighters,** which were much smaller than transports, were used to transfer goods from a larger vessel anchored offshore to ports. **Tugs,** small steam- and later oil-powered vessels, were capable of pushing and "tugging" larger vessels into constricted areas. When all-metal construction was used to build large ocean-going transports, tugs and lighters became increasingly popular.

Defensive Items

Fortifications played a major role in the outcome of the American Civil War. Many of the techniques of seventeenth and eighteenth century siege warfare were employed to drive the Confederates from their **citadels,** the strong fortifications that command the approaches to a city. Southern citadels or fortifications were often surrounded and besieged, or **invested.** The trenches that surround the fortification were called **lines of investment.** Trenches are also referred to as **intrenchments,** or **entrenchments.** A trench that connects two other trenches is called a **communication trench.** Any construction or digging that is intended to afford protection in the midst of battle is termed a **battlework. Revetments,** which are walls of dirt, stone, or sandbags built to protect weapons or other materiel, are a form of battlework. One technique employed in the Civil War to negate enemy fortifications was the attempted **breach** of the enemy line by **mining**—that is, by digging a shaft under the enemy line, placing a large explosive in the shaft, and attempting to explode the charge to break the line. The most publicized attempt made by the Union to mine a Confederate position occurred during the siege of Petersburg in 1864; the attempt ended in failure. An underground attempt to break out from an invested position was referred to as **countermining.**

Among the less elaborate items that supported defensive positions during the Civil War were abatis, *chevaux-de-frise,* rifle pits, bomb pits, demilunes, and bombproofs. **Abatis** were rows of felled trees from which the smaller branches had been removed, while the remaining branches had been sharpened to a point. Abatis were employed as **obstacles** (objects that impede enemy movement). The *chevaux-de-frise, chevaux-de-frize,* or (singular) *cheval-de-frise,* were six- to eight-foot timbers from which long spikes or spears projected. The term literally translates from the French into "horse(s) of Friesland," where the device was first employed. **Rifle pits** were holes or short trenches about four feet long and three feet deep with earth piled up in front of them to form a parapet. They provided cover for two men, and today would be called **foxholes.** Logs, called **head logs** because they protected the head, were sometimes placed across sandbags on the parapet of the rifle pit to form a loophole. A **demilune** (half moon), also called a **ravelin,** was a work constructed to cover the entrance, the bastion, or the curtains of a fort. A **bombproof** was a shelter or building, either wholly or mostly underground, that afforded protection from bombs and shells.

Communications

Although the means of exchanging the information essential to the functioning of a large army took on new dimensions in the early nineteenth century, the old, proven use of messengers, or **couriers,** remained the most reliable. **Signal guns,** which fired smoking projectiles, were effective when visibility was good. **Pyrotechnics,** or multi-colored fireworks, were also used for communication, with the same limitations as signal guns. **Semaphores,** which were similar in concept to the mechanical telegraph used by Napoleon, were also limited by visibility. **Heliographs,** which reflected sunlight from mirrors, could send coded messages; however, they required both good visibility and sunshine to operate. The first reasonably reliable means of all-weather communication, other than messengers, was the **telegraph, electric telegraph,** or **field telegraph,** which used an electric impulse to send codes over a wire. The telegraphs used during the Civil War were a product of Samuel F.B. Morse's laboratory. The most commonly used code, consisting of combinations of dots (short electrical impulses) and dashes (longer electrical impulses) that represented letters of the alphabet, was called the **Morse Code.** Today, the Morse Code remains one of the principal

international codes used in **telecommunications,** a term that refers to any long-distance communications. Balloons, too, assisted in communication because they could be readily seen from remote distances. The balloons used in the Civil War sometimes carried a **telegraph key,** a device that allowed the transmission of the electrical impulse and trailed the wire necessary to send the impulses to their destination. The balloon's principal task was reconnaissance, but when the information gained from observation could not be communicated to the commanders who would benefit from the information, the balloon served little purpose.

Operations

Some of the terms that affected operations in the Civil War necessarily involved the highest officials of the states, the Union, and the Confederacy. **Secession,** the act of withdrawing from an organization, was a political decision without parallel for some states. Other state leaders attempted to maintain **neutrality,** a condition in which no favoritism toward either side is to be shown. One problem common to all the neutral states was the need to maintain **internal security,** a term that refers to the condition of prevailing law and order and all attempts to prevent hostile actions against life and property. When internal security was threatened, governments relied on **police power**—their inherent power to exercise reasonable control over persons and property within their jurisdiction. When states or governments **surrendered,** or yielded to the requirements and wishes of their antagonist, **armies of occupation,** or troops in effective control of enemy territory, were sent to insure that law and order existed and that surrender terms were implemented.

Command and Control

The term **command and control** refers to the means by which authority is exercised by a properly designated commander over assigned forces in the accomplishment of a given mission. Command and control is generally exercised through written orders. When a commander feels that a subordinate has not fully comprehended the meaning of a written or spoken instruction, the commander often gives a **direct order,** which is generally a verbal order that does not pass through staff officers or intermediate commanders. A direct order nearly always includes the warning, "This is a direct order!" Orders that emanate from the highest level of command and contain general instructions for other high-level commanders or government officials are called **war orders. A letter of instruction** is an order from a superior commander that gives information on broad aims, policies, and strategic plans for operations in large areas over a considerable period of time; it is like the operations order that is used at lower echelons of command. A **directive** establishes policy or orders a specific action. A **warning order** is a preliminary notice of a more detailed order or action that is to follow. Its purpose is to allow subordinates additional time to plan and prepare for some coming event. A **general order** usually concerns policy or administrative matters that apply or are of interest to all members of a command. General orders also refer to the instructions that govern the duties of **sentries,** or guards, in general. **Special orders** are administrative or policy instructions that affect only individuals or small groups. Special orders also define the specific instructions given to a sentry on a particular post. A **summons** is an order that announces either a meeting or an appointment to command. **Dispatches** include all the written communications that are sent from one location to another location. Orders that are cancelled or altered by subsequent orders are said to be **countermanded.**

Intelligence

A reliable **intelligence source,** that is, any agency, device or individual used in the collection of information, is essential to successful operations. The most reliable intelligence is called **hard intelligence.** It is often obtained from a **patrol,** or an intelligence source that is a detachment of land, sea, or air forces dispatched by a larger unit to gain information about the enemy (a **reconnaissance patrol)** or to destroy or harass the enemy (a **combat patrol).** To accomplish their reconnaissance and combat missions, patrols generally rely on **infiltration,** that is, on the process of surreptitiously moving through areas occupied by enemy forces. Members of reconnaissance patrols, or individuals or vehicles whose principal function is reconnaissance, are called **scouts.** Scouts are also involved in **counterintelligence,** which refers to the measures taken to destroy the effectiveness of enemy intelligence sources and to protect information from espionage, personnel from subversion, and installations or materiel from sabotage.

Intelligence is essential to the effective use of snipers, sharpshooters, and bushwhackers. In the Civil War, **snipers** were skilled riflemen who shot at exposed enemy troops. **Sharpshooters** were members of elite regiments, who during the Civil War were only incidentally armed with the Sharps rifle. The term can also refer to an individual skilled in **marksmanship,** or shooting, or to a marksmanship rating below **expert** and just above **marksman.** Before undertaking any marksmanship training, soldiers must **zero, zero in,** or **zero their sights,** terms that refer to the procedure of aligning the sights of a rifle with the trajectory of the bullet at a specified distance. Under ideal conditions, a bullet will strike the center of a target at the specified distance if the rifle is properly zeroed. During the Civil War, the term **bushwhacker** was applied to those men who claimed to be noncombatants when in the presence of a superior force and who to outward appearance pursued peaceful avocations, but who did not hesitate to become combatants when an opportunity arose to slay soldiers from concealed positions.

Tactics

The basic tactics of the Civil War were little different from those of the Napoleonic era. ⁶Attacks were made by successive lines, each two ranks deep, or in columns. Each leading regiment would deploy one or more of its companies as skirmishers to cover its advance. The diagram on page 90 shows the formation used by Major General William Henry French's third division of the II Corps during the Battle of Fredericksburg in December 1862. Three regiments were used as skirmishers, and the brigades of Kimball, Andrew, and Palmer were formed with their regiments on line. Similar formations were used by the Confederates. Against well defended positions, divisions often attacked with columns rather than lines of regiments. Late in the war, the increased range and accuracy of the rifled musket led infantry to attack increasingly by rushes, to rely on cover during the rush, and to attack in "looser" formations that presented less of a target to an opponent firing from well prepared positions.

Logistics

In the Civil War era, logistics was affected by the great distances over which **stores** (a term that refers both to supplies and to the place where supplies are kept) had to be transported, by the large number of soldiers that had to be supplied and moved, by the increasing variety and usage rate of **ordnance stores** (weapons and ammunition), and by the improved methods of transportation. Numerous stores, or **supply bases,** were established. The large supply base located at the rear of each major unit was called the **main supply base.** Those supply bases located near the front were called **forward supply bases.** Trains, wagon trains, or **supply trains** carried the stores from the place of purchase to the units. The **commissary** was the place where food supplies, which were also called subsistance supplies, provisions, provender, or rations, were obtained. **Subsistance stores,** or **subsistance supplies,** consisted principally of rations; however, they also included other items used by the individual, including candy and toilet articles. The term **ration** generally referred to food, but

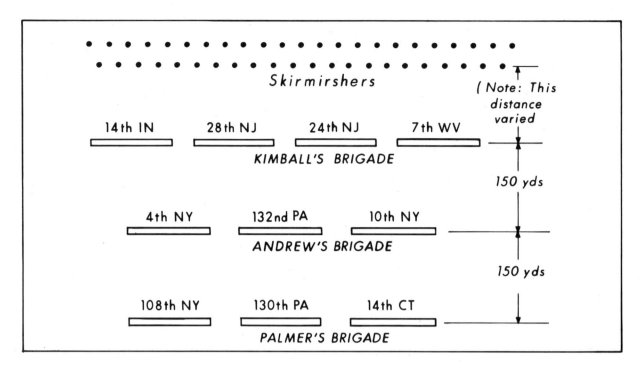

A Typical Attack Formation Used by a Division During the Civil War

specifically, it was the amount of food and drink authorized for one person or one animal per day. The ration for one horse, for example, was called its **forage,** and consisted of 14 pounds of hay and 12 pounds of grain. Any coarse food fed to domestic animals was called **fodder.** The term **provisions** usually referred to food, but could also be applied to any needed supply. **Provender** was another term that applied generally to food; specifically, it applied to the dry food fed to animals. When food supplies were scarce, the commissary officer often ordered **half-rations,** meaning that each soldier or animal received only one-half of his authorized subsistance supplies. Half-rations and **hardtack,** which was a hard biscuit or bread made of flour and water without salt, were common but generally unpopular with Civil War soldiers. Empty **haversacks,** or coarse linen bags in which soldiers carried their rations, were as much of a disappointment to individual soldiers as empty **war chests** (the funds accumulated to finance a war) were to political leaders.

Selected Bibliography

The number of books on the American Civil War is tremendous and increases monthly. The titles in this bibliography are representative, and many contain bibliographies that will in turn assist the inquiring student. A work that ably puts the Civil War into the perspective of the one hundred years from 1815 to 1914 is William McElwee's *The Art of War, Waterloo to Mons.* The chief military events of the decades after 1815 and before the Civil War are expertly covered in Glyndon van Deusen's *The Jacksonian Era, 1828-1848* and Jack K. Bauer's *The Mexican War, 1846-1848.*

Dictionaries and other excellent reference works on the Civil War include Mark M. Boatner's *Civil War Dictionary;* Colonel Henry Lee Scott's *Military Dictionary* (published in 1861 and reprinted in 1968); Francis A. Lord's *They Fought for the Union,* which is a detailed account of the circumstances surrounding Union soldiers; and Brigadier General Vincent J. Esposito's *West Point Atlas of American Wars,* the first volume of which presents narratives and campaign maps of American wars prior to 1900. The most complete reference work on the Civil War is the *War of the Rebellion: Official Records of the Union and Confederate Armies,* generally called simply the *O.R.* Compiled under the direction of the Secretary of War in the decades after the struggle, it contains orders, correspondence, maps, reports, and other source material. This monumental collection consists of four series, the longest of which has 53 volumes, which are often bound in several parts each. A shorter (four-volume) collection of observations by participants in the war is entitled *Battles and Leaders of the Civil War.*

Among the best general histories of the war is Shelby Foote's three-volume *The Civil War: A Narrative.* More general in its coverage is the excellent one-volume history by J.G. Randall and David Donald, entitled *The Civil War and Reconstruction.* A more recent one-volume study, and one that brings an outsider's perspective to the interpretation of the war, is Peter Parish's *The American Civil War.* A popularly oriented single volume that focuses on the Army of the Potomac is Bruce Catton's *This Hallowed Ground.* His *Grant Moves South* deals with the decisive campaign in the Eastern Theater, while *A Stillness at Appomattox* focuses on the desperate fighting in Virginia during the last year of the war. Perspectives of the individuals who directed the war can be found in T. Harry Williams' *Lincoln and His Generals,* Douglas Southall Freeman's four-volume *R.E. Lee: A Biography,* and Freeman's three-volume *Lee's Lieutenants.* Other comprehensive series that deal with the problems of high command in the war are K.P. Williams' *Lincoln Finds A General: A Military Study of the Civil War* (five volumes) and Allan Nevins' *Ordeal of the Union* (eight volumes). The first two volumes of Nevins' study are entitled *Ordeal of the Union,* the next two are entitled *The Emergence of Lincoln,* and the final four are entitled *The War for the Union.* An excellent work on the war, on the theory of war, and on Thomas J. "Stonewall" Jackson was written by the British soldier, teacher, and writer, George Frederick Robert Henderson and is entitled *Stonewall Jackson and the American Civil War.* An excellent volume on the war from a southern perspective is Clement Eaton's *A History of the Southern Confederacy.* An assessment of the reasons for the northern victory is found in a series of essays edited by David Donald and entitled *Why the North Won the Civil War.* Among the best novels about the war are Margaret Mitchell's *Gone with the Wind,* Stephen Crane's *Red Badge of Courage,* and Michael Shaara's *The Killer Angels* (important reading for anyone contemplating a visit to the Gettysburg Battlefield).

Doctrine for the Civil War era army can be found in the *Revised United States Army Regulations of 1861,* and in Egbert L. Viele's *Handbook for Active Service.* The authority for field fortifications was Dennis Hart Mahan's *Field Fortifications.* Arthur L. Wagner's *Organization and Tactics* and John K. Mahon's *The Army Lineage Book* are also useful sources for information regarding American doctrine of the late nineteenth century.

Notes

[1]United States War Department, *Revised United States Army Regulations of 1861* (Washington, D.C., 1863), p. 9.

[2]Francis A. Lord, *They Fought for the Union* (Harrisburg, PA, 1960), p. 55.

[3]John K. Mahon, "History of the Organization of United States Infantry," in *The Army Lineage Book,* II (Washington, D.C., Infantry, 1953) as cited in Mark M. Boatner, *Military Customs and Traditions* (New York, 1956), pp. 88–89.

[4]Lord, *They Fought for the Union,* p. 61.

[5]*Ibid.,* p. 69.

[6]The description of Civil War tactics is taken from Arthur L. Wagner, *Organization and Tactics* (London, 1895).

To the Great War, 7
the First
World War

The impact of the American Civil War on the military institutions of Western civilization was slight in comparison with the impact of the Prussian example from the **Wars of German Unification.*** Of brief duration, the Wars of German Unification were characterized by maneuver as a prelude to battle and by battles of decision fought in the open field. The prolonged trench warfare characteristic of the latter stages of the American Civil War seemed an anomaly. Moreover, as the diplomats of the late nineteenth century turned to **alliances,** which are agreements between nations intended to further common interest and to guarantee the commitment of war resources against mutual enemies in case of war, the armies of the West turned to Prussia to discover the precepts that form the foundation of military success. The search for such precepts occurred in the midst of profound political, social, and technological change. Soldiers and governments attempted to control the pace of change and to exploit the changes themselves. Their efforts, however, failed miserably; the coming holocaust that would destroy the youth committed to fight for the principal European powers and would bring deprivation and suffering to the societies represented by the powers was a product of man's inability to foresee the imminent—even in the light of recent follies.

Major Themes of the Age

The major themes of any age occur in response to the themes of earlier ages. The Industrial Revolution, which began in the eighteenth century, continued to spread its influence to new areas and new states. In turn, industrialization spawned **im-**

*This term includes Prussia's war with Denmark (1864), the Seven Weeks' War between Austria and Prussia (1866), and the Franco-Prussian War (1870-1871). The Seven Weeks' War is also referred to by German scholars as the *Bruderkrieg* (Brother's War).

perialism, which was the policy of extending a nation's control over remote territories for one or more purposes: to acquire raw materials, to support commerce; to Christianize, or civilize, people whose culture was more primitive than that of the imperialists; or to reap the glory associated with territorial, spiritual, and economic growth. The policy by which an imperialist power extends and maintains its control over other areas and people is called **colonialism.** The Industrial Revolution, imperialism, and colonialism were largely responsible for the extension of nationalism in the late nineteenth century and for the popularity of socialism, communism, and Marxism. **Socialism** refers to those policies and practices by which collective or government ownership of all means of production is advocated and in which the government controls all distribution of goods. Socialism differs fundamentally from **capitalism,** which advocates private ownership and free enterprise, and from **liberalism,** which seeks to minimize governmental controls over individuals. Like socialism, **communism** advocates the collective or government ownership of means of production and distribution of goods; unlike socialism, it advocates violent means to achieve government ownership and control. Communist ideology is based on the writings of Karl Marx, the most significant work being Marx's *Communist Manifesto,* which was written in concert with Friedrich Engels in 1848. **Marxism** connotes a belief in the struggle between economic and social classes as the fundamental force in history. It also emphasizes that the increasing concentration of industrial control in the capitalist class and the resulting intensification of class antagonism brought about by the misery of the working class will lead to a revolutionary seizure of power by the worker, or **proletariat,** and to the establishment of a classless society.

Along with the political movements of the era, two major social movements attracted the interest and attention of the masses—Darwinism and Victorianism. **Darwinism** was the belief in the theory advanced by Charles Darwin in his *Origin*

1865	⚔ Königgrätz: July 3, 1866	I Prusso-Danish War February 1864-August 1864 I Seven Weeks' War June 1866-July 1866
1870	† Death of Jomini: March 22, 1869	I Franco-Prussian War July 1870-May 1871
1875	⚔ Little Big Horn: June 25, 1876	
1880	† Treaty of Berlin: July 13, 1878 † Three Emperor's Alliance: June 1881	I Russo-Turkish War April 1877-March 1878
1885		I Serbo-Bulgarian War November 1885-March 1886
1890	† Bismarck dismissed: March 16, 1890 ⚔ Wounded Knee: December 20, 1890	
1895	† Franco-Russian Alliance: January 4, 1894	
1900	I Boxer Rebellion June 1905-September 1901	I Spanish American War April 1898-December 1898 Boer War October 1899-May 1904
1905	I Revolution in Russia January 1905-December 1905	I Russo-Japanese War February 1904-September 1905
1910	† Agadir Incident: July 1911	
1915	⚔ The Marne: August 1914 ⚔ Verdun: February-December 1916	The Great War August 1914-November 1918 Anglo-Irish Civil War April 1916-December 1921
1920	† Treaty of Brest-Litovsk: March 3, 1918 † Armistice: November 11, 1918	

From the American Civil War Through the Great War

of Species (1859), which maintained that organisms tend to produce varying characteristics in their offspring and that those best adapted to the environment would be the ones to survive. The application of this theory to human societies was known as **social Darwinism. Victorianism** was the acceptance and practice of the middle-class social standards that prevailed in Great Britain during the reign of Queen Victoria (1825–1901). Victorianism connoted solid comfort and prosperity for the middle class, a rising standard of living for workers, increased social welfare programs, high standards of morality, and a firm belief that what was British was best—for the world. The British Navy ruled the seas, the British pound dominated commerce, and British industry was second to none.

Growth was another characteristic of the decades between the American Civil War and World War I. Populations in Western nations increased markedly, and cities, especially, expanded rapidly as industry attracted masses of workers from farms and villages. Growth was apparent in not only the measurable areas of population, commerce, and industry, but also areas such as knowledge and military professionalism. The belief that man has the capability to understand all things in terms of natural laws was formalized in a movement known as **positivism.** The growth of military professionalism was evidenced by the opening of schools for advanced study in the art and science of war for commissioned officers, the appearance of numerous journals that dealt almost exclusively with military topics, and the efforts of nearly every major world power to improve the organization of its highest headquarters and to better its methods of training and recruitment.

In the arts, **romanticism** flourished. Its emphasis on feelings, imagination, and emotions arose in response to the order and reason that had characterized the eighteenth century. Romanticism fit the age, for on the surface, the belief that man's progress, growth, and stature was never greater prevailed. To those who had influence, it was an age of high hopes, noble purposes, and great expectations.

Beneath the surface, however, rumblings of socialists, Marxists, militarists, and Frenchmen who were intent on *revanche* (the widespread attitude that France should seek revenge from Germany because of the humiliations suffered by France in the Franco-Prussian War) threatened to upset the halcyon age of feather beds, steampowered trains, ocean cruises, and international peace. **Militarism,** which is the placing of martial beliefs, characteristics, and activities in a dominant position in the hierarchy of national priorities, was especially prevalent in Germany after the crowning of William II in 1888 and master diplomat Otto von Bismarck's subsequent dismissal in 1890. Bismarck had fabricated an elaborate alliance system during his tenure as Chancellor of the German Empire, but that system paradoxically became a threat to peace and a major cause of the pervasive and total nature of the Great War.

The alliances led to two armed camps: the **Triple Entente,** the name given to the "understandings" rather than formal treaties that existed between Great Britain, France, and Russia; and the **Central Powers,** the name given to Germany and its allies. The Central Powers originally consisted of Germany, Austria-Hungary, also referred to as the **Dual Monarchy,** and Italy. Italy chose not to declare war in 1914, and on May 23, 1915 joined the entente powers. Turkey and Bulgaria joined the Central Powers in October 1914 and October 1915 respectively.

The great historical forces of the late nineteenth and early twentieth centuries all found expression in the Great War. Commerce, nationalism, technology, and industrial growth contributed to both the arms race and the **naval race,** the part of the arms race that focused on naval construction. Romanticism and militarism brought euphoria to the preparation for the opening battles of the war. Marxism destroyed the cohesiveness of Russia more quickly and decisively than German arms. Colonialism spawned colonies that became the objects of conflict, and their raw materials fed the mills that turned out the tools of total war. In the end, military professionalism brought victory to the Allies, for without the training, discipline, and courage of professional soldiers, the German war machine might well have prevailed in 1918, the year of decision.

Participants in the Profession of Arms

The number of individuals in full-time military service, or called to **active duty,** during the Great War exceeded the number called in any previous modern war. **Reservists,** those who in times of peace principally pursue other than military occupations, filled the ranks of the mass armies of the great powers. The rank structure of the American Army differed little from the rank structure in use since the eighteenth century, except that (1) the system of brevetting had been eliminated, (2) a new rank was established for John "Black Jack" Pershing, the commander of the American Expeditionary Force, and (3) the increasing complexity of the technical aspects of managing men and materiel increased the number and roles of **warrant officers,** who were appointed by **warrant,** or written document.* A warrant officer is a highly

---

*Originally, the appointments were made by a senior commissioned officer; today, they are made by the Secretary of the Army.

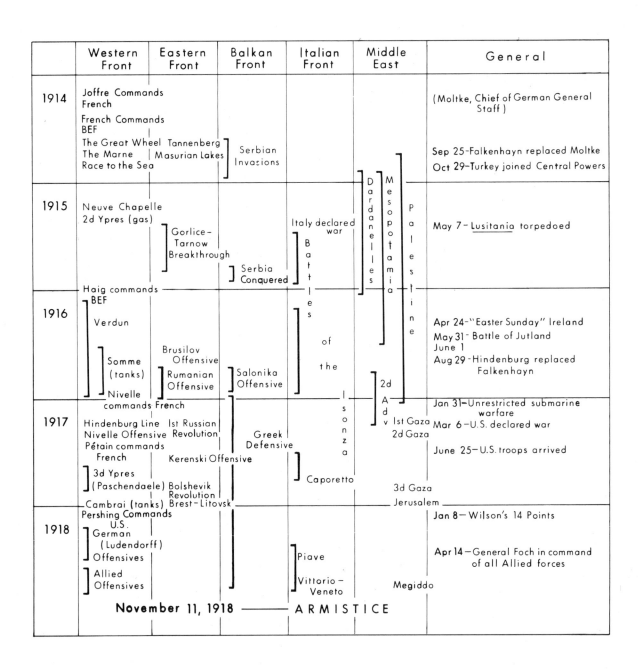

Principal Campaigns, Battles, and Command Changes of the Great War

skilled technician who fills those positions above the enlisted level that are too specialized to permit the effective use of a broadly trained, branch-qualified commissioned officer. The warrant officer's rank and precedence are below those of second lieutenants and above those of cadets.

The equivalent ranks of the American, British, French, and German armies during World War I are listed on page 97, but, in many cases, the equivalents are only approximations, since organizations and customs of the different armies varied considerably.[1]

American	British	French	German
Commander-in-Chief	Commander in Chief	généralissime	Oberbefehlshaber Generalissimo
Chief of Staff of the Army	—	chef d'état major de l'armée	Generalstabs-chef
General of the Army	Field Marshal	maréchal (de France)	Feldmarschall; Feldherr
General	General	général	General; General oberst
Lieutenant General	Lieutenant General	lieutenant général; général de corps d'armée	Generalleutnant
Major General	Major General	général de division	Generalmajor
Brigadier General	Brigadier (brigade commander)	général de brigade	
Colonel	Colonel	colonel	Oberst
—	Second in command	commandant en second	nachst-Kommandierende
Lieutenant Colonel	Lieutenant colonel	lieutenant-colonel	Oberstaleutnant
Major (Infantry)	Major (Infantry)	chef de bataillon; commandant	Major
Major (Cavalry)	Major (Cavalry)	chef d'escadron; commandant	Oberstwachtmeister
Captain (Infantry)	Captain (Infantry)	capitaine	Hauptmann
Captain (Cavalry)	Captain (Cavalry)	capitaine	Rittmeister
First Lieutenant	Lieutenant	lieutenant	Leutnant; Oberleutnant
Second Lieutenant	Second Lieutenant	sous lieutenant; lieutenant en second	Unterleutnant

Table of Equivalent Officer Ranks

The term **colonel general** is used to refer to the officers of some foreign armies whose equivalent rank in the United States Army would be a full general. The highest ranking generals of any army are referred to as the **high command,** and the senior officer in a **service,** such as the army, navy, or air force, of a nation is referred to as the **service chief.** Any senior military official is colloquially referred to as a **brass hat,** while his civilian counterpart is called a **silk hat** or a **frocked coat.** Formed in 1904, the **Committee of Imperial Defense (CID)** was a body of soldiers and politicians whose function was—and remains—to direct and coordinate Great Britain's national policy. One influential member of the CID is the **Chief of the Imperial General Staff (CIGS),** who is the senior member of the British military forces.

The nature of the war and new technological developments brought new functional titles to participants in the wars of the late nineteenth and early twentieth centuries. For example, long-range artillery and radio and wire communications contributed to the creation of the **forward observer,** an artilleryman who positioned himself where he could watch artillery rounds land and then transmit to the gunners corrections necessary to bring the rounds onto their target. The airplane allowed forward observers to direct artillery fire from the air, and other **aerial observers** used the plane as a platform from which to conduct reconnaissance. Those who flew the early aircraft were known as **aviators** or **pilots.** Today, the term aviator denotes a far greater involvement in flight requirements and capabilities than simply those necessary to pilot an aircraft. One of the principal objectives of World War I pilots was to shoot down enemy aircraft. A pilot who scored five

American Non Commissioned Officers	British Non Commissioned Officers	French Sous-Officers	German Unteroffiziere
Sergeant Major	Highest non commissioned officer	adjudant	Feldwebel
Sergeant Major (Infantry)	Sergeant Major (Infantry)	sergent major	Feldwebel
Sergeant Major (Cavalry)	Sergeant Major (Cavalry)	maréchal des logis chef	Feldwebel
Sergeant (Infantry)	Sergeant (Infantry)	sergent	Sergent
Sergeant (Cavalry)	Sergeant (Cavalry)	maréchal des logis	Wachtmeister
Corporal (Infantry)	Corporal (Infantry)	corporal	Korporal
Corporal (Cavalry)	Corporal (Cavalry)	brigadier	Korporal
Lance Corporal	Lance Corporal	tourrier	Gefreiter
Private	Private	soldat	Gemeiner

Table of Equivalent Ranks of Soldiers

kills (the destruction of enemy aircraft) was called an **ace,** a term that had formerly been applied to an exceptionally gallant cavalryman. Another newly designated title that resulted from technological innovation was **cryptographer,** an individual who encoded messages to be sent over wire or via radio.

The **litter** (stretcher) bearers of the medical service, or **sanitary service,** who had to recover the dead and seriously wounded, were nicknamed **body snatchers.** Other terms that reflected the peculiar humor of the trenches were coined relative to the injuries and death that reduced the number of **line effectives,** that is, the number of fit soldiers occupying the trenches. The number of infantry in the trenches was sometimes referred to as **bayonet strength,** and the number of cavalry available for combat was referred to as **saber strength.** The casualties suffered by a force were known as **wastage.** Among the wastage were the *grand blessés* (a French term meaning seriously wounded) and those with **blighty wounds** (an injury serious enough to return a soldier to **blighty,** the Indian term for "over the sea" that was used by British soldiers to refer to their home over the sea, the British Isles). A soldier who was killed was said to be **hanging on the wire,** a reference to the barbed wire that stretched across no man's land; to have **bought the farm,** a term that was revived with the same meaning during the American involvement in Vietnam; or to have simply **gone west,** an allusion to the setting sun or, in the case of American soldiers, to the westward journey home for burial.

Because of the great numbers involved in World War I, the age, experience, and training of different groups of soldiers varied widely. The most carefully selected and highly trained of the new soldiers were called **first line troops,** and the first line troops that the Germans prepared for mobile and decisive advances were called **storm troops.** Belgian troops who were generally too old to fight among the regulars but were experienced and motivated fought a guerrilla war after the occupation of Belgium by the Germans; these men were called *franc tireurs.* The same term described the French soldiers who had fought the Germans under similar circumstances during the Franco-Prussian War. Veteran troops in Great Britain became a part of the **constabulary troops,** or armed police forces, that were organized on military lines but were used domestically to maintain the peace.

Many participants in the Great War were known by the nation that they represented. **Magyars** were the people of Hungary and the Hungarian contingent of the Austrian Army. Those Magyars who were a part of the Hungarian National Guard were known as *honveds.* The **Uhlans** were lancers from Tatar (today a province of the U.S.S.R.), who served the Prussians in the Franco-Prussian War and served Ger-

many in the Great War. The **Senegalese** were black colonial troops in the service of the French; originally referring to blacks from Senegal, a state in West Africa, the term generally pertained to all black French troops, regardless of their place of origin. Native colonial troops fighting in Africa were called *askaris.*

Nicknames were popular in the Great War. By way of illustration, a **Holy Joe** was a chaplain or a highly religious individual. The **Old Contemptibles** were the British, stemming from Kaiser Wilhelm's (Emperor William II, sometimes called **Kaiser Bill**) reference to the British Expeditionary Force as a "a contemptible little army." A British soldier was also referred to as a **Tommy,** because "Thomas Atkins" was the fictitious name used as a sample on official blank forms used by the British Army. "Tommy" was made famous in Rudyard Kipling's ballads about the British Army. The common French soldier was called a *poilu,* which translates into "hairy" and referred to the dark hair that distinguished the French soldier from most of his British and American allies. The derogatory term **frog** was also used by the English-speaking allies to refer to their French compatriots. It was a shortened version of **frog eater,** which referred to the French custom of eating frogs, a practice regarded with disgust by the English. The Western allies had a variety of nicknames for German soldiers: the *Boche,* a favorite term of the French, which translates literally into "hard-headed ones"; **Fritz;** the **Hun; Heine; Jerry;** and **Hans Wurtz,** which was applied especially to German infantry. Members of the **ANZAC,** or Australian-New Zealand Army Corps, were referred to as **diggers** because of their penchant for using the entrenching tool. The Turkish soldier was known as a **Johnny,** while the Russians were given the nickname **Ivan.** Americans were generally referred to as **Yanks,** or **Yankees,** a term that was applied to New England colonists in the seventeenth and eighteenth centuries, to northerners in the American Civil War, and to American soldiers in general by the early twentieth century. "Yankee" possibly came from the American Indians' attempts to pronounce the French word *"l'anglais"* (the English) in referring to English settlers. The term **sammy** was also applied to Americans in World War I. The nickname **doughboy** was applied to Americans, but more accurately to any infantryman. In the British Navy, a doughboy was a kind of a dumpling, and because the large buttons on late nineteenth century infantryman's tunics resembled dumplings, or doughboys, the name was given to the wearers of the tunics. Another explanation suggests that the adobe clay that infantrymen of the American Southwest marched through and that covered their uniforms with "dobe" gave them the nickname "dobes" or "doughboys."[2] Regardless of the origins, the term is used without derision and is today widely accepted.

The 50 years after the end of the American Civil War constituted an era in which professional military institutions flourished in Western nations.³ The School of Application for Infantry and Cavalry was established at Fort Leavenworth, Kansas in 1878, and the Army War College was founded in Washington, D.C. in 1902. Professional journals, too, were established, and the level of training of commissioned officers was enhanced. **Service schools,** which were practical schools of instruction in branch-related skills, were founded in many nations. One such institution was the school for coastal artillery at Fort Monroe, Virginia. **Officer training schools,** or **officer candidate schools,** were established to provide a source of trained officers for the mass armies that characterized the Great War. An extensive series of officer candidate schools was established in the United States prior to its entering the war, and because the first of these schools was located in Plattsburg, New York, the system was referred to as the **Plattsburg Movement.**

At the war schools, practical instruction was stressed. **War games,** which are training exercises that simulate war on a map (sometimes simply called **map exercises**) or sand table, and **terrain rides,** which are exercises in which participants simulate the conditions of war by traveling over large areas and considering the manner in which the area could be organized for war, were favorite methods of practical instruction. **Terrain exercises without troops,** or **command post exercises (CPXs)** as they have been more recently known, were also methods of practical instruction used at service schools and staff colleges. **Training centers,** or places where practical instruction took place—especially the instruction of newly recruited soldiers—were established by every army. **Cadres,** a term that refers to the key group of officers and enlisted men necessary to perform essential duties in the formation, administration, and training of additional men to fill their units, were often returned to training centers to rebuild decimated units. **Training teams,** which were small groups of specialists who traveled from unit to unit to inform personnel of new methods and equipment, often assisted the cadres. Improvements in technology, the size of armies, and the extraordinary casualty rates combined to make training one of the principal support activities of the war.

The war was initially greeted with an air of excitement, but the horrors and routine of trench warfare and the disease and drudgery that characterized most operations quickly brought morale to a low ebb. To enhance morale, armies at different times—and when the situation allowed—used a liberal leave policy. **Permissions,** the French word for leaves of absences or passes, allowed soldiers a brief respite from the trenches, and the policy of *roulement,* or rotation of frontline troops to areas of rest, relaxation, and retraining, also provided a wel-

come change from the drudgeries of combat. Soldiers who stayed too long in the trenches often suffered from **shell shock,** a condition in which an individual acted erratically, often hysterically, because of his repeated exposure to enemy attack. While on permission or while recovering from psychological or physical wounds, soldiers were known to console themselves by partaking of the services of prostitutes. Frequent **short arm inspections,** which were inspections for venereal diseases conducted with no regard for privacy and named by analogy to rifle inspections, were conducted in the interest of the general health of the troops.

Awards and **decorations** represented another area in which the services and nation recognized the importance of morale in war. In ancient times, the **laurel** was the symbol of victory. The first **decoration** (the appellation for a distinctively designed mark of honor denoting heroism or meritorious service or achievement) authorized in the United States was the "figure of a heart in purple cloth or silk, edged with narrow lace or binding."⁴ George Washington issued the order authorizing the **purple heart** from his Newburgh, New York headquarters on August 7, 1782; but although the order authorizing the decoration was never abolished, the decoration fell into disuse and subsequent oblivion until its revival in 1932, when it was authorized for wear by those injured as a result of enemy action. In 1861, Congress created the second decoration, the **Medal of Honor,** and for almost 40 years, during which nearly 1,600 medals of honor were awarded, it was the sole American military decoration awarded. After the Spanish-American War, new decorations were authorized, and in January 1918, President Wilson established the **Distinguished Service Cross (DSC),** which was awarded for extraordinary heroism not meriting the Medal of Honor, and the **Distinguished Service Medal (DSM),** which was given in recognition of extraordinary meritorious service in a duty of great responsibility. At the conclusion of the Great War, the **Victory Medal** was authorized for all who served on active duty in either the Army or Navy during the war.

By the time of the Great War, other nations had established far more elaborate systems of awards and decorations than the United States. The highest of the British decorations was the **Victoria Cross,** or **V.C.,** which was named for Queen Victoria and first awarded in 1857. The highest French decoration in the Great War was the *croix de guerre,* the "cross of war." Germany's highest award for valor was the *Pour le Mérite* (for merit). Established by Frederick the Great, it was nicknamed the **Blue Max. Iron crosses,** the symbol of the German Empire, were also awarded to German soldiers for valor during the Great War.

The impact of the system of awards and decorations on the morale of the soldier is difficult to assess, but there is little doubt that, during the Great War, the humor and indomitable spirit that many men possess prevailed. For example, when the British Expeditionary Force (BEF) escaped from the overwhelming German strength massed at Mons, Belgium in August 1914, the troops attributed their survival to **"the angel of Mons,"** a mythical power that saved the BEF and whose recollection comforted British soldiers for the duration of the war. Complaints were heard from soldiers when all was not well, and French, British, and German frontline troops each coined terms that expressed their contempt for the orders and instructions that emanated from higher headquarters and led not only to minor inconvenience but also to unnecessary slaughter. The French referred to the efforts of their higher headquarters as *Système D,* from the verb *se débrouiller,* meaning "to unravel." The British referred to their higher command's war efforts as **muddling through;** the German equivalent was *notbehelf,* which translates into "makeshift."

The frontline troops, however, caused their share of problems for the high command. **Mutiny** (the willful refusal to obey or observe authority) was especially prevalent among certain French units in 1917. Other problems existed for high ranking commanders. For example, a French officer who failed to achieve what was expected of him was *Limoged,* that is, sent to Limoge, a garrison with a reputation as a holding area for ineffective, displaced officers. The British term for such an assignment originated in the Boer War. Their term was **Stellenbosched,** which was an anglicized version of German words meaning "put into place by being botched up." Another British term, **degommered,** was an anglicized version of the French word meaning "to become unstuck." The British soldier anglicized other foreign words to suit his phonetic abilities and his brand of humor. Among the most widely known were the nicknames applied to Ypres, Belgium, the place where the British Expeditionary Force was decisively engaged on three occasions during the war. The popular terms for Ypres included **Wipers, Eeps,** and **Weeps.** The participants in the profession of arms during the Great War may have brought more humor into their war than soldiers of earlier wars, but humor in this world-wide clash was often no more than an attempt to temporarily escape from the horrors that, by all human standards, seemed impossible to survive.

Principal Organizations

Technological developments and the increased size of armies in the era of the Great War contributed to the creation of new organizations to manage the war. Combatant organizations, however, generally used the terms of earlier periods to describe the units ranging in size from platoon to corps. As in the past, the common organizational progression was: platoon, company, battalion, regiment, division, and corps.

Because of the number of national armies committed to the war on the side of the Allies, a war policy coordinating body called the **Inter-Allied Supreme War Council,** and more commonly referred to as the **Supreme War Council** or **SWC,** was established in November 1917. Before the SWC was established, each of the Allied nations determined its military policy and objectives in national **war councils, military cabinets,** or **war cabinets,** which were policy bodies composed of the nation's most powerful individuals, including the prime minister or chancellor, the heads of the services, the secretary of state or minister of foreign affairs, and other key governmental officials. In 1915, the British war council was referred to as the **Dardanelles Committee** to emphasize the importance that the British war planners then placed on the campaign in the Dardanelles. The members of the war councils or war cabinets were nearly always members of the nation's **cabinet,** the group of individuals who advised the chief of state on issues of national importance and established national policies and priorities. When national policy indicated that closer relations with another country would be beneficial, a **military mission,** that is, a group of experienced soldiers sent to advise the military forces of the other country, was often formed.

During major wars, it was not unusual in the interest of unity and domestic harmony to establish a **coalition cabinet,** which was a cabinet whose members were chosen from the major political parties, rather than the party with a majority of members in the national legislative assembly.

War policies were often determined at the cabinet or council level, but **war plans,** which consisted of the military means of implementing war policy, were the principal concern of the nation's highest military headquarters. This headquarters was referred to as the *Stavka* in the Russian Army; the **General Headquarters,** or **GHQ,** in the British and American Armies; the *Grand Quartier Général (GQG)* in the French Army; and the *Oberstehheeres Leitung (OHL)* in the German Army. When the Kaiser of Germany personally worked from a field headquarters during the war, it was referred to as the *Grosses Haupt-Quartier,* or *GHQ.*

The British troops who landed in France in August 1914 were known as the **British Expeditionary Force,** or **BEF.** All troops and units that subsequently served with the original contingent of the BEF on the Western Front were also considered to be a part of the BEF. Similarly, all the troops and units from the United States that went to France in support of operations on the Western Front during the Great War made up the **American Expeditionary Force,** or **AEF.** Some chauvinistic Americans believed that AEF referred to those sent to the Western Front "After England Failed."

The term "front" assumed new meanings during World War I. In addition to referring to the portion of an army nearest the enemy or the direction in which a unit is facing, a front referred to the entire area in which conflicting forces opposed each other, such as the Western Front. In the Russian Army, a front referred to the large organization usually composed of four armies. Allied **field armies,** which were organizations composed of from two to seven corps, were also combined into large organizations, like the Russian fronts, and called **army groups.** On the German side, corps within the field armies were on occasion organized into **corps groups.**

Each corps in the United States Army during World War I consisted of a headquarters; two or more infantry divisions; one or more cavalry brigades or a cavalry division; one field artillery brigade; one telegraph battalion; one field signal battalion; ammunition, supply, and engineer units; and **sanitary trains,** which were composed of ambulance companies, field hospital companies, camp infirmaries, a medical supply unit, and reserve medical supplies.[5] The **ambulance companies,** or **ambulance service,** established **regimental dressing stations,** which were simply dry and sheltered places where wounded were taken for treatment. The ambulance companies transported the more seriously wounded to higher echelons of the **hospital organization,** that is, to field hospitals and base hospitals. Each ambulance company was authorized 12 motor ambulances, 3 trucks, 3 motorcycles, and a touring car. The **field hospital** was located as close to the fighting as possible, and generally consisted of tents or other temporary buildings. A **base hospital** usually occupied permanent buildings, if they were available. A base hospital was assigned to each division.

Several specific corps in the Great War were unique in their organization and function. The *Woyrsch Korps* was an unnumbered German reserve corps named after its commander. It often stiffened the Austro-Hungarian sectors on the Eastern Front or filled the gap between regular German troops and Austro-Hungarian troops. The **Desert Mounted Corps,** or **DMC,** was an Australian cavalry force active in the campaign in Palestine in 1917 and 1918. The British **Royal Tank Corps** was an organization formed in 1917 to study and utilize the shock characteristics of the tank. The portion of Great Britain's military establishment that included the aviation elements of the army was called the **Royal Flying Corps.** When this corps was combined in 1918 with the Navy's aviation elements, referred to as the **Royal Naval Air Service,** the **Royal Air Force** was born. Prior to its incorporation into the Royal Air Force, or **RAF,** an organization initially known as the **British Bomber Wing,** and subsequently called the **Independent Air Force,** existed to perform long-range bombing missions. In the United States services, the **air force,** or **air service,** which is the term applied to a nation's military organization that is primarily responsible for air warfare, was organized into **aero squadrons** composed of two **aero companies,** each of which had four aircraft. Aero squadrons, wire companies, radio companies, and telegraph battalions were all a part of the **Signal Corps,** the branch of the Army responsible for the communication of information and, during the Civil War and World War I, for the collection of information as well. The German air force was also organized into squadrons. The **Flying Circus** was the name given by the Allies to the squadron commanded by Baron Manfred von Richthofen; in turn, he was nicknamed the **Red Baron** because he painted the aircraft from which he commanded bright red to facilitate his being seen by his own troops.

The typical infantry division of World War I consisted of two or three brigades of infantry, one or two regiments of field artillery, a squadron to a regiment of cavalry, a battalion to a regiment of engineers, one or more companies of signal troops, ambulance companies, field hospitals, and ammunition and supply units. At full strength, the infantry division numbered about 20,000 men. Foreign cavalry divisions numbered from 3,600 to 5,400 men, while the American cavalry division was much smaller, having a war strength of about 1,500. **Investment divisions** were divisions especially trained for siege operations.

The brigade was the largest unit composed wholly of one arm. In most European armies, the brigade was made up of two regiments or six battalions. The British brigade had four battalions, and the United States brigade, commanded by a brigadier general, was composed of three regiments.

The regiment in foreign armies generally had three battalions, although the Russians and Austrians had four. The American infantry regiment consisted of a headquarters company, a **machinegun company** (with four machineguns), a supply company, and three infantry battalions. The **headquarters company** was the unit to which staff and some support troops were assigned. In World War I, it was made up of supply personnel; cooks; stable personnel; color sergeants; a first sergeant; a sergeant major; and a 28-man **regimental band,** which, in addition to its responsibility for producing music of inspiration and sounding bugle calls, assisted the medical or sanitary personnel with the evacuation of wounded. Nicknames continued to be a colorful part of the traditions of the British regiments. The *Greys* were a regiment raised in Scotland in 1781. Known initially as the Royal Regiment of Scots Dragoons, they acquired their nickname because they were mounted on grey horses. The **Flames** were the Dorsetshire Regiment; they had burned New London, Connecticut during the American Revolutionary War. The Germans referred to the kilted Scottish regiments as the **Ladies from Hell.** During the Great War, black soldiers were assigned to **colored regiments.** John Pershing, who has commanded a colored regiment prior to his assignment as commander of the AEF, was nicknamed "**Black Jack**" Pershing.

The United States infantry battalion was made up of 4 companies, had a war strength of 14 officers and 600 men, and was commanded by a major. In the German Army in the last years of the war, selected individuals (storm troops) were organized into specialized **assault battalions,** called the *Sturmabteilung* or a *Sturmbattaillon.* Cavalry units that were mounted on bicycles or motorcycles were organized into **cyclist battalions** and nicknamed the **pneumatic cavalry.** The company in most armies had 4 or 5 officers and 200 to 250 men, but the American company's war strength was 3 officers and 150 men. **Frontier companies** were the organizations responsible for reconnaissance and security along international borders.

Companies were generally composed of four platoons, which in turn were made up of **squads** of eight men each, or **sections,** which were organizations larger than squads. When two individuals worked at a common task, they were referred to as a **team.** Two such organizations of World War I were the **wireless interceptor teams,** whose task was to gain information from enemy radio broadcasts, and the **burial teams,** whose task was to recover and properly dispose of the dead. Small groups that are not part of a permanent organization and are given a specific duty are referred to as **parties. A wiring party** is tasked with laying wire. **Recruiting parties** are often sent to personnel depots or to certain areas to enlist soldiers for the unit they represent.

During the Great War, armies were recruited through the two methods used by all major powers since the Napoleonic era: (1) conscription, draft, or **compulsory service,** and (2) voluntary enlistment. In Germany, every 17-year-old male was considered to be in the military service. From that age until March 31 of the year in which he had his twentieth birthday, he was a member of the **first contingent of untrained** *Landsturm,* and could volunteer to serve in the active army. After March 31 of the year of his twentieth birthday, he could be drafted into the active army. If he was not drafted, he remained in the *Landsturm.* If drafted into the cavalry or horse artillery, he served in the active army for three years, followed by four years in the reserve. If drafted into any other arm of the service, he was in the active army for two years, with five years in the reserve. Following his active and reserve service of seven years, he was placed in the *Landwehr,* a second reserve where he remained until March 31 of his thirty-ninth year. From the *Landwehr,* the individual was transferred to the **second contingent of the** *Landsturm,* generally called simply the *Landsturm,* until his forty-fifth birthday. Those who were not drafted in their

twentieth year were placed in the **Ersatz reserve** where they remained until they were 32 years of age. The Ersatz reserve received no military training except during war, when they were at once put in training and assigned to active army **Ersatz units.** The *Landwehr* was thus a **second line reserve,** that is, recently trained but not active, and the *Landsturm* was a **third line reserve,** that is, qualified but either untrained, self-trained, or not recently trained. Similar systems prevailed in other countries, where second line reserve forces were often called the militia, the **National Guard,** or, in Russia, the *Opolchenie.* In Great Britain, the third line reserve was often self-trained and was called the **Home Guard,** the **home island defense force,** the **Territorial Army,** or simply **Territorials.** The largely self-trained and self-organized forces that fought the **Bolsheviks,** the Communists who gained control of Russia in 1917, were called **white forces,** while the German soldiers who fought communism in Germany after the armistice of 1918 were organized into *Freikorps,* which translates into "free corps" and refers to the independent nature of the units.

The Technology of the Age

Industrialization, continued commercial growth, and an unyielding belief in man's ability to unlock the secrets of his universe contributed to significant improvements upon the tools of war during the decades after the American Civil War. Internal-combustion engines, smokeless powder, iron warships, and heavier-than-air craft were just a few of the developments that brought profound changes to the methods and practice of war.

Striking Weapons

Even though the American Indians continued to use knives, tomahawks, and lances in their struggle against the encroachment of Western civilization, striking weapons continued to decline in importance as weapons of war. Cavalry relied on the saber for close fighting throughout the period from the American Civil War to the Great War, and the infantry rifles of the period continued to mount the bayonet. A **trench knife,** which was a knife with a short pointed blade used in hand-to-hand fighting, was popular in the Great War, but the effect of rapid-firing weapons dominated the outcome of every battle of significance. Knives, daggers, and sabers were more significant in the campaigns in the Middle East and colonial empires than on the major European fronts, but even among more primitively armed forces, gunpowder weapons predominated.

Missile Weapons

The ability to make more effective gunpowder weapons was greatly enhanced by advances in chemical, metallurgical, and mechanical engineering. A French chemist, Paul Vielle, developed a satisfactory **smokeless powder** in 1884, and even though the powder was far from smokeless—its smoke was grayish-white rather than blackish-gray—it was far harder to see from a distance and produced significantly less residue than black powder. The latter characteristic was especially important in the development of effective rapid-firing, or **quick-firing weapons.**

Prior to the adoption of quick-firing shoulder arms, major powers continued their efforts to develop an effective breech-loading rifle. An improvement over the Dreyse needlegun, and the first of the famous Mauser designed weapons, was the **German Mauser, Model 1871.**

The 1871 model was a single-shot, black-powder weapon, but in 1884, a nine-round tubular **magazine,** the part of the firearm that holds ammunition ready for chambering, was added. The **Mauser Model 71/84,** the name of the modified piece, was the principal weapon used during the 1916 Irish Rebellion. Because the guns were landed at Howth, near Dublin, in a well publicized gunrunning expedition, the Irish call the Mauser 71/84 a **Howth rifle.**

In 1886, the French adopted the first Standard shoulder arm to use the new smokeless powder. Known as the **French 8mm rifle M1886** or, by the French, as the *Lebel mam'se-le,* it was, like the Mauser, a **magazine rifle,** and was used in both World War I and World War II.

The first smokeless-powder small-caliber rifle adopted by the United States Army was the **Krag-Jorgenson, Model 1892.** It was a bolt-action rifle and had a five-round magazine. The German Army adopted smokeless powder in its five-shot magazine rifle known as the **German Mauser—Model 1898.** After 1935, when modifications changed its designation to the **98K,** the 1898 Mauser remained the standard bolt-operated rifle of the Germany Army throughout World War II.

In 1903, the United States Army adopted a five-round magazine rifle known as the **Springfield, Model 1903.** Manufactured at the Springfield Arsenal, the '03 was a copy of the 1898 Mauser, and its German designer, Peter Paul Mauser, brought suit against the United States for patent infringement. The case was dismissed, however, when the United States entered World War I. The Springfield was used

German Mauser—Model 1871–11 -mm

United States Rifle— Model 1892–Caliber .30

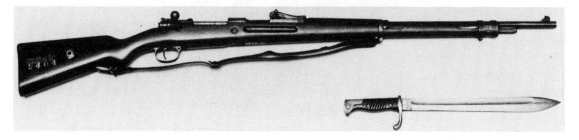

German Mauser—Model 1898–7.92 -mm

United States Rifle–Model 1903–Caliber .30 (Springfield)

again in World War II until sufficient M-1 Garands were available, and with telescopic sights it was employed as a sniper rifle throughout World War II and the Korean War. Each of the magazine rifles discussed could fire only one round each time it was cocked, loaded, and discharged.

In 1851, Frenchmen Faschamp and Montigny began developing a gun that could produce a greater volume of fire than the single-shot, single-chamber, single-barrel weapons characteristic of the period. By 1869, they had perfected a one-ton gun, called the *mitrailleuse,* with 35 barrels, 35 chambers, and 35 firing pins that could fire about 370 rounds in one minute. In 1862, Dr. R.J. Gatling of the United States developed a gun, known as the **Gatling gun,** that also produced a great volume of fire. His piece, however, utilized 10 barrels that rotated past a common chamber and common firing pin. It was capable of rapid fire because it reduced the problem of overheating; but because of its size and weight, it was not a popular weapon. It was, however, the forerunner of modern automatic weapons. **Automatic** refers to firearms that employ either gas pressure or force of recoil and mechanical spring action to both eject the empty cartridge case after firing and load the next cartridge from the magazine. A **semiautomatic weapon** requires that the trigger be pulled for each shot fired, while a **fully automatic weapon**

fires repeatedly as long as pressure on the trigger is maintained and ammunition is fed into the chamber. An **automatic weapon** is a small-caliber weapon that fires either semi- or fully automatically.

A **machinegun** is a fully automatic, rapid-fire small arm. The first machinegun produced was the **Maxim heavy machinegun, Model 1893.** It was invented by Hiram Maxim, an American citizen, who, when unable to sell his product to the United States Government, sold it to the English and the Germans. The **German Maxim, Model 08-15** was similar in design to Maxim's earlier models, but was the first machinegun that could be handled by one man. The designation 08-15 refers to the year the model was introduced, 1908, and the year of its modification, 1915. The weapon was water-cooled to control barrel heating. With the hoses detached, it could fire approximately 800 rounds before the water in the jacket would start to boil. Another water-cooled gun was the **Browning machine gun, Model 1917.** It was adopted by the United States Army in 1917, and later air-cooled models were used throughout World War II and the Korean War. It was an excellent weapon for defensive purposes.

Because the new machineguns were very heavy, they contributed far more to the defense than the offense. To gain

Gatling Gun–Model 1862–Caliber .58

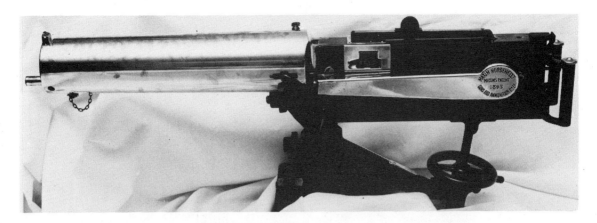

Maxim Heavy Machinegun—Model 1893

German Maxim—Model 08-15—7.92 -mm

Browning Machinegun—M1917A1—Caliber .30

French Light Machinegun—Model 1915—8 -mm

volume of fire in the attack, light, automatic weapons were employed. These included **light machineguns** (which by most definitions weighed less than 30 pounds); **submachine guns** (which were also called **machine pistols** because the German term for submachinegun was *Maschinen-pistole*); and **light assault guns** (which were lighter than the light machineguns and often fired semiautomatically). However, these early light, automatic weapons were generally incapable of overcoming the volume of fire from the heavier guns aligned against them. The first automatic weapon light enough to accompany assaulting infantry was the **French light machinegun, Model 1915.** Nicknamed the *Chauchat,* which literally means "hot cat," it was poor in workmanship and design, had a brutal recoil, was inaccurate, and malfunctioned frequently. In 1911, an American Army lieutenant colonel, Isaac N. Lewis, developed a gas-operated, air-cooled light machinegun that weighed about 25 pounds. Called the **Lewis machinegun,** it was first adopted by the British in 1915, and later by the United States after its entry into the war. The gun, which was the first to be fired from an aircraft, was also the most effective ground antiaircraft gun used in World War I. The German Army adopted a **7.92-mm parabellum mach-**inegun as its standard aircraft-mounted weapon. The German light assault gun in 1918 was called the **MP** (*Maschinen-pistole*) **18,** the **9-mm machine pistol,** the **German 9-mm parabellum submachine gun MP 18,** or simply the **German MP.** The gun weighed only 11½ pounds and used a pistol magazine designed for the Luger pistol.

The **Luger automatic pistol** was one of the most popular sidearms developed prior to the Great War. The earliest Luger model was called the **parabellum pistol Model 1900,** and its caliber was 7.65 mm. Successor models were the official weapon of German services from 1902 until 1938 and had a 9-mm bore. The automatic pistol adopted by the United States Army in 1911, the **U.S. Pistol, M1911,** with slight modifications was the standard sidearm of the Army for 75 years. A .45-caliber weapon, it was known simply as the **forty-five** to generations of soldiers.

With the advent of heavier-than-air craft, a new family of weapons was developed. Capable of doing enough damage to render aircraft unflyable, they were called **antiaircraft weapons, archies,** or simply **AA guns.** Prior to the production of guns designed specifically for antiaircraft fire, artillery pieces like the **British 13-pounder gun** were mounted on

Lewis Machinegun—Model 1915

German MP—Model 18-1

revolving pedestals and employed against aircraft. The Germans developed an antiaircraft cannon called the *Flugabwehrkanone,* and from its abbreviation, "Fl-a-k," the term **flak** was coined by pilots to refer to the shells fired from any antiaircraft weapon. One of the popular Allied antiaircraft guns was the **Vickers-Maxim 1-pounder automatic machine cannon.** It was nicknamed the **pom-pom gun** because of its distinctive sound.

Just as quick-firing small arms dominated the infantry organizations of World War I, **quick-firing artillery** represented the most advanced product of artillery technology. Smokeless powder contributed to the development of such artillery, but of greater importance was the adoption of **hydraulic buffer systems,** or **recoil systems,** which consisted of pistons in oil-filled cylinders that absorbed the recoil of the gun tube as it slid back on its carriage and returned the gun to its firing position. Successive rounds could thus be fired without having to **re-lay,** or re-aim. The first gun to employ such a system was the **French 75-mm gun.**

Also known as the *soixante-quinze* (French for "seventy-five"), the **75,** the **Josephine** (a reference to Napoleon's wife, Josephine), and the **75-mm rapid-fire field gun,** it fired a 13.5 pound projectile to a maximum range of 12,780 yards at a rate of up to 40 rounds per minute. Larger recoil guns were developed during the war, including the **British 60-pounder,** nicknamed the **Long Tom,** a term that has subsequently been applied to any long-barrelled artillery; the **German 15-inch gun,** nicknamed the **Jack Johnson** because of the heavy black smoke of its shells that reminded soldiers of Jack Johnson, the black heavyweight boxing champion of the world; **320-mm railroad artillery,** converted naval guns that were too heavy to move from their railroad carriages; and the famous German siege weapon, the **Paris gun,** which sent 250-pound shells into Paris from a distance of 75 miles. The Paris gun was made by attaching three large naval guns end-to-end and inserting a 100-foot-long tube of 8-inch diameter into the bore.

Effective direct fire weapons were also employed during

French 75-mm Gun—Model 1897

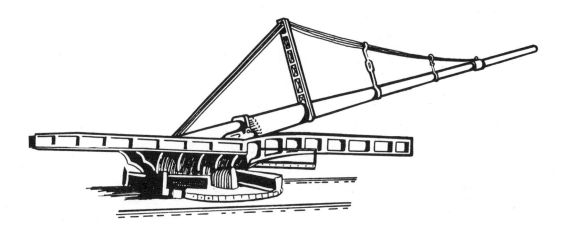

The Paris Gun

the war. The **German 88-mm flat trajectory gun** was nicknamed the **Whiz Bang** because of the high velocity of its shell, a characteristic of all direct fire weapons. The **British 37-mm direct fire cannon** was nicknamed the **Pound Wonder** because of the devastating effect of its relatively small one-pound projectile.

Very different from the flat-trajectory direct fire weapons were the mortars and howitzers, which also played an important role in the trench and siege warfare of the Great War. One of the smallest of the mortars was the **Stokes mortar** or **Stokes trench mortar.** Named after its inventor, Sir Wilfred S. Stokes, it had a 3-inch bore and fired an 11-pound projectile. The term **trench mortar** referred to any mortar small enough to be moved readily into and out of a trench. A popular German trench mortar was called the *Minewerfer,* or **mine thrower,** since mines were defined during World War I as explosives placed under or projectiles fired from under the surface of the ground. Among the largest of the howitzers employed during the Great War was the **German 420-mm siege howitzer,** known to the Allies as **Big Bertha** and, to the British particularly, as the **Wipers Express** because it was first encountered at Ypres.

Hand grenades, which had been popular missile weapons in the seventeenth and eighteenth centuries, were not used during the Napoleonic era, but were reintroduced in the last half of the nineteenth century. They proved especially useful in assaulting enemy trench lines in World War I. The popular English hand grenade was called the **Mills bomb.** Named after its inventor, Sir William Mills, it weighed about one and a half pounds. The German Army placed a grenade on a

short stick, which made the weapon easier to throw; such grenades were called **potato mashers** because of their resemblance to the common kitchen tool. Offensive grenades were referred to as **concussion bombs,** while defensive grenades were more generally referred to as **shrapnel bombs.**

Ammunition and Other Implements of Destruction

A few fundamental changes occurred in the types of ammunition employed in the Great War. Solid shot was supplanted by improved shell ammunition, especially **high explosive shell,** which contained chemical compounds such as **trinitrotolulene** (abbreviated **TNT**); **ammonal,** a mixture of ammonium nitrate, TNT, and flaked or powdered aluminum; **cordite,** a mixture of **guncotton** (cotton treated with nitric or sulphuric acid), **nitroglycerin** (a colorless liquid high explosive), and **mineral jelly** (a petroleum extract); and **lyddite,** which is mostly piric acid and was used in shells manufactured in Lydd, England. High explosives are characterized by extremely rapid detonation and a powerful disruptive or shattering effect. Some high explosive shells produced a very black smoke. The shells of the German 5.9-inch howitzer were called *Kohlenhasten,* or **coal boxes,** because of their black smoke, while the Germans called Allied heavy artillery shells *Schwarze Marias* (Black Marys). Many shells used in the Great War had an **impact fuze,** which detonated the shell's explosive charge when the item struck any hard object, such as the ground, a building, or any military hardware. Shells with impact fuzes were called **Daisy**

Cutters because of their devastating effect on daisies as well as everything else near the point of impact. In addition to high explosives, smoke was often used in artillery munitions. **Smoke** is a composition that when burned creates a thick vapor that prevents enemy observation of certain areas or activities.

The role of chemicals in warfare extended to a new dimension with the introduction of **gas,** or **war gas,** which is any chemical substance that produces poisonous or irritant effects on the human body. Gasses are classified according to their effects on the body. Common classifications include **asphyxiating** (affecting the repiratory system), blistering, blood, choking, nerve, tear, and vomiting agents. A **lethal gas** is one that is used with the intention of killing its victims. A common blistering agent used during the Great War was **mustard gas,** nicknamed **Yperite** because it was used at Ypres. Gas was dispensed directly from the canisters; from the **Levens projector,** a device used to hurl the canisters before dispersing their contents; and from grenades, like **tear gas grenades,** which contained an irritant that caused a painful watering of the eyes. It was not considered wise to dispense lethal gas from grenades, since friendly troops would be too close to the released gas; a slight change of wind direction would create risks that might not be entirely negated by **respirator masks** or **gas masks,** which were worn over the mouth and nostrils and had a **canister** of chemicals to filter or neutralize toxic agents.

Chemicals also figured prominently in the **flame thrower, flame projector,** or *Flamenwerfer.* Invented by the Germans, and nicknamed the **blowtorch,** it was an offensive weapon that used compressed nitrogen to project thickened, ignited fuel onto human and materiel targets. Chemicals also were responsible for the increased use and effectiveness of mines; **booby traps** (concealed explosives that are set to explode when an unsuspecting person touches off its firing mechanism by stepping on, lifting, or moving what is thought to be a harmless object); and **Bangalore torpedoes** (long tubes filled with explosives that can be used to cut a path through barbed wire or to detonate buried mines).

Mobility

During the era of the Great War, mobility was enhanced tremendously by the development of the internal-combustion engine. A relatively small internal-combustion engine could produce enough power to move not only itself, but also a significant load of people or goods over roads and through the air. A **truck** is any wheeled contrivance, including the wheel at the top of a flagpole. However, the term is generally applied to a vehicle, such as a **gasoline-powered truck,** that has an internal-combustion engine and can carry a cargo of goods, people, or animals on its own **chassis,** or frame. A truck with low or open sides and sometimes a canvas cover was called a **lorry** by British soldiers in World War I. The term **ambulance,** which comes from the Latin verb *ambulare,* "to walk," referred to a moving (or walking) hospital, to those who manned such a hospital, and—by World War I—to the trucks that moved the sick and wounded to the hospital.

The internal-combustion engine had an even more profound effect on mobility in the air than on mobility on the ground. The ability to move an airfoil through the air generated the lift essential to the development of heavier-than-air craft, often called **airplanes.** Balloons had been available to military forces since the late eighteenth century, and **dirigibles,** which were lighter-than-air craft that had their own motive power and could be steered, seemed the logical successor to the balloons used in the American Civil War. But like the **kite balloons,** which were used for artillery observation and were also known as **sausages** or **tethered artillery observation balloons,** dirigibles were limited in their usefulness by their slow speed, ponderous size, and vulnerability to rapid-fire weapons. **Zeppelins,** the large, rigid dirigibles designed and produced by Count Ferdinand von Zeppelin, were used to bomb France and especially England during the Great War.

The airplane was faster, smaller, and far less vulnerable to ground fire than the lighter-than-air craft. **Tactical aviation,** which refers to air capabilities that directly contribute to the outcome of ground combat, increased in importance when tasks like **strafing,** or the raking of a body of troops with fire at close range from a low-flying aircraft, were added to the former lighter-than-air tasks of observation and limited communication. An additional task for the airplane developed in the sky, for aircraft soon became the targets of other aircraft, leading to **dogfights,** a term that describes air-to-air combat between fighter aircraft. Long-range bombing added still another dimension to warfare, and the fledgling air services of the period found the need to develop both **fighter aircraft** (for ground support and air-to-air combat missions) and **bombers** (for carrying heavier loads over greater distances than fighters).

Among the first of the military aircraft used during the war by the Allies was the **Sopwith Tabloid,** a British single-seat, single-engine **biplane** (two wings) that was first flown in 1913 and had a speed of 92 miles per hour. The Sopwith Tabloid was intended to be a scout aircraft, but the war in the air persuasively and quickly converted scouts into fighters. The most famous of the Sopwith fighter aircraft produced during the war was the **Sopwith Camel,** a single-seat, single-engine biplane first flown in December 1916. It had a speed of about 115 miles per hour and was armed with two **Vickers**

Sopwith Camel

Fokker D-VII

machineguns, which were **synchronized,** meaning that they could fire "through" the propeller. It was called the Camel because of the humps over its nose guns.

The most popular French fighter of the war was the **SPAD 13.** SPAD is an abbreviation for *Société Pour Aviation et ses Dérivés,* the Organization for Aviation and Related Activities. A single-seat, single-engine biplane with a top speed of 137 miles per hour, the SPAD, like the Sopwith Camel, was armed with twin synchronized Vickers.

For a time, **Hansa-Brandenburg** craft were the best of the German fighters, but in October 1917, the Germans introduced a **triplane** (three wings), that brought Fokker designs to undisputed preeminence. Called the **Fokker DR-1,** it was the plane that the Red Baron was flying when he was shot down in April 1918.

The most famous German fighter aircraft of the war was the **Fokker D-VII.** Introduced in April 1918, it was a single-seat, single-engine biplane that had a speed of 118 miles per hour and twin Spandau machineguns.

The most famous of the bombers of the World War I era were the British Handley-Page and the German Gotha. The **Handley-Page** had a four-man crew and weighed over four tons when empty. Its top speed was just over 80 miles an hour. Originally powered by two 275-horsepower Rolls Royce engines, the Model 0/400, manufactured in the United States, had 400-horsepower Liberty engines. The German **Gotha** was a **pusher,** that is, its propellers were behind the wings and pushed rather than pulled the aircraft through the air. Powered by two Mercedes 260-horsepower engines, it had a crew of three and a top speed of 78 miles per hour.

The internal-combustion engine also profoundly affected ground combat. As the stalemate of the trenches extended

SPAD 13

Fokker DR-1

Gotha G-V

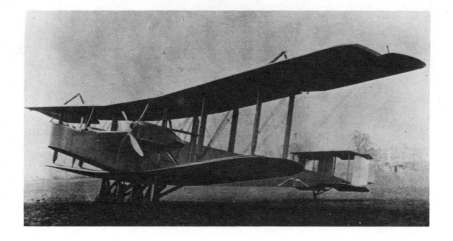

Handley-Page 0/400

British Tank, Mark IV

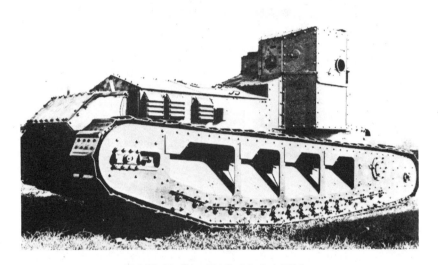

British Medium Tank, Mark A Whippet

French Tank, St. Chamond

weeks of suffering into months of suffering without either side making decisive gains, men of talent used their intellect and imagination to devise a means of breaking the stalemate. Among such men was Colonel E.D. Swinton, a British officer who suggested "a caterpillar tractor machine gun destroyer for the frontal attack of prepared positions."[6] Swinton's idea was supported by Winston Churchill, then First Lord of the Admiralty. Following design work by a joint Admiralty and War Office Committee, a company that produced water tanks used to carry water from purification stations to the trenches received the contract to manufacture the "tractors." In order to maintain secrecy, the new vehicles were said to be a special type of mobile water tank designed for the horrible roads and terrain that the Russians had to cope with on the Eastern Front. The first "tanks" shipped—their machineguns were shipped separately and installed after their arrival in France—were labeled in Russian, "To Petrograd—Handle with Care." The ruse apparently worked, for the Germans were surprised when the strange vehicles appeared during the Battle of the Somme in September 1916. Several official names, such as **land ship, land cruiser, land destroyer,** and **trench tractor,** were suggested, but the War Office chose to retain the term **tank** for the new combat vehicle. The French adopted the term *char d'assaut,* literally, an "assault carriage," for both armored track vehicles and armored wheeled vehicles, which in English are called **armored cars.** The Germans called the tank a *Sturmpanzerkraftwagen* (an armored, powered assault vehicle) or a *Schutzengrabenvernichtigungsautomobil* (a powered vehicle for the destruction of trench positions). *Panzer,* a shortened form of the first term, was soon adopted for common usage.

The first British tank of note was the **Mark IV.** Weighing 30.2 tons, it had a crew of 8, a range of 15 miles, and a speed of 3.75 miles per hour. The early British Mark-series tanks were designated **males** when they were armed with both cannon and machineguns, and **females** when they were armed with machineguns alone. In 1918, the **Mark V Heavy Tank** was widely used by the British. Today, a **heavy tank** is one that weighs between 56 and 82 tons, but early heavy tanks weighed far less. A **light tank** today weighs up to 25 tons, while **medium tanks** fill the gap between the lights and the heavies. The British **Whippet** or **Mark A** tank was called a medium tank and weighed only 15.7 tons. It had a crew of 3, a range of 40 miles, and a speed of 8.3 miles per hour. It was armed with four machineguns in the **turret,** which is the structure on forts, warships, airplanes, and tanks that contains one or more guns.

The most popular French light tank of the First World War was the **Renault FT.** It had a crew of 2, a speed of 6 miles per hour, weighed 7 tons, and mounted either a machinegun or a 37-mm gun in its turret. The first tank to mount a field gun was the French **St. Chamond.** It had a crew of 9, a range of 37 miles, and a speed of 5 miles per hour. Weighing 25.3 tons, the St. Chamond was armed with the French 75 field gun and four machineguns. Its frontal armor was sloped to help deflect bullets.

In spite of the advances associated with internal-combustion engines, steam engines and railroads remained the principal movers of bulk quantities of men and supplies. Differences in gauges,* which is the distance between the rails, plagued the logistical efforts on the Eastern Front. Nevertheless, the mass supplies needed to sustain the forces of the war generally found their way to the **railheads,** those areas where masses of supplies and personnel were placed on and taken from trains. Bulk cargo, like tanks, traveled on **flatcars,** which were freight cars that had no permanent sides or covering. People, animals, and loose cargo traveled in **boxcars,** which had permanent sides and roofs. World War I boxcars were nicknamed **sidedoor Pullmans** (after the comfortable coaches designed by American industrialist G.M. Pullman in the late nineteenth century); or **Forty and Eights,** *Hommes-Forty,* and *Chevaux 8-Hommes 40* (after the French designation on the side of the car that specified a maximum load of 8 horses or 40 men).

Naval Objects

Technology changed the nature of sea warfare as well. The **main battle fleets,** a term that referred to the organization composed of the major combat ships of the two opposing powers, fought only one major engagement and seldom journeyed far from their home bases.† The aggressive vessels of the war were the **submarines,** or **U-boats** (from the German *Unterseeboot,* or "under-the-sea boat"). Designed for operations under the water, these warships' primary mission was to locate and destroy ships of all kinds, including other submarines. When submarines were not stalking the waters for unsuspecting commercial vessels, they were often refitting in **submarine pens,** piers constructed to provide protection from attack via land, sea, or air.

To oppose the threat of German-U-boats, a variety of countermeasures were mounted. **Barrages,** which in naval terminology refers to an antisubmarine barrier of sea mines, were established along suspected submarine routes. **Minesweepers** were used to locate and remove or neutralize the

---

*A **standard gauge** (also spelled gage) rail system measures four feet, eight and a half inches between the rails. A **narrow gauge system** refers to any of a variety of measures that are less than standard gauge.

†The German main battle fleet was called the **High Seas Fleet,** while the British main battle fleet was referred to as the **Grand Fleet.**

Submarine—USS *Holland*

underwater mines. **Q-boats, decoy ships,** or **trap ships,** which were armed vessels disguised to look like unarmed commercial vessels or **trawlers** (fishing vessels), lured the U-boats to a close range before dropping false bulwarks and firing on the U-boats with their suddenly exposed guns. Steampower, iron hulls, and the improvement of naval guns all contributed to the construction of larger and larger armed vessels. **Torpedo boats,** which were boats that carried a torpedo propelled by compressed air, were used in shallow waters as an antiship weapon. Vessels larger than torpedo boats that often served as scout boats were called **destroyers.** Fast, with both offensive and defensive capabilities, destroyers were the principal anti-U-boat vessel in the fleet. The generations of vessels that succeeded the frigates of earlier centuries were called **cruisers.** A variety of cruiser "families" resulted, including **armored cruisers, semiarmored cruisers,** and **light cruisers.** A **battle cruiser** was a ship with the size and armaments of a battleship but without the heavy armor protection of the latter. All of the battleships extant when the HMS *Dreadnought** was launched in 1906 were referred to as **predreadnoughts.** They generally mounted four big guns and a variety of other armaments. The **dreadnoughts** were the family of battleships that followed the general design of the HMS *Dreadnought,* the first of the all-big-gun battleships that made every other existing battleship obsolete. The *Dreadnought* had ten 12-inch guns and could steam at 21 **knots** (nautical miles per hour, or 1.1508 statute, or land, miles per hour).

Landings on hostile shores, too, brought new and different vessels to the inventories of modern navies. Any small boat

with an inboard or an outboard motor is called a **motor boat.** Motor boats were called **guides** when they were used in directing the movements of other vessels to shore during amphibious operations. **Tows** were motor boats that pulled **barges,** flat, unpowered vessels often used during amphibious operations to transport people and materiel. A self-propelled barge-like vessel designed to carry coal was called a **collier.** This vessel proved to be an effective means of carrying infantrymen ashore during amphibious operations. The collier design suggested the design used in vessels called **landing craft, infantry (LCI)** which were principal personnel carriers during amphibious operations in World War II.

Defensive Items

The increased lethality of gunpowder weapons brought concomitant changes in the protective items worn by soldiers. Some body armor was reintroduced, and the armies of every major power on the Western Front adopted steel helmets to protect the soldier's head as he peered from the trenches and sought to advance against the rain of bullets from enemy rifles and automatic weapons. The German Army was among the first to revert to metal head coverings; its late nineteenth century helmet had a spike, or *Pickel* (pickax), protruding from the top, and was called the *Pickelhaube.* Later in World War I, the *Pickelhaube* was replaced by a new German-designed helmet called the *Stahlhelm* ("steel helmet"), which protected the ears and neck. The round appearance of the British helmet earned it the nickname **battle-bowler;** it was also referred to as a **tin hat.** Later versions of steel helmets in the American Army have been referred to as **steel pots.**

Armies that did not fight on the Western Front did not adopt steel headgear, but generally relied on **camouflage,** the

*The designation **HMS** stands for "His/Her (depending on whether a king or a queen is on the throne) Majesty's Ship," and is used with the title of a capital ship in the British Royal Navy. The designation **USS** stands for "United States Ship," and is used to identify capital ships of the United States Navy.

Battleship—HMS *Dreadnought*

disguising of personnel or materiel in an attempt to conceal identity and location from the enemy. The Austrians used *feldgrau,* or field gray, as their uniform and *Feldkappe,* or field cap, color. In spite of the increasing lethality and danger of head wounds, tradition prevailed in some units. The Turkish Army, for example, continued to wear the **fez,** a brimless, cone-shaped hat with a flat top and, usually, a tassel. Many French units retained the **kepi,** a close-fitting hat with a round flat top that slopes toward the front and has a small visor. During the American Civil War, armies often wore kepis, referring to them as **forage caps.**

Group protective items during the Great War were engineering masterpieces—not from an esthetic point of view, but in terms of magnitude. Multiple lines of trenches stretched from Switzerland to the English Channel on both sides of **no man's land,** the area of 30 to 300 yards between friendly and enemy trenches. The trench line closest to the enemy was called the **frontline ditch,** while lines to the rear that contained mortars, artillery, and infantry reserves were called **support line trenches.** A separate trench occupied by reserves was called the **reserve trench line.** The trenches that connected the different lines were called **traverses.** Narrow trenches, often in their initial stages of construction, were called **saps,** and saps that were opened with explosives were called **Russian saps.** Any spot along a trench that had been especially prepared as a firing position was called an **emplacement.** An emplacement could consist of a simple **fire step,** which was a step cut into the forward wall of the trench on

which a rifleman could stand to fire over the lip of the trench, or be as elaborate as a **pillbox,** which is a small, low fortification that houses machineguns and antitank weapons and is usually constructed of concrete, steel, or sandbags. Because of the water that accumulated in trenches, **duckboards,** or timber walkways, were constructed in the trenches; apparently, they reminded someone of the boards that led from the edge of ponds to duck houses. When dirt was thrown from a trench, it formed a parapet in front of the trench and a *parados* when thrown to the rear of the trench. The *parados* protected the trench from attack from the rear and also confused the enemy concerning the location of the occupied trench. When ammunition and **trench stores,** which consisted of items commonly used around trenches, were **stockpiled,** or accumulated, near the front, they were often placed in **dugouts,** which were underground shelters. The trench stores included barbed wire, entrenching tools, **fascines** (bundles of sticks used as obstacles), and **screw pickets** (twisted stakes upon which barbed wire was stretched). Underground storage shelters associated with elaborate fortifications were called **casemates.** Dugouts and casemates often served as a World War I soldier's quarters. In the rear, soldiers often slept in **Nisson huts,** which were semicylindrical buildings of wood and steel.

Elaborate fortifications were as much a part of the Great War as they had been of previous wars, and technology had its impact on fortifications just as it had an impact on other items of war. **Disappearing cupolas** were installed on Belgian fortifications, and **turreted gunpits,** which were permanently constructed emplacements that had large-caliber guns mounted in above-the-ground turrets, were used along the major defensive lines. Each of these lines was called a *Stellung* by the Germans. The elaborate **Wotan Stellung,** located in the Drocourt-Queant area, was known to the Allies as the **Switch Line;** the **Siegfried Stellung** was known as the **Hindenburg Line.** The true extent of sophisticated enemy positions was often hidden from ground observation—even with the aid of **periscopes** (optical instruments that provide a raised line of vision when direct vision is impractical) and **field glasses,** or **binoculars,** which provide a magnified view of distant objects and activities.

Communications

Electronic communications, the application of electricity to communications, had a profound effect on the communications capabilities of early twentieth century armies. Some mechanical semaphores remained in use, and **flares,** which were pyrotechnical devices that emitted a distinctive color of smoke, continued to serve as signaling devices. A large-caliber pistol, the **Very pistol,** was developed to fire flares. Nevertheless, communications on the modern battlefield relied increasingly on the efforts of electrical engineers.

The **field radio, wireless radio,** or simply **radio,** gave commanders a means of sending and receiving essential information to and from remote locations without having to lay wire, as telephone and telegraph systems required. The radios of World War I, however, were cumbersome and relatively unreliable, with each set requiring as many as 20 men to insure its successful use. Accordingly, radios were located primarily at division and corps headquarters.

Because of the ease with which enemy radios could receive friendly **radio traffic,** a term that refers to all the information broadcast by radio transmitters, **encryption systems** utilized codes to confuse and negate enemy intelligence efforts. Messages that were not encrypted were said to be sent **in the clear.** The term **radio net** referred to all the radios that operated on a specified, single **frequency,** which is the wave length or wave lengths of the signal that carries the transmitted information. **Shortwave radios** are those that use frequencies of 60 meters or less.

Disease

A subject that receives all too little attention in military textbooks is that of disease. During World War I, men used technology to combat the spread and effect of disease, but in the

trenches and in regions where sanitary standards were far below those to which the Western soldier was accustomed, disease far outweighed battle casualties as the more effective wartime killer and incapacitator. **Trench foot,** a painful condition of the feet resembling frostbite, was caused by the combined effect of low temperature and moisture on the feet; its prevalence among the inhabitants of the trenches of the Western Front gave the condition its name and exacted a heavy toll of noncombatant casualties. Unsanitary conditions led to diseases like typhoid fever, enteric fever, dysentery, and diarrhea. **Malaria,** a disease characterized by fever and chills, and yellow fever were transmitted by mosquitoes, and were especially prevalent in tropical and semi-tropical regions. **Field sanitation,** which refers to all the measures taken to combat infectious diseases, failed to prevent the often epidemic spread of disease. On both inactive and active fronts, disease casualties often outnumbered battle casualties by a ratio of 10 to 1.

Operations

To a greater extent than in any previous war, the key operational decisions of World War I, at least on the Allied side, were made by civilians. And because it was a world war, it was also a **coalition war,** that is, a war in which the vital interests of the several Allied Powers merged into the common goal of defeating the Central Powers. Yet the inherent difficulties of waging effective coalition warfare plagued both sides throughout the war. While some members of the alliance were content with **saber rattling,** a term that refers to the threatened use of force, other nations insisted on the *offensive à l'outrance,* an extreme belief in the power and efficiency of the offensive. In addition to differences in the level of violence to be applied, the Allied governments held differing views on the overall strategy for the war effort. Even within the British Cabinet, disagreements arose over whether to use a **blue-water** or **Eastern strategy** (one that placed emphasis on the resources destined for theaters like the Grecian Theater and the Middle Eastern Theaters), or a **continental** or **Western strategy** (one that gave priority of resources to the Western Front). When the Western Front turned to **stalemate,** a condition in which neither side is able to gain a decisive victory, some politicians favored an end to the **strategic attrition,** a term that refers to the slow and steady deterioration of the national resources available to wage war. To resolve the conflict, they were willing to accept the *status quo ante,* a return to the political and social conditions and institutions that existed before the war began. Other states were anxious to acquire **buffer territories,** areas or belts that would provide protection against surprise attacks by neighboring states. When the United States entered the war, acceptable terms of surrender increasingly included the connotations of **unconditional surrender,** a surrender in which the victors dictate all terms of settlement. Unconditional surrender usually means that the defeated will have to pay **reparations,** fees levied against the losers to pay for damages caused by the war. Reparations often include payments to offset a nation's **war debt,** the money spent to finance the war. Revenue from **war bonds,** promissory notes issued by the warring powers, pay for just a fraction of the total war debt.

Other key decisions that were made at the highest levels of the warring nations' political systems included the decision by the Germans to adopt a policy of **unrestricted submarine warfare,** which meant that any vessel at sea was an appropriate target for submarine attack. Similarly, the Russian High Command, as in the Napoleonic 1812 Campaign, adopted a **scorched earth policy,** which refers to the systematic destruction, usually by burning, of all resources that could be of benefit to an advancing army. For all participants in the war, operational realities dictated that **air superiority,** the ability of one side's air forces to conduct significant air and ground operations in a given area for a reasonable period of time without serious disruption from the opposing air force, be attainable at any given time. Resources that were to be expended for **strategic bombing,** the bombing of militarily related targets (such as factories, railheads, and communications centers) deep in the enemy rear, was a matter of considerable military and civilian debate.

Even though many decisions that affected the war effort were made by high ranking civilians, the feelings and opinions on the **home front,** a term that refers to the society behind the politicians and the armies, could not be ignored. Until the spirit on the home front was broken, **evacuations** (the surrendering of a given area by removing all military forces from the area), **drawdowns** (reductions in the number of personnel in a unit), and **deactivation** (the elimination of entire units) were unrealistic alternatives as far as national leaders were concerned.

For senior military commanders in World War I, frustrations were caused not only by political requirements but also by operational realities. Those who had sought a war of movement, or **open warfare** (a war in which armies have the capability of maneuvering over relatively vast areas), found themselves faced by the stalemate of **trench warfare** (a form of **position warfare** in which mobility in the vicinity of the enemy is restricted to a subterranean world).

Intelligence was difficult to obtain in trench warfare because living things could not readily cross the no man's land that separated the opposing sides. **Map reconnaissance,** the detailed study of a good terrain map, provided as clear a look at an area as many commanders could obtain before launching an attack. At night, patrols often penetrated the forbidden zones, but such **ground reconnaissance** efforts were extremely limited in the scope and range of their activity.[7] **Aerial reconnaissance** supplemented ground-based sources, but many useful and often essential details concerning the enemy were concealed by camouflage and deception. The days of cavalry **sweeps,** a rapid, on-the-ground look at an area to gain useful and essential military information, had passed.

Tactics

For the combat units that fought in the Great War, contemporary doctrine often failed to reflect the reality of the war they were fighting. Trench warfare had not been foreseen as a dominant form of fighting, and the idea of a **continuous front** (an enemy front that has no flanks) was foreign to the leaders of the prewar forces. Attention was focused on the offense. Concepts as basic as the **elastic defense,** a **flexible defense system** devised by the Germans in which multiple defensive zones were prepared to deceive the enemy concerning the area that was occupied in strength and to make rearward positions available should withdrawal be necessary, were a product of wartime thinking rather than prewar thinking. Even the attack differed fundamentally from prewar ideas concerning the attack. **Assembly areas,** locations where troops prepare for operations, were often in the frontline trenches, and **shakedown marches,** a trial deployment that allows commanders to witness the state of preparedness of their units, generally could not be attempted by units that attacked from their defensive positions. The daily routine was one of boredom and misery. **Stand-to,** which referred to the period just before dawn—or any other time when an attack was likely and all members of the command stood at the ready in their firing positions—alternated interminably with **stand-downs,** when only a part of the unit remained in its prepared firing positions.

On only rare occasions were the troops in the trenches preparing for the arrival of **Z-hour** or **Zero Hour, Z-day** or **Zero day,** terms indicating that there were zero hours or zero days until the next major attack. For the frontline troops, zero hour was the time to go **over the top,** to leave the relative safety of the underground world and to become a part of the **hammerhead,** the leading echelons of the assaulting waves. **Assault fire,** which is fire delivered at close range at a highly sustained rate from weapons pointed at small areas rather than aimed at targets and suspected targets, often commenced as soon as the infantry left its trenches.

The war brought new methods to artillerymen as well. Artillerymen learned of the increased importance of **preparatory fires, preparations,** or simply **preps,** which are supporting fires that prepare, or **soften,** an area by disrupting or destroying the attacked unit's **tactical integrity,** a term that means that a unit has an adequate number of people doing the tasks that insure the unit's effective performance. For example, when key individuals in a unit's chain of command are victims of preparatory fires and no one can effectively take their place, or when an infantry unit has lost most of its riflemen, the unit's tactical integrity has been destroyed. The new recoil systems on artillery pieces also affected the manner in which artillery was employed. The modern guns could be accurately **registered,** or **laid on a target,** that is, fired until their rounds landed at the desired location. Elevation and deflection were recorded, and when the same elevation and deflection were applied at a later time, the gun crews could accurately and quickly send rounds against the registered point. Their actions were coordinated by a **fire direction center,** which computed the range and direction to targets and communicated this information to the gun crews. It was also relatively easy to shift the fires to targets near any point of registration. When artillery is firing at a target and friendly troops enter the target area, the fire is said to be **masked,** and the artillery must be shifted to other targets to preclude injury to friendly personnel. Troops that are under artillery fire from either friendly or enemy fires warn persons near them by shouting "**Arrival**" or "**Incoming.**"*

When ammunition became plentiful, new techniques of fire support by artillery were utilized. **Rolling barrages** were intense fires that shifted when advancing friendly infantry masked the barrage. Extremely intense barrage fires, sometimes described as a curtain of fires, were referred to as **hurricane barrages, typhoon barrages,** or **drum fires** (because the sound of impacting rounds resembled the frantic beating on a drum). When the German artillery shifted rapidly from counter-battery fire to barrage fire and rolling barrage fire, the technique was known as the *Feuerwalz* (firewaltz) or **fire dance.**

But new artillery techniques and technology failed as miserably as every other technological or operational attempt to break the stalemate on the Western Front. Units that were broken either **consolidated** (regrouped) or were reinforced in time to preclude a decisive advance by the attacking force.

*The author's understanding of the Vietnamese language was profitably increased one quiet afternoon in September 1967 when he heard the Vietnamese whom he was "advising" suddenly run about excitedly while shouting *"Phao-kich."* A Vietnamese soldier, who saw the author's inactivity as a sign of stupidity rather than courage, physically pulled him into a muddy water-filled bunker just as a rain of mortar rounds landed within yards of their position. Given a moment to reflect, the author recalled being taught that *"Phao"* means "related to artillery" and *"kich"* means a "surprise attack"; *"Phao-kich"* translated nicely into "Incoming."

Hot pursuit, a term that refers to a pursuit while in sight of or contact with the enemy, became a term for airmen, but was rarely applied to ground operations. Because of the decisiveness and the high cost of attacks, the ground soldier tended to prefer the "**Chinese attack,**" an operation in which artillery prepped but the infantry only faked its advance. On occasion, the infantry became involved in **clearing operations;** sometimes called **mopping up,** these operations were intended to eliminate pockets of resistance that remained after a successful advance. More often, the operations of the Great War were characterized by disappointment and defeat for the side attempting to advance.

Amphibious Operations

Among the **peripheral operations,** those operations conducted on other than the principal fronts of World War I, were a number of planned and executed and planned but not executed **amphibious operations,** which are attacks launched from the sea by naval and landing forces against a shore that is controlled and often occupied by a belligerent.[8] Amphibious operations include raids, demonstrations, and withdrawals, but the term usually refers to an **amphibious landing,** or **amphibious assault,** which is a landing by a considerable force that intends to remain ashore until its combat missions are accomplished. Amphibious operations are usually **joint operations,** a term that refers to operations that involve units from at least two services. Most amphibious operations in the present century's wars have involved the Army and the Navy, in spite of the fact that amphibious operations are within the specific charter of the United States Marine Corps. When an amphibious operation is conducted solely by the Navy and Marines, it is not a joint operation but a **uniservice operation** being mounted under naval doctrine.

In general, amphibious operations are conducted to provide mobility to land combat forces. Specifically, they are conducted to obtain a lodgment area at the start of a land campaign; to seize an advanced base for naval, air, or logistical operations; to deny the use of the seized area to the enemy; to maneuver land combat forces preparatory to continuing an existing land campaign; to create deception; to gain information; to destroy installations or forces; or to make a show of force. Amphibious operations are termed **joint campaigns** when naval gunfire supports the advance or when an air force provides close air support. The units and personnel who participate in the amphibious operations compose the **amphibious task force.** Many factors are involved in the planning of an amphibious operation, and innumerable **staff studies,** which are examinations of specific factors involved in an operation or activity by a member of a commander's staff, assist the commander in selecting his concept. **Backward planning,** which refers to the planning of the total operation in the reverse order from which it is to be executed, is used in amphibious operations. The land campaign is planned before the landing is planned, and the landing is planned before the naval loading plans are outlined.

Landing areas are selected in accordance with a variety of criteria established by both the naval and ground forces. The **landing area** includes the beach, the approaches to the beach, the **transport areas** (where larger ships transfer troops and cargo to lighters or landing craft), the **fire support areas** (where naval gunfire ships operate), the air utilized by closely supporting aircraft, and the land up to the initial objective, which is called the **toehold. Control ships** guide the lighters along **boat lanes,** which lead to designated landing beaches. Navy gunships do not remain in support of landings for extended periods of time, because once located they are vulnerable targets. Their effectiveness lies largely in their mobility.

Once landing areas are selected, the landing plans can be completed. The **beachhead** is the area extending inland from the water's edge that, once secured, permits the continuous landing of troops and materiel and provides the maneuver space required for further operations ashore. This area is designated in the planning by the amphibious task force commander. The development of an effective logistical system ashore is intially the responsibility of special task organizations known as **shore parties.** The provide all combat service support until the normal supply units, known during World War I as the **Service of Supply,** or **SOS,** came ashore.

Logistics

Even more than in previous wars, the procurement of **"beans and bullets,"** the stuff that sustained the armies, placed astronomical demands on the logisticians and their supply systems. Rapid-fire guns consumed ammunition faster, and mass armies required food in greater quantities. The **required supply rate,** which is the rate at which goods must be supplied

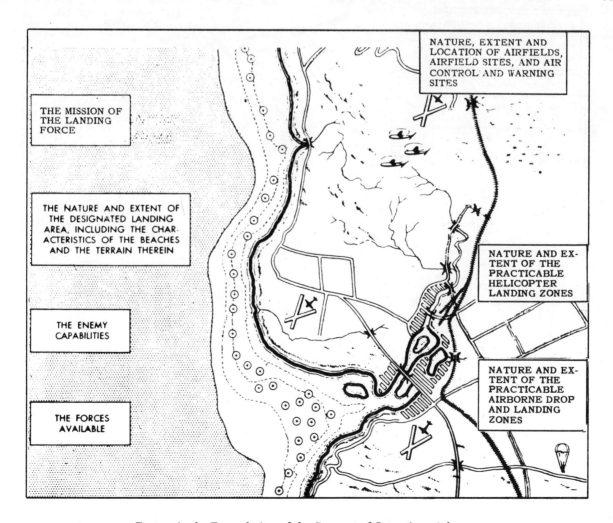

THE MISSION OF THE LANDING FORCE

THE NATURE AND EXTENT OF THE DESIGNATED LANDING AREA, INCLUDING THE CHARACTERISTICS OF THE BEACHES AND THE TERRAIN THEREIN

THE ENEMY CAPABILITIES

THE FORCES AVAILABLE

NATURE, EXTENT AND LOCATION OF AIRFIELDS, AIRFIELD SITES, AND AIR CONTROL AND WARNING SITES

NATURE AND EXTENT OF THE PRACTICABLE HELICOPTER LANDING ZONES

NATURE AND EXTENT OF THE PRACTICABLE AIRBORNE DROP AND LANDING ZONES

Factors in the Formulation of the Concept of Operations Ashore

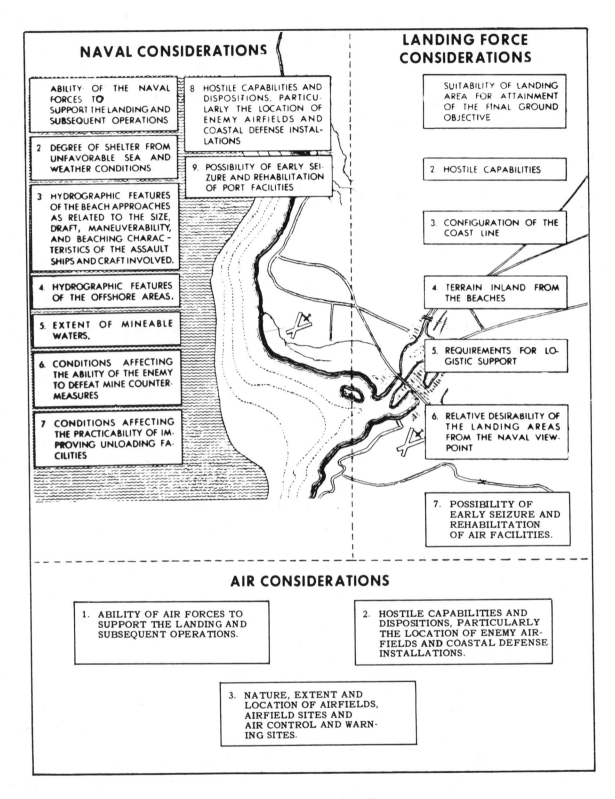

Factors in the Selection of Landing Areas

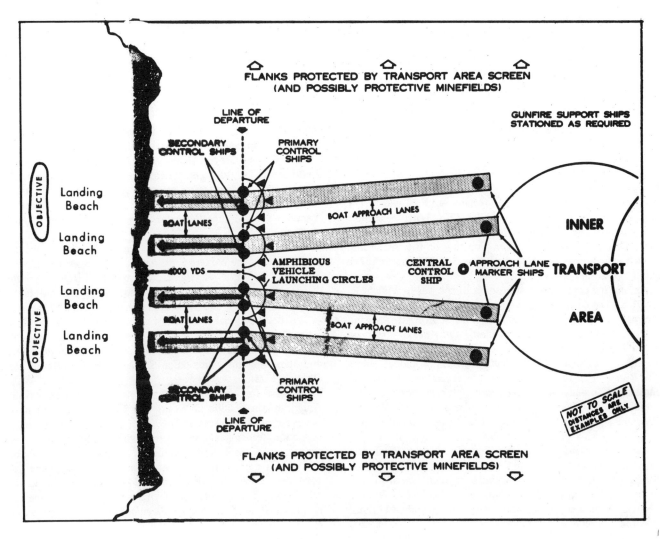

Schematic of the Landing Area

to keep pace with normal consumption, often far exceeded the **available rate of supply,** which is the rate at which goods are being supplied. When depots were empty, proper functioning became impossible for the **regulating stations,** supply facilities that assigned the goods coming from depots to **dumps** (the name given to supply facilities nearest the front). Considerable engineering effort was required to keep the **main supply routes (MSRs)** capable of carrying the necessary traffic. Because of the great variety of goods, like items were grouped into **classes of supply.** For example, Class I included those items consumed at a uniform rate; therefore, these items required a relatively uniform **rate of supply.** Food and forage were Class I supplies; the French referred to them as *revitaillment.* Like the rations of earlier wars, a soldier's food was often an object of verbal abuse. When fresh food was unavailable, troops had to turn to canned rations, referred to as the **iron ration** and the **last hope.**

Selected Bibliography

An understanding of the European wars of the twentieth century can be considerably enhanced by examining the European political and military events of the late nineteenth century. Such books as Otto Pflanze's *Bismarck and the Development of Germany* and Michael Howard's *The Franco-Prussian War* are among the best on Germany. A popular satire on the British experience in the Boer War is found in tank pioneer E.D. Swinton's *The Defense of Duffer's Drift.* A brief examination of the political roots of the First World War is included in Joachim Remak's *The Origins of World War I 1871–1914,* but the classic works on the origins of the war remain A.J.P. Taylor's *The Struggle for Mastery in Europe 1848–1918,* Luigi Albertini's three-volume *The Origins of the War of 1914,* and Sidney B. Fay's *Origins of the World War.* Focusing on the British and French entente that preceded the war is S.R. Williamson's *The Politics of Grand Strategy.*

Among the comprehensive histories of the war is William R. Griffiths' *The Great War.* If you wish to follow the operations of the war on detailed maps, another good source is the second volume of Vincent J. Esposito's *West Point Atlas of American Wars.* Readable, popular, and generally available in paperback are S.L.A. Marshall's *World War I* and Basil H. Liddell Hart's *The Real War, 1914–1918.* A more concise treatment is found in James E. Edmonds' *A Short History of World War I,* and for detail in a pleasant though often biased manner, Winston Churchill's five-volume *The World Crisis* is recommended.

Specific events of the Great War are the topic of excellent works like Gerhard Ritter's *The Schlieffen Plan: Critique of a Myth* and Barbara Tuchman's popular *The Guns of August.* Alexander Solzhenitsyn has presented a poignant view of Russian society and the war in his novel *August 1914.* Critics of Winston Churchill should enjoy the interpretations in Trumbull Higgins' *Winston Churchill and the Dardanelles.* Alistair Horne expertly captures the intensity and horror of Verdun in his *The Price of Glory: Verdun 1916,* while a German participant's view of the war on the Eastern Front is found in Max von Hoffman's *The War of*

Lost Opportunities. American participation in the war has no finer treatment than Edward M. Coffman's *The War to End All Wars: The American Experience in World War I.*

Biographies also offer unique perspectives of the war. Correlli Barnett's *The Swordbearers: Supreme Command in the First World War* presents sketches of two Allied leaders (General Philippe Pétain and Admiral Sir John Jellicoe) and two German leaders (Colonel General Helmuth von Moltke and General Erich Ludendorff). A full-volume treatment of Ludendorff is D.J. Goodspeed's *Ludendorff: Genius of World War I.* The most comprehensive and current work on General Pershing is Frank E. Vandiver's two-volume *Black Jack: The Life and Times of John J. Pershing.*

Two excellent first-person stories from the trenches on the opposing sides of the Western Front are Robert Graves' *Goodbye to All That!* and Eric M. Remarque's novel, *All Quiet on the Western Front,* the tale of a German squad that experiences the futility, frustration, and physical and mental trauma of the war.

The best dictionary of World War I terms is Edward S. Farrow's *A Dictionary of Military Terms* (1918). The inimitable British slang of the war is thoroughly covered in Edward Fraser and John Gibbons' compilation *Soldier and Sailor Words and Phrases* (1925).

Doctrine for the United States Army applicable to the American participation in World War I is found in the United States Army *Field Service Regulations* (1914). Six changes to the regulations were made in the course of the war, but none of these changes was particularly significant. More detailed fighting instructions were written for infantrymen in the American Expeditionary Force; they were published in Paris in 1918 under the title *Infantry Drill Regulations (Provisional).* British doctrine was promulgated in *Field Service Regulations Part I Operations* (1909). Modern doctrine on amphibious operations is detailed in FM 31-12, *Army Forces in Amphibious Operations (The Army Landing Force).*

Notes

[1] E.G.A. Beckwith, *Military Expressions in English, French, and German* (London, 1914), pp. 35–37; *Dictionnaire militaire* (Paris, 1894), pp. 349–350.

[2] See Elbridge Colby, *Army Talk: A Familiar Dictionary of Soldier Speech* (Princeton, 1942) for further explanations of the origins of "doughboy."

[3] Samuel P. Huntington, *The Soldier and the State* (Cambridge, MA, 1957), Chapter 9.

[4] Robert W. Wyllie, *Orders, Decorations and Insignia, Military and Civil* (New York, 1921), p. 7.

[5] Details of both United States and foreign army organizations in this section rely heavily on G.J. Fiebeger, *Army Organization*

(West Point, 1916).

[6] Edward Fraser and John Gibbons, *Soldier and Sailor Words and Phrases* (London, 1925), p. 276.

[7] See Robert Graves, *Goodbye to All That!* (New York, ca. 1930) for a well-written, first-person report on patrols in no man's land.

[8] The discussion of amphibious operations relies heavily on FM 31-12, *Army Forces in Amphibious Operations,* March 1961. Current amphibious doctrine, rather than World War I amphibious doctrine, which differed in definition and detail but not in concept, is discussed.

From the Great War to the Second World War

8

The Great War was called the war to end all wars. No sane official in the months and years immediately after the armistice of November 1918 would have predicted that in just over two decades an even greater war would affect the lives of more people, change the destinies of more governments, and shatter the illusions of more leaders than had World War I. The Second World War was more total than any previous modern war, and the changes that occurred in the midst of the war were more extensive than the changes that had occurred in centuries of earlier history. While foot power and horsepower were prevalent throughout the war, jet aircraft and high altitude bombers and missiles provided more sophisticated forms of wartime mobility. Cavalrymen carried swords as a legitimate weapon of war in the early campaigns; but nuclear weapons brought an impersonal extreme of destruction before the war was ended. **Pacifists,** who believed that man's injury of man as a means of resolving differences between nations and societies should be outlawed, campaigned in the name of humanity on the eve of the war. Nevertheless, over 25 million **servicemen**—soldiers, sailors, and airmen who wear the uniform of their country—died during the war.[1] **Disarmament,** which refers to multinational agreements to reduce the number or types of weapons or weapon systems to a specified level, and **rapprochement,** which refers to multinational desires to work in harmony, succumbed to more powerful forces. People's lives became dominated by arms races, **aggrandizement** (a nation's desire to grow in size, wealth, or influence), and **genocide** (the mass extermination of peoples of a certain race, color, or creed). **Appeasement,** the policy of granting concessions in the interest of peace, led to **protracted war,** a war that is longer and more intense than anticipated. Contrasts had never been as extreme as they were for the generations that lived through and remembered the era from the **Roaring Twenties,** a decade of euphoria, frivolity, and dreams, to the disillusionment and horror of the late thirties and early forties.

Major Themes of the Age

In the wake of the armistice that ended the first great war, **idealism,** a belief that civilization is capable of reaching perfection, was briefly apparent, but reality turned idealists to more mundane beliefs. Many turned to **cynicism,** which maintains that human conduct is motivated wholly by self-interest. Others became **fascists,** who believe that progress and perfectability can come only from a centralized, autocratic, national regime with severely nationalistic policies, highly regimented industry and commerce, rigid censorship, and forceful suppression of opposition. The National Socialist Workers' party of the newly formed German Republic was among the most ardent fascist parties of the day. To many Americans, as well as others who were to carry on the fight against Germany, the party's shortened title, **Nazi,** became synonymous with hatred, evil, cruelty, and madness. Idealism was also shattered by economic hardship. The **Great Depression**—a term that refers to the years after the crash of the stock market in October 1929, during which unemployment was high and money scarce—turned people from idealism to the ever-popular American themes of **antimilitarism,** which denies that any value exists in military institutions, and **isolationism,** the belief that a community, especially a national community, should concern itself primarily with its own affairs and hence avoid commitments, obligations, and intercourse with external organizations or nations.

Change and adaptation, action and reaction were so frenzied that the age seemed foreign even to those who were a part of it. To the professional soldiers who were trying to understand the Great War in terms of technology and tactics, **mechanization**—the change from manpower and animal power to machine power—was recognized as an irreversible trend. But few outside of Germany, and few even within Germany, conceived of the devastation that could be wrought by

1920	† Treaty of Versailles: June 28, 1919.
	† League of Nations: November 1920.
1925	† Beer Hall Putsch: November 8-11, 1923.
	† Locarno Agreements: October 5-16, 1925.
	† General Strike in Great Britain: May 1, 1926.
1930	† Pact of Paris (Kellogg-Briand): October 29, 1928.
	† Stock Market Crash: October 29, 1929.
	† Mukden Incident: September 19, 1931.
	† Hitler became Chancellor: January 30, 1933.
	† German withdrawal from League: October 14, 1933.
	† Blood Purge in Germany: June 30, 1934.
1935	† German repudiation of Versailles: March 16, 1935.
	† German reoccupation of Rhineland: March 7, 1936.
	† Rome-Berlin Axis: October 27, 1936.
	† Annexation of Austria: March 12-13, 1938.
	† Munich Agreements: September 29, 1938.
	⚔ Czechoslovakia overrun: March 10-16, 1939.
	† German-Russian Non-aggression Pact: August 23, 1939.
1940	⚔ Invasion of Poland: September 1, 1939.

Anglo-Irish Civil War
April 1916–December 1921

Paris Peace Conference
January–June 1919

Washington Naval Conference
1921–1922

Russian Civil War
September 1921–October 1922

London Naval Conference
1930

World Disarmament Conference
1932–1934

Abyssinian War
October 1935–May 1936

Spanish Civil War
July 1936–April 1939

Sino-Japanese War
July 1937–August 1945

WAR IN EUROPE   and   THE PACIFIC

Between the World Wars

the mechanized and orchestrated air and land forces that combined to usher in the age of *blitzkrieg,* or **lightning war,** in 1939. The French term for *blitzkrieg* was *attaque brusque,* but their having a term to describe the method fell far short of being prepared either to use it successfully or to stop the German **juggernaut,** an inexorable force that initially crushed everything in its path.

Participants in the Profession of Arms

The totality of World War II forced politicians and soldiers to face common problems with a frequency to which neither group was accustomed. In foreign armies, political and military problems were often resolved by the fusing of political and military authority. Adolf Hitler, as his self-proclaimed title *Führer* (leader) suggests, was indisputedly in command of both the German state and the German armed forces. Benito Mussolini, known as the *Duce* (leader), also had full military and political power. In the Union of the Soviet Socialist Republics, **political commissars** were, and are, assigned to military units to insure that commanders adhere to the principles and policies of the predominantly civilian, but absolutely totalitarian, government. In Japan, military leaders relied on the ancient Japanese traditions of the *shogun,* a political leader whose powers rivaled those of the emperor, and the *daimyo,* a powerful feudal baron. Both *shogun* and *daimyo* were served by *samurai,* the warriors who had the power of life and death over commoners. The military leaders of Japan were cabinet members who wielded the political power in the *shogun* and *daimyo* tradition. The professional soldier, in turn, saw himself in the superior position of the *samurai.* The policies of soldiers-turned-politicians were as disastrous to Japan as the policies of politicians-turned-soldiers were to Germany and Italy.

In the United States, the President served concomitantly as the nation's political leader and as Commander-in-Chief of the armed forces. During World War II, the term commander-in-chief was often applied to other extremely high command positions, and when used in this way, it was abbreviated as **CINC.** Because of the tremendous responsibility associated with the highest levels of military command in World War II, new ranks were created. In the United States Army, a five-star rank, **general of the army,** was designated. The Navy added an equivalent five-star rank, **fleet admiral.** The Germans named *Reichsmarschalls,* literally "marshals of the empire," and senior German naval commanders were called *Grossadmiralen,* or **grand admirals.** A German rear admiral was a *Konteradmiral.* A German rank

with no direct English equivalent was the *Generaloberst,* which referred to the senior general, as in the phrase *"Generaloberst der Waffen SS"* ("senior general of the SS"). The British referred to their senior air force officers as **air marshals** (equivalent to lieutenant generals), while the British officer in the Royal Air Force with the equivalent rank of an army colonel was a **group captain.**

Several new designations for enlisted men became prevalent during World War II. During World War I, the senior non commissioned officer in a company, who was responsible for company administration, had sometimes been referred to as the **first duty sergeant.** By World War II, this individual was called simply the **first sergeant.** A **lance corporal** was generally a private appointed to perform the duties of a corporal temporarily; but in the United States Marine Corps, a lance corporal ranked just below the lowest non commissioned officer (corporal) and above a private first class. As navies grew in size and diversity, the number of enlisted designations also increased. The lowest ranking enlisted personnel were called **seamen.** Senior to seamen were three classes of **petty officers.** The highest naval enlisted rank was the **chief petty officer (CPO).**

The new dimensions of war brought new tasks to **soldiers** (a term that can refer to any serviceman, but as the air and sea components of warfare became more specialized, was increasingly used to refer strictly to the ground-oriented service); **sailors** (personnel of the sea component of warfare); and **airmen** (personnel of the air component of warfare). The term **flight officer** referred to anyone in charge of a number of aircraft; but in the British Royal Air Force, it was an officer rank equivalent to that of first lieutenant in the Army. In the United States Army Air Forces, a flight officer was a rank equivalent to a junior warrant officer. A **control officer** was an officer who regulated the pace of certain military activities, such as the rate of a march or convoy or the rate at which planes took off from an airfield or carrier. Soldiers trained to make assault landings from the air are called **airborne soldiers.** The term **airborne** sometimes refers to those landed in gliders and aircraft; hence, airborne soldiers are to be distinguished from **parachute troops,** or **paratroopers,** who jump from aircraft with a parachute. At other times, "airborne" refers only to parachute troops and **glider troops,** those who are trained to make assault landings in unpowered aircraft. **Pathfinders** are individuals who mark the way for larger groups of either aircraft or airborne troops. *Panzer* **forces** are those troops that are a part of German tank or other armored units. *Panzer* means armored. A **commando** is a soldier especially trained in night scout and reconnaissance work, hand-to-hand fighting, surprise raids, and extraordinary small-unit missions. The commandos were originally a part of the British armed services, but the word has become a

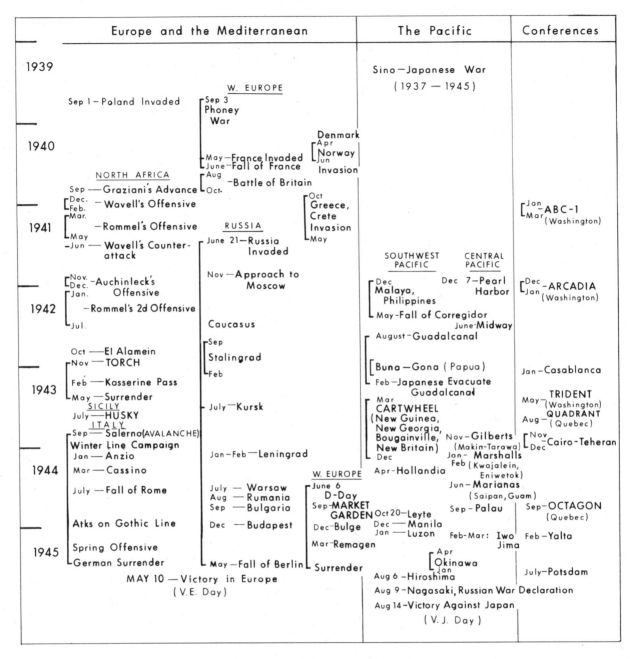

Principal Events of the Second World War

general term for any troops so trained. In the American Army, such troops are called **rangers**. The common soldier, a *Landser* in the German Army, was referred to as a **G.I.** in the American World War II Army. Before the war, the abbreviation stood only for galvanized iron. During the war, the all too common yellow soap used for washing dishes and clothes was officially called "government issue soap," or "G.I. soap." Soldiers extended the use of the term to anything official that came in bulk, such as G.I. food and G.I. haircuts. Eventually, the term was applied to the soldiers themselves.

Like the Army, the World War II Navy enlarged its vocabulary to encompass new terms. The term **skipper**, which referred to the commander of any ship, even was adopted by the Army, where it referred to any company commander. **Frogman** was a nickname for an **underwater demolition team** member. The title **deck officer** applied to all naval officers, except engineer and staff corps officers, who assisted in the navigation of a ship or in the working of cargo in port. The term **deckhand** applied to all seamen who assisted in the navigation, loading, and unloading of ships. A **coxswain** was an enlisted man in charge of a ship's boat, such as a lifeboat. The officer who was in charge of a **watch**, or a specific section of a ship, such as the engine room or main battery, was called the **watch officer.*** The **officer of the deck** was in charge of the whole ship and all of its watches.

In the United States Army and many other armies, the **officer of the day** during a 24-hour tour of duty was the officer who represented the commanding officer in routine matters pertaining to the preservation of order, the protection of property, and the performance of guard. The **staff duty officer** represented the commander and his staff in matters that required immediate attention when the commander and his staff were absent from their headquarters. Since World War I, the staffs of American units commanded by general officers have been made up of **G-sections:** a **G-1** (responsible for personnel), a **G-2** (responsible for intelligence), a **G-3** (responsible for operations and training), and a **G-4** (responsible for supply and services).

During the Second World War, many paramilitary forces operated against and in support of the armies. Guerrillas were active as in earlier wars, but the irregulars known as the **underground,** or the **resistance,** were a new breed. They were organized in strict secrecy in occupied or totalitarian countries to conduct espionage and sabotage, maintain communications with sympathetic governments, and assist their cause and liberation forces in any way possible. In the Pacific, the war effort was aided by another special group: the **coast watchers.** Their mission was to observe and report the air and naval activities of the enemy in a given area.

To enhance the abilities of officers and enlisted men, education remained a principal concern throughout the war. From the war **colleges,** which were the schools of learning for senior field grade officers, to **branch schools,** where newly commissioned officers and enlisted men learned the rudiments of their particular arm, to **maneuvers,** during which entire units functioned in a warlike environment that lacked only a hostile enemy, emphasis was placed on learning and doing those things essential to the defeat of the enemy. In the Japanese Army, leaders relied less on education and more on tradition and spirit when military results fell short of expectations. For example, the *bushido* was the code of the warrior, and it emphasized loyalty, benevolence, bravery, self-control, and the valuing of honor above life. For soldiers imbued with the *bushido, kamikaze* ("divine wind") missions —the suicidal attacks by Japanese planes and small boats against Allied ships and other high-value objects—were fitting, proper, and expected. Similarly, *hara-kiri,* also spelled *hari-kari,* which is suicide by disembowelment, was an accepted and not uncommon practice among disenchanted Japanese soldiers.

---

*The term "watch" also refers to the specific time periods, such as, morning watch (4:00 a.m. to 8:00 a.m.) and forenoon watch (8:00 a.m. to noon).

Western armies also relied, although to a lesser extent, on custom and spirit to motivate soldiers. The practice of recognizing bravery and achievement by the presentation of awards and decorations was carried forward from earlier times. In the Germany Army, the Iron Cross, which had been first authorized by Frederick William III of Prussia in 1813 and revived during the Franco-Prussian War and World War I, was reauthorized by Adolf Hitler in 1939. Eight grades of the Iron Cross were awarded during the war: second class; first class; Knight's cross; Knight's Cross with oak leaves; Knight's cross with oak leaves and swords; Knight's Cross with oak leaves, swords, and diamonds; Knight's Cross with golden oak leaves, swords, and diamonds; and the Grand Cross. In the United States, the **Distinguished Flying Cross** and the **Navy Cross** were established for air and sea heroism comparable to that signified by the Army's Distinguished Service Cross. The **Legion of Merit** was established during the war in four degrees by the United States, primarily for award to Allied servicemen. The Legion of Merit without degree was conferred on American forces. When the second award of a decoration was made to an American serviceman, a bronze **oak leaf cluster** was added to the decoration. Five awards of the same decoration merited a silver oak leaf cluster. In addition to individual awards, unit awards were conferred during World War II. The Army was authorized to award the **Distinguished Unit Citation** beginning in 1942; the Navy and Marine Corps were authorized to award the comparable **Presidential Unit Citation** in the same year. No nation has authorized more awards and decorations than the United States.

Principal Organizations

The size of armies and the complexity of the tasks that soldiers performed during World War II led inexorably to new military organizations that were inherently more complex and more diverse than organizations of earlier wars. Before the war, the United States Army had been organized into three major components. **Army Field Forces** were responsible for the training of ground forces; **Army Air Forces,** which succeeded the Air Service as a separate arm (not a separate service) for aviation, were responsible for the training of air forces; and the **Services of Supply** (later designated the **Army Service Forces**) were responsible for logistical planning and for the execution of logistical plans by **service and support troops.** When army units were sent overseas, they generally became a part of a **unified command,** which was an organization in which a **joint force** (a force made up of two or more services, such as Army and Navy) operated as a single unit under the long-term command of a duly appointed senior officer. This officer exercised control through the commanders of service components assigned or attached to his joint force, or through the commander of task forces that he himself formed. The term **assigned** refers to the placement of units or personnel in an organization where such placement is relatively permanent. A unit that is assigned to a larger organization by the authority of Army **Tables of Organization and Equipment,** or **TOEs,** is said to be an **organic** unit. The weapons and equipment authorized by these tables are referred to as **organic weapons** and **organic equipment.** When a unit is **attached,** it is temporarily placed under a larger organization that is responsible for the attached unit's rations, quarters, supply, administration, training, and operations. The attached unit's **parent organization** retains responsibility for the promotion and transfer of personnel. Assignment and attachment are **command relationships.** A third command relationship commonly encountered in the United States Army is **operational control,** in which a unit is temporarily placed under the authority of another organization for the purpose of accomplishing a specific task or mission. Administration, discipline, internal organization, training, promotion, and transfer of personnel in a unit placed in operational control remain the responsibility of the unit's parent organization.

The German military establishment was known as the *Wehrmacht* (*Wehr* meaning "defense" and *Macht* meaning "power" or "might"). Among its components were the *Heer,* or Army; the *Luftwaffe* (*Luft* meaning "air" and *Waffe* meaning "arm" or "weapon"), or Air Force; and the *Kriegsmarine* (*Krieg* meaning "war" and *Marine* meaning "navy and merchant fleets"), or Navy. The headquarters that directed the German military establishment was called the *Ober Kommando-Wehrmacht* (*ober* meaning "over" and *Kommando* meaning "authority"). Hence, it was the authority over the *Wehrmacht,* or *OKW.*

As in previous large wars, both area, or geographic, organizations and unit organizations were prevalent. In China, for example, the largest geographic organization was the **war area;** 12 such areas were designated. The rear part of the theater of operations, known as the **communications zone,** also was an area organization, as were the **corps areas**

designated in the United States in the 1920s. **Military installations,** a term that refers to any land or improvements thereon devoted to military purposes, also are area or geographic commands.

As in World War I, army groups were the largest unit organizations employed on the Allied side. Armies, or field armies, were subordinate to the army groups. The Japanese equivalent of the army was the **area army.** Ironically, the area army was not an area, or geographic, organization, for it was not limited to a specific region. Allied armies were made up of corps, while in the Japanese area army, **armies** were the next subordinate organization. In other words, the Japanese army was the equivalent of the Allied corps. In Russia, those armies that were extensively trained for offensive action were called **shock armies;** they were characterized by high morale and strict discipline.

Many types of divisions were organized during World War II. The **square division,** which was composed of two brigades of two regiments each, was used early in the war by the Japanese. However, most divisions employed in the war were **triangular divisions,** which were smaller and more flexible than the 28,000-man square division. There were no brigades in the triangular division; each of three regiments reported directly to the division headquarters. The triangular division, like most United States fighting organizations, derived its logistical support from a **support command.**

In the German Army, distinctions were made between field divisions and static divisions. The **field division** was well trained and capable of conducting operations in the field over long periods of time. Field divisions were often either mechanized or motorized, but many pure infantry field divisions were also deployed. **Mechanized units** were organized and trained to be transported by and fight from armored and armored motor vehicles. **Motorized units** were equipped with enough motor vehicles to transport simultaneously all assigned personnel, weapons, and equipment. The **static division** had little or no organic transportation, was often composed of large numbers of inexperienced troops, and could not deploy for field operations over long periods of time. German divisions that were extensively trained in movement by air were known as **airlanding divisions.** Divisions trained, equipped, and employed to defend against a seaborne attack

British and American	Russian	Japanese	German
army group	front	—	army group
army or field army	army	area army	army
—	—	—	corps group
corps	corps	army	corps
division	division	division	division

Equivalent Large Unit Organizations of World War II Powers

were called **coastal divisions.** A division whose principle weapon system was composed of organic armored vehicles was called a *panzer* **division.** Elite units in many armies have been called **guards units,** or grenadiers; elite armored forces in the German Army were called *panzer* **grenadiers.** A separate elite Nazi party force that fought for Germany was called the *Schutzstaffel* (protection group), or **SS.** The infamous *Waffen SS* **divisions** were respected and feared, both for their fanaticism and for their military skills.

World War II units smaller than divisions were known by a great variety of titles. A brigade augmented by nonorganic forces was sometimes called a **brigade group.** In the Japanese Army, two or more infantry regiments, which in most armies formed a brigade, were called an **infantry group.** A **mixed force,** such as a **mixed brigade,** was a unit that was composed of forces from two or more arms. A **tailored unit** (a unit that was not a TOE unit) similar to a brigade and designated only in United States armored divisions was called a **combat command,** or **CC.** An infantry regiment that was reinforced by a battalion of artillery and an engineer company was called a **regimental combat team,** or **RCT.** RCTs were employed throughout the Pacific theater. **Mountain units,** especially **mountain regiments,** were trained in such techniques as mountain climbing and skiing. They were equipped with the special equipment, including light artillery, needed to facilitate the conduct of military operations in mountainous terrain. The **1st Special Service Force,** with its insignia of the crossed arrows of the Indian Scouts, was organized in three regiments of two battalions each. Its Canadian and American soldiers were trained in demolition, rock climbing, amphibious assault, skiing, and airborne techniques; they were the antecedents of the present-day United States Army Special Forces.[2] For resupply, mountain troops often had to rely on **muleskinner units,** units whose principal means of transportation was the reliable army mule. (A **muleskinner,** or **muleteer,** was one who led or drove a mule or muleteam.) **Ranger battalions** were composed of soldiers trained to make surprise attacks on enemy territory, generally in the face of formidable obstacles. **Ranger companies** often operated independently of their parent battalions. **Sapper companies** were composed of engineers who specialized in the digging of trenches, tunnels, and underground fortifications. Their functions later included the laying and neutralizing of minefields. The British still refer to all military engineers as sappers. American infantry battalions in the triangular division were composed of three **rifle companies** and a **heavy weapons company.** Each rifle company had three rifle platoons and a **weapons platoon.** The weapons units employed mortars, machineguns, and, later in the war, recoilless rifles.

Varying considerably in size, according to their assigned mission, *ad hoc* organizations were used frequently in World War II. A **flying column,** for example, was a strong and highly mobile military force that operated at a considerable distance from the main body. A **strike force** was any force organized for an offensive mission; its operations were generally characterized by rapidity and shock action. Japanese special attack units, called **takko units,** became suicide units as the war waned in the Pacific.

The increased size and new missions of air forces necessitated the creation of new aviation organizations. An **air group** consisted of a headquarters and two or more squadrons, and was roughly equivalent to a ground regiment. A **combat air group** was composed of either fighter or bomber aircraft. The **air transport fleet** was made up of all cargo and troop-carrying aircraft.

The increased role of airpower also affected naval organizations, for the principal fighting organization of the United States Navy became the **carrier task force.** The assemblage of all the vessels in a carrier task force and all necessary supporting combat vessels was referred to as the **carrier group.** When two or more submarines conducted a coordinated attack against the same target, they constituted a **wolfpack.**

During World War II, civilians were organized to support the military needs of the combatant nations. In the United States, the **organized reserve** (those civilians who were trained for duty in war) was called upon to serve on active duty. Similarly, the **National Guard** (those citizen-soldiers responsive to the separate states in peacetime) was called into Federal service by the President and with approval from Congress. Officers were obtained in large numbers from the **Reserve Officers Training Corps (ROTC),** an organization in which military training was and is given to students at civilian educational institutions. In occupied and totalitarian countries, civilians were often forced into **labor battalions,** or **labor camps,** where rigorous manual labor, such as digging extensive trench systems, was demanded. Labor battalions were sometimes made up of prisoners, whose journey to a place of work or incarceration was often called a **death march** because of the dire conditions that brought death to so many along the route.

The Technology of the Age

The totality of the Second World War gave impetus to invention and discovery. The economics and attitudes that had prevailed in the early years after the Great War had slowed military research, development, and procurement; hence, except for the aggressor nations, Western military forces began

United States Rifle M-1—Garand—Caliber .30

the second great war with arms and equipment that were generally obsolete. The outbreak of the war, however, marked the beginning of a period of technological growth unequalled in the annals of history.

Striking Weapons

Swords continued to be carried for ceremonial purposes, and late in the war they were tokens of surrender for some and suicide instruments for others. The *samurai* **swords** of the ancient Japanese warriors became symbols of the tragedy that marked the end of war for the Japanese people. Bayonets and knives were issued to nearly all combatants, but were more often used for the opening of cans and idle whittling than for warlike purposes. The striking weapon remained dormant as a war instrument of consequence.

Missile Weapons

A tremendous variety of missile weapons were used by the armies of the Second World War. The basic rifle of the American infantry was the **M-1 Garand .30 caliber,** a gas-operated, eight-round semiautomatic. Developed in the late 1920s by John Garand, it was officially adopted by the Army in 1936 and remained the standard shoulder arm for over 20 years.[3] Another popular infantry weapon was the **M-1 carbine.** Smaller and lighter than the Garand rifle, it had a 15-round magazine and was carried by airborne troops. Toward the end of the war, the night effectiveness of both carbines and rifles was enhanced by the development of the **sniperscope,** an electronic device that used infrared rays to illuminate targets. The basic British infantry weapon was the **Lee Enfield .303,** which had a range of 2,000 yards and a 10-round magazine. The Soviets relied on 7.62-mm rifles, machineguns, and submachineguns as their principal infantry weapons. One of the most popular was the **PPSh-41 submachinegun;** by the late 1940s, over five million had been produced. The German infantry used the 7.9-mm **KAR-98K**

rifle, which had a five-round magazine. The Japanese basic infantry weapon was the **Model 38 6.5-mm rifle.**

Automatic weapons and machineguns served a variety of functions during the Second World War. In addition to their use by infantrymen, machineguns, especially, were installed in tank turrets, mounted in aircraft, and built into casemates and other defensive positions. Although the United States Army had adopted a light assault gun designed by American John Browning late in World War I, it was not put into production. Its successor, the **Browning automatic rifle 1918A1 (BAR),** was adopted in 1937. The 1918A1 had a tripod and a detachable 20-round box magazine. Modified versions of the A1 were the standard automatic rifles in American infantry squads of both World War II and the Korean War. The American infantry also used a Browning **light machinegun M1919A6** during World War II; it was a successor of the first air-cooled Browning, the M1919, which was a tank-mounted weapon. The **Thompson submachinegun,** which had a 30-round magazine, was also widely used by American ground forces. The British Army had a reliable light machinegun called the **Bren Gun - Mk I.** Developed during the 1930s from a Czechoslovakian model, the Bren Gun could fire single shots or short bursts. In 1942, the Germans adopted a light machinegun that was the first machinegun made from stamped, rather than machined, parts. Designated the **MG-42,** it had a locking system that allowed extremely fast barrel changing, and fired at a rate of 900 to 1,200 rounds per minute from either a metallic belt or 50-round drum. German paratroopers were armed with the **Schmeisser submachinegun,** which ironically was not developed by the noted gunsmith Hugo Schmeisser. The first weapon to be made entirely of steel and plastic, it fired the 9-mm parabellum cartridge and was officially designated the **MP-38** (*Maschinenpistole* 38). Modifications of the MP-38 led to the **MP-38/40** and the **MP-40.** Over a million MP-40s were produced between 1940 and 1945. The standard Japanese machinegun of World War II was the **Type II light machinegun.** Developed from the French Hotchkiss M1914 by General Nambu, it was adopted in 1922, and, like the M-38 rifle, used ammunition

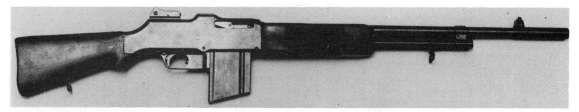

Browning Automatic Rifle—Model 1918—Caliber .30

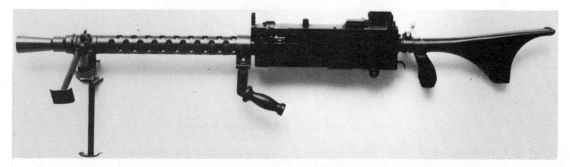

Browning Light Machinegun—M1919A6—Caliber .30

British Bren Gun-Mk I—Caliber .303

German MG-42 Machinegun—7.92 -mm

Japanese Type II Light Machinegun—6.5-mm

that was packaged in five-round stripper clips. The Scandinavian and Swiss Armies relied heavily on the **Suomi submachine gun** for automatic weapon fire. This weapon, which fired the 9-mm parabellum cartridge from either drum- or box-type magazines, was adopted by the Finnish Army in 1931. Popular heavy machineguns were the Allies' **Browning .50 caliber M-2** and the **Soviet 12.7mm Degtyarevs,** the latter of which could fire 600 rounds per minute. Most aircraft-mounted machineguns, like the **Browning .303,** and most **coaxial machineguns** (guns mechanically linked to fire in the same direction as a main tank gun), were in the .30-caliber range.

Many types of **antitank weapons,** that is, weapons used primarily against tanks, were used during the Second World War. Among these weapons were **assault guns,** which were either self-propelled or tank-mounted guns used for direct fire against point targets. **Bofors**—any of a variety of antitank, antiaircraft, and **antiboat guns** in the 37-mm to 57-mm range—were assault guns manufactured by the Bofors Armament Works in Sweden. The most popular of the Bofors was the 40-mm gun used by the United States Army and Navy, the British, and, to a lesser extent, the Russians and Germans. The Germans mounted an 88-mm gun on a tank chassis to create an effective antitank gun known as the **Elephant** or the **Ferdinand assault gun.** A hand-held antitank weapon, the **2.36-inch rocket launcher,** was developed by the United States. Nicknamed the **bazooka,** the German copy was known as the *Panzerfaust* (literally, the "armored fist"). As the armor on tank turrets and hulls was increasingly thickened, larger caliber antitank weapons were needed and developed. Shortly after the war, the 2.36-inch bazooka was replaced by the **3.5-inch rocket launcher,** or **super bazooka.** During the war, it was replaced by **recoilless rifles,** which were light tubes from which **recoil** (the reaction force created by the force imparted to the projectile) was eliminated by the

controlled, rearward escape of propellent gases through openings in the breechblock. The **57-mm recoilless rifle** was a shoulder-fired weapon; the **75-mm recoilless rifle** was fired from a portable tripod or from a light vehicle.

Necessity often dictated that World War II artillery be employed as antitank or antiaircraft weapons. One of the best weapons in this regard was the **German 88,** the same 88-mm gun that was accurate in direct fire roles. The German 88 had a range of 12,000 yards when employed as a field gun, and could fire eight rounds per minute to an altitude of 35,000 feet when employed as an antiaircraft gun. Included in the vast arsenal of German artillery pieces were the huge **600-mm self-propelled howitzer,** which fired a 3,746-pound projectile to a range of 7,300 yards or a 4,840-pound projectile to a range of 4,900 yards, and the **800-mm railroad gun,** nicknamed the **Gustav.** The Gustav was the largest railroad gun ever used. With a crew of 500 (commanded by a colonel), it fired a five-ton high explosive projectile to a range of 29 miles. The principal artillery pieces of the United States Army were the **105-mm howitzer,** the **155-mm howitzer,** and the **155-mm gun,** the last of which had a range in excess of 25,000 yards. The United States had used captured German 105s in World War I, and, as World War II was beginning, produced a variety of more sophisticated models with such innovations as pneumatic tires and a hydropneumatic recoil system. Numerous **self-propelled guns** were also produced during the war. These guns were mounted on motorized carriages—usually a tank chassis. Among them was the **Gun motor carriage M7,** a Sherman tank chassis mounting a 105-mm howitzer and a .50-caliber machinegun. Because of the pulpit-like turret that housed the .50, it was nicknamed the **Priest.** Japanese artillery consisted principally of **105-mm howitzers** and **150-mm howitzers.** Howitzers, because of their high angle of fire, were well suited to the jungle warfare that typified much of Japan's fighting.

United States 105-mm Howitzer—Model 101A1

Like artillery, mortars of a wide variety were used during World War II. The smallest was a Japanese model called a **knee mortar** because of the mistaken belief that it could be fired with its base resting on the knee. Officially, it was the **Japanese 50-mm grenade launcher.** The Japanese also employed the **320-mm spigot mortar,** which propelled a warhead (320-mm) larger than the bore of the mortar by means of a closed tube (or **spigot**) attached to the warhead.

United States forces used the **60-mm mortar,** which had a range of about 2,000 yards; the **81-mm mortar,** some models of which could attain a range of nearly 4,000 yards; and the **4.2-inch mortar,** which had been designed to fire chemical rounds. When gas warfare was outlawed, the 4.2 was used to fire conventional rounds to a range of 4,500 yards.

Rockets, which in previous eras had made little impact on the conduct of war, became formidable weapons in World

United States Gun Motor Carriage M7

War II. The Soviet **Katyusha rocket launcher** was a system of eight rails that could launch **82-mm rockets** or **132-mm rockets** to a range of nearly two miles. The American **4.5-inch multiple rocket launcher** consisted of a cluster of 25 tubes, each of which could fire a 39-pound rocket to a range of 6,000 yards.

The Germans, however, were the masters of rocket development. Their *Vergeltungswaffen* (revenge, retaliation, or vengeance weapons), or **V-weapons,** were the most advanced rockets that had been developed up to that time, and they marked the beginning of the space age. The **V-1,** the first of the V-weapons, was called the **flying bomb,** the **pilotless aircraft,** or the **buzz bomb.** From launch sites in France, more than 8,000 of these 400-mph jets, each with a one-ton warhead of high explosives, were launched against London on courses determined at **plotting stations.** The **V-2** was a **ballistic missile**—a missile that does not rely on aerodynamic surfaces, or **airfoils,** to produce lift. The V-2 generated lift through thrust. Weighing 13 tons, it carried a one-ton warhead, had a top speed of mach 5 (five times the speed of sound), and could climb to an altitude of 116 miles.

Ammunition and Other Implements of Destruction

As with **ordnance**—which includes all military weapons, ammunition, explosives, combat vehicles, battle materiel, and maintenance supplies—ammunition types and amounts increased dramatically during the Second World War.

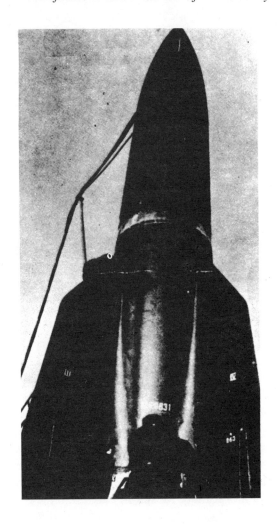

German V-2 Ballistic Missile

German V-1 Flying Bomb

Ammunition parks, which were places where ammunition was stored and distributed, were often unable to furnish the authorized **units of fire** (the units of measure for ammunition supply expressed in rounds per weapon per day). The variety of munitions, too, complicated the World War II ordnance situation. New types of **bursting rounds** (projectiles with an explosive charge designed to produce demolition, fragmentation, or chemical action) were introduced. Some were **semifixed ammunition**—ammunition in which the cartridge case can be readily detached from the projectile so that the charge in the cartridge case can be adjusted to obtain a desired range. Many bursting rounds used the new **variable time (VT) fuze,** or **proximity fuze,** which was designed to detonate a shell when activated by an external influence in the close vicinity of the target. **Armor piercing rounds,** which had a hardened steel casing designed to penetrate armor plate, were constantly improved during the war. **Armor piercing bombs** were also developed. Smoke generating devices (such as the simple **smoke pot**) and **illumination devices** (shells that contained a parachute and a brightly burning magnesium compound designed to light an area of suspected enemy activity) also were among the ordnance stores of World War II. "**Corkscrews**" (a slang term referring to explosives); **satchel charges** (blocks of explosives taped together and having a fuse and a rope or wire for carrying and attaching them to targets, such as bunkers or tanks); **shaped charges** (explosives shaped with a hollow that caused the blast to be directed at a desired point); and **time bombs** (explosive devices with a mechanism or fuse programmed to detonate the device at a predetermined time) were other popular ordnance items used in the Second World War. **Land mines,** which are explosives placed and generally concealed in the path of the enemy to impede his movement or to deny him access to an area, were also widely used.

The aerial dimension of World War II brought still other ordnance items into the inventory. The Japanese experimented with a **balloon bomb,** a bomb attached to a paper balloon that drifted across the Pacific at an altitude of about five miles. Solar-powered batteries released the bomb at a predetermined time, presumably when the balloon was over the United States. Out of the 1,000 balloons sent aloft, nearly 300 were recovered. Damage caused by the bombs was negligible. Enjoying greater success than the balloon bombs were the **fire bombs,** which contained an incendiary mixture (often a gasoline gel) that spread and was ignited on impact. The Allies used firebombs against cities in both Japan and Germany. **Napalm,** which is made from napthenic and palmitic acids, is a powder used to thicken gasoline for use in flamethrowers and firebombs. The firebomb containing napalm is often referred to simply as napalm, an effective incendiary weapon. One of the most efficient weapons for use against personnel targets was the **fragmentation bomb.** Usually, it was a thin steel tube filled with explosives. A square wire, which was spirally wound on the tube, was the principal source of fragments. **Glide bombs** were bombs fitted with airfoils to provide lift; hence, they could be released at a considerable distance from their target. Glide bombs that were remotely-controlled by radio and had control surfaces, like the rudders and elevators of aircraft, were used extensively during the war in Vietnam, and were nicknamed **smart bombs.**

Instruments of destruction developed during the war were more lethal, more sinister, and more sophisticated than weapons of any previous era. The **atomic bomb,** or **A-bomb,** was the weapon that had the single greatest impact on the war and on civilization in general. Moreover, it was the only totally new weapon in terms of its source of energy. The A-bomb's energy came from the natural splitting of uranium atoms. Only two bombs were produced during the war. The first, nicknamed the **Thin Man** or the **Little Boy,** weighed 9,700 pounds and had the force of 20,000 pounds of TNT. The Thin Man was dropped over Hiroshima, Japan on August 6, 1945. Its descent was retarded by a parachute, causing it to explode 800 feet above the ground. Over 78,000 Japanese were killed, nearly 40,000 were injured, and over 13,000 were missing as a result of the blast. The second bomb, nicknamed the **Fat Man,** was dropped over Nagasaki, Japan on August 9, 1945. Today, the world's major powers have a wide variety of **atomic weapons** (weapons that utilize atomic energy as their explosive source). The portion of an atomic projectile that carries the source of the atomic energy, usually U-235 atoms, is called the **warhead.** The term warhead also applies to any projectile armed with substances —be they biological, chemical, or explosive—intended to inflict damage. **Biological weapons** are weapons that use living organisms to inflict damage. **Bacteriological weapons,** for example, use bacteria to inflict damage through disease. **Radioactive poisons,** which can be used to contaminate water and food, provide another insidious, lethal, and effective means of incapacitating enemy personnel.

Mobility

The means that the fighting man used to gain mobility during the Second World War ranged from crude wagons, like the Russian *panje,* to jet-powered aircraft. The **diesel engine,** an internal-combustion engine in which air is compressed to a temperature sufficiently high to ignite fuel injected directly into a cylinder, provided power for heavy land and sea vehicles. **Four-wheel drive,** a system that supplies power to all four of a vehicle's wheels, enhanced the mobility of small vehicles in

Hiroshima "Little Boy"

unstable soil, mud, and snow. The variety of wheeled vehicles increased dramatically, ranging from two-wheeled motorcycles to trucks of all sizes. Although these were several more classifications, according to both size and use, the workhorses among the trucks were the one-quarter ton utility vehicle, commonly known as the **jeep;** the reliable **one and one-half ton truck;** and the **two and one-half ton truck,** known as the **deuce and a half** or the **twice and one-half.** The weight in the designation refers to the approximate weight carrying capacity of the vehicle. A two and a half ton **amphibious truck** used to ferry or land cargo or troops was called a **DUKW,** or, more commonly, a **duck.**

The aircraft that saw service in the Second World War were dramatically greater in number and type than those craft used in World War I, and new aerial terms kept pace with the burgeoning aircraft manufacturing industry. The term **airfield,** which at one time referred to a reasonably level and smooth field where planes could take off and land, later referred to the hard-surfaced runway area of an **airport,** which was an expanse of land where buildings, maintenance,

Nagasaki "Fat Man"

and passenger facilities existed along with an airfield. An **airbase** is generally a military facility that supports the activities associated with a contiguous airfield. An **airdrome** is chiefly a British term for an airbase. An **airstrip** is generally a primitive or unimproved landing area, especially one that is constructed in forward areas by military engineers.

Performance criteria of aircraft increased as each side tried to gain technological superiority over its opponent. **Range** is determined by the number of hours that an aircraft's fuel allows it to stay in the air, while **cruising speed** is affected by altitude, gross weight, and power used. Hence, **endurance,** which is measured at top speed and expressed in hours and minutes, is often used in lieu of range as a measure of a plane's performance. **Loiter time** is the maximum time that a plane can stay airborne; fuel consumption rates are minimized in determining loiter time. Loiter time, endurance, and range can all be increased by adding additional fuel tanks. The auxiliary fuel tanks that can be jettisoned when empty are called **drop tanks.** Because of the weight they add, however, drop tanks cut down on the **combat bombload,** the measure of the maximum weight of bombs that an aircraft can carry. Other important performance criteria include top speed, rate of climb, and service ceiling. The **service ceiling** is the highest altitude at which a rate of climb of 100 feet per minute can be attained. The **absolute ceiling** is the highest a plane can go, but it is a less meaningful measure of a plane's performance than service ceiling.

Aircraft have been classified in a variety of ways. Sometimes, the number of wings has been the criterion. Thus, a **monoplane** has one wing; a **biplane,** two; and a **triplane,** three. Aircraft also have been distinguished by the number and configuration of seats. For instance, there is the **single-seater,** which has one seat, and the **tandem,** which has two seats front to back. Planes are also classified according to their source of thrust. Using this method, there is the **propeller plane,** or **prop,** and the **jet,** which uses compressed, superheated air to move it through the atmosphere. Planes also have been categorized according to their function. Early aircraft, used for observation purposes, were called **reconnaissance planes.** When pilots began to drop grenades, which were at one time called bombs, their planes were called **reconnaissance bombers.** When the bombing became more important than the observation, the planes were called simply **bombers.** To rid the sky of the nuisance created by the bombers, designers attached machineguns to the wings or **cowling,** the covering over the engines, and the planes were known as **fighters.** To protect themselves against the fighters, the early bombers added machineguns to their arsenal and became known as **fighter-bombers.**

Better bombers meant heavier loads, and better fighters needed to be faster and more maneuverable. Hence, two new families of aircraft developed. Within the bomber family, those planes designed to carry light bomb loads for short distances were called **light bombers.** Light bombers were generally used against small targets during daylight, and those bombers assigned to daylight bombing missions were also known as **day bombers.** Bombers that generally flew higher and farther than light bombers, but that were slower because of heavier loads, were called **heavy bombers.** They were generally assigned to night bombing missions against area targets, and those bombers assigned to night bombing missions were called **night bombers.** The term **medium bomber** was applied to bombers that were intermediate in size. They were used for both day and night bombing. When ground forces needed a bomber that was effective against small targets located near the front, they called on the **dive bomber,** which the Navy had developed as an antiship weapon. To distinguish dive bombers from other bombers, the latter group was known as **level bombers.** The largest of the bombers in use in a given period of time were known as **line bombers.** Those that were capable of dropping torpedoes from the air were called **torpedo bombers.** Some bombers were used to strike targets—such as factories, supply depots, and ports—the destruction of which would have a long-term effect on the conduct of the war rather than an immediate effect on the outcome of a single battle. These were called **strategic bombers.**

Fighter aircraft, too, had a variety of subcategories. Those designed with a high rate of climb, good maneuverability, and short range in order to stop longer-range enemy bombers were called **interceptors.** Those fighters that had sufficient range or endurance to accompany bombers and sufficient maneuverability to be competitive with interceptors were called **escort fighters.** A **pursuit plane** was a small fighter having a high speed and a limited radius of action. A short-range fighter designed for high speed interception of enemy aircraft in conditions of good visibility and daylight was known as a **day fighter.**

Naval aviation, which encompasses all aircraft under the control of the Navy, used some unique terms to classify their aircraft. **Seaplanes** were planes that could land on and take off from water by utilizing floats or a flat-bottomed fuselage. Those seaplanes with floats were called **float planes;** those with flat-bottomed fuselages were called **flying boats.**

Until 1919, aircraft manufacturers assigned designations to their own aircraft, but afterwards the United States Army Air Service established a standard classification system: "P" designated pursuit and "B" designated bomber. The Navy used its own system of "F" for fighter, "A" for attack, and "P" for patrol. Other Navy designations applied such abbreviations as "TB" to torpedo bombers and "PB" to patrol bombers. The numeral in the Navy designations pro-

Mitsubishi A6M Type Zero

vided the model number, while a letter identified the manufacturer. Planes manufactured by Grumman, for example, used the letter "F"; those manufactured by Consolidated used the letter "Y". Hence the **PBY** was a patrol bomber manufactured by Consolidated. After June 11, 1948, all United States Air Force combat aircraft were designated either "F" for fighter or "B" for bomber.[4] Designations for reconnaissance aircraft were prefixed by the letter "R."

One of the first low-wing (the wing was attached low on the fuselage) monoplanes ordered by the United States Army was the Boeing **P-26.** It was developed because older biplanes, which had filled the pursuit role in World War I,

were unable to keep pace with newer, single-wing bombers. Its top speed was 234 miles per hour, and its wing was supported from the fuselage by wire cables. In 1936, the Army bought the Seversky **P-35,** a low-wing pursuit plane that could carry up to 300 pounds of bombs. Its top speed was 259 miles per hour. The first mass-produced low-wing fighter that had an engine behind the pilot was the Bell **Airacobra,** or **P-39.** A variety of models were produced from 1939 until 1944.

In general, the Airacobra was a disappointment because it was inferior to the Japanese Zero. Formally called the **Mitsubishi Type O,** and nicknamed the **Zero** or **Zeke,** the Zero

Curtis P-40

North American P-51 Mustang

was Japan's greatest fighter. Its top speed was 351 miles per hour, and it had a service ceiling of 33,000 feet. It was armed with one 7.7-mm machinegun and one 12.7-mm machinegun in the fuselage, and four 20-mm cannon in the wings.

The Curtis **Hawk 75A** was the first American fighter used in the Second World War. Its maximum speed was in excess of 300 miles per hour, and its service ceiling was 32,800 feet. Although a dependable aircraft, it was soon replaced by more advanced designs, like the long line of Curtis **P-40s,** which had a maximum speed of 357 miles per hour. The P-40's service ceiling was 50 feet lower than the Hawk's, but it could climb over 700 feet per minute faster than the Hawk. P-40s delivered to Great Britain were designated **Tomahawks.** Later model P-40s were called **Warhawks** in the United States and **Kittyhawks** in Great Britain. The North American **P-51 Mustang** was built to replace the P-40 at the request of the

British. It was the finest American fighter of the war. The P-51D had a top speed of 437 miles per hour and was armed with six .50-caliber machineguns. The Republic **P-47 Thunderbolt,** which had a top speed of 429 miles per hour and was armed with eight .50-caliber machineguns, was the largest and heaviest single-seat, piston-engine fighter ever built. It could also carry ten 5-inch rockets. Over 15,660 Thunderbolts of various models were produced during the war.

The **Messerschmitt Bf-109** was Germany's standard fighter for over a decade. First used in the Spanish Civil War, it had a top speed of 354 miles per hour and was armed with two 20-mm cannon and two 7.9-mm machineguns. During the Battle of Britain in the fall of 1940, however, the Messerschmitt 109 met its match in the British **Supermarine Spitfire,** which had evolved from a 1931 racer. The Spitfire

Republic P-47 Thunderbolt

Messerschmitt Bf-109

was the only Allied fighter in continuous production throughout the war. The prototype first flew in March 1936, and the last model, a Mark 22, was delivered in 1947. It was armed with eight .303-caliber Browning machineguns, and later models attained speeds in excess of 450 miles per hour.

In addition to the Messerschmitt 109, other prominent German fighters included the Focke-Wulfe 190, the Messerschmitt Bf-110, and the Messerschmitt 262. The **Focke-Wulfe Fw-190** first saw combat in 1941, at which time it was superior to every Allied fighter then in service. Fortunately for the Allies, an Fw-190A landed intact in Britain in 1942, and many of its features were copied. It had a top speed of 408 miles per hour and was armed with four 20-mm cannon and two 13-mm machineguns. The **Messerschmitt**

Bf-110 was a twin-engine airplane that was used as a long-range day fighter, a night fighter, and an escort fighter. First flown in the 1930s, various versions of the aircraft were produced throughout World War II. The **Messerschmitt 262** was the first military jet-propelled combat aircraft, and the only jet to see extensive service in World War II. It could climb to 38,400 feet with a rocket booster in four and a half minutes, and had a top speed of 540 miles per hour at that altitude. The Me-262 was armed with four 30-mm cannon and could carry twenty-four 50-mm rockets.

The United States Navy developed fighter aircraft that were capable of landing and taking off from aircraft carriers. Because storage space on carriers was limited, naval aircraft had to be compact—with either small wings or a means of

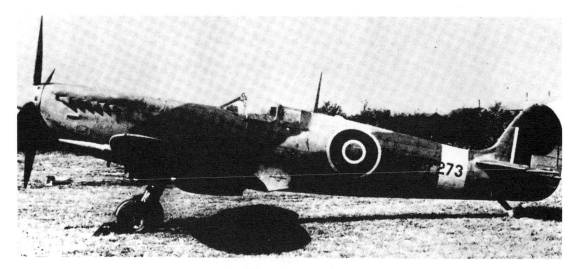

Vickers Supermarine Spitfire

Focke-Wulfe Fw-190A

folding large ones. The first monoplane fighter used by the Navy was the barrel-shaped **Brewster Buffalo.** Constantly outperformed by the Zero, the Buffalo was fortunately replaced early in the war by Grumman **Wildcats.** The Wildcats, like the Buffalo, were slower than the Zero, but had better climb and maneuverability because of their large, low wings. Their pilots were generally better trained than the Japanese, and their tactical doctrine was well-suited to the capabilities of the aircraft. By 1943, Grumman had produced its last Wildcat, and the Grumman plants turned to the more advanced Hellcat. The Grumman **F6F Hellcat** was first flown in 1942. Its speed was 386 miles per hour, and it was armed with six .50-caliber machineguns and either two 1,000-pound bombs or six 5-inch rockets. The Vought **F4U Corsair** was superior to its contemporary naval aircraft, but its pilots had difficulty landing on carriers because of poor visibility from the cockpit. Nevertheless, the plane, which was the first American fighter to exceed 400 miles per hour in level flight, was used extensively by the Marines.

Early in the war, the British used a few biplanes, including the Albacore and the Gladiator. The **Albacore** was a torpedo bomber, and the **Gladiator,** first flown in 1934, was a land-based fighter armed with four .303-caliber Browning machineguns. The **Hawker Hurricane** and the **Beaufighter,** both monoplanes, played decisive roles in the Battle of Britain. The former, which was the first combat aircraft to exceed 300 miles per hour in level flight, was armed with eight .303-caliber Browning machineguns. The Beaufighter was a night fighter; with the aid of radar, it effectively stopped the *Luftwaffe's* night air raids directed against England in 1940

Messerschmitt Me-262 Schwalbe

Yakovlev Yak-3

Boeing B-17F Flying Fortress

Consolidated B-24H Liberator

and 1941. Another British night fighter was the **Bolton Paul Defiant.** It had a power-operated turret that housed four .303-caliber Brownings behind the cockpit.

The French relied early in the war on their **Morane-Saulnier M.S. 406,** a single-engine, single-seat, low-wing monoplane that flew at speeds in excess of 300 miles per hour, and the **Dewoitine D. 520,** a single-seat, low-wing fighter-interceptor that went into large-scale production in 1940. The D. 520 had a maximum speed of 329 miles per hour and was armed with one 20-mm Hispano-Suiza cannon and four 7.5-mm machineguns. Russia produced several of its own fighters, and also used many American models provided through lend-lease. Among the early Soviet fighters were the I-15 and I-16; "I" is the abbreviation for *Istrebital,* which means fighter. These fighters were superseded by the **MiGs,** a designation that credited the designers of the aircraft (Artem Mikoyan and Mikhail Gurevich) and the **Yaks,** which were designed by Aleksander Yakovlev, who copied the Spitfire and Me-109.

A great variety of bombers, too, were used during World War II. In the late 1930s, the Douglas Corporation produced a versatile design known as the **A-20** (the "A" designation meant attack or light bombardment), or the **Havoc.** Later models of the twin-engine A-20 were adapted to night fighting and redesignated as **P-70s.** In 1934, the United States Army issued specifications for a multirange bomber with a bomb load and range that were twice as great as existing models. The result was the **B-17,** known popularly as the **Flying Fortress.** The top speed of the B-17 was about 300 miles

per hour, and it typically carried a bomb load of 4,000 to 5,000 pounds over operating ranges averaging 1,400 miles. The development of the **B-24,** or **Liberator,** began in 1939, and before the war had ended, more B-24s were produced in the United States than any other bomber. Its maximum speed was 300 miles per hour, and its range with a 5,000-pound bomb load was 1,700 miles. It was often used for clandestine transport missions. The first **B-25** flew in August 1940. It was a twin-engine bomber with squared-off lines that made production easier. Its range was 1,300 miles with a 3,000-pound bomb load. The **B-29,** or **Superfortress,** was the largest of the World War II bomber fleet. This four-engine heavy bomber, which was first flown in 1942, was used exclusively against Japan. Its raids culminated in the single-plane attack on Hiroshima by the *Enola Gay,* which dropped the Thin Man. The Superfortress had a top speed of 357 miles per hour and an armament of twelve .50-caliber machineguns and one 20-mm cannon. It could carry a 20,000-pound bomb load for short distances. In all American heavy bombers used during the war, bombing accuracy was greatly improved by the **Norden bombsight,** a gyroscopically stabilized bombsight that computed the correct bombing angle and course necessary to get a bomber and its bombs to their targets.

As with fighters, the United States Navy had its own family of bombers. One of the most successful of these was a flying boat known as the **Catalina PBY,** the **Catalina,** or simply the **PBY.** This twin-engine patrol bomber was first flown in 1935 and was in continuous production for 10 years. It could carry four 1,000-pound bombs, twelve 100-pound bombs, four

Boeing B-29 Superfortress

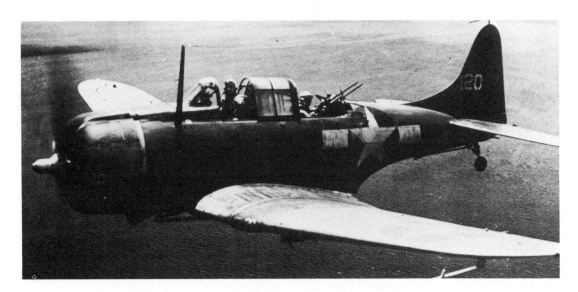

Douglass SBD Dauntless

650-pound depth bombs, or two torpedoes. Its greatest contribution to the war effort, however, was made in its capacity as a patrol plane. The principal torpedo bomber used by the Navy during the war was the single-engine Grumman **Avenger TBF,** which replaced the Douglas **TBD-1 Devastator.** The standard Navy dive bomber was the Douglas **Dauntless,** which was probably the best dive bomber of the war. During the Battle of Midway, Dauntlesses were credited with the sinking of four Japanese carriers. Later models of the Dauntless had a top speed in excess of 250 miles per hour, were armed with two .30-caliber and two .50-caliber machineguns, and could carry bomb loads of up to 1,000 pounds.

Although they generally relied on designs and shipments of American manufacturers, other Allied nations did develop a few bombers. In 1936, the French developed and produced a single-engine torpedo bomber and reconnaissance, twin-float seaplane. Known as the **Latécoère,** it was armed with three 7.5-mm machineguns and carried one 1,477-pound torpedo, three depth charges, or two 330-pound bombs. The **Martin 167** bomber was developed in the United States, but when no purchase contracts were offered by the Government, France purchased a number of the aircraft. After the fall of France in June 1940, the Martins were diverted to Great Britain where they were known as the **Maryland I.** One of the earliest and best British torpedo bombers to be used in the war was the **Swordfish.** Although primarily a carrier-based aircraft, some twin-float models were catapulted from British cruisers and battleships. After flying, the plane landed on the water near its mother ship, and was returned to the deck by a large crane. The Swordfish was replaced by the Albacore, also a two-wing torpedo bomber.

The British **Lancaster** carried a heavier bomb load than

any other bomber in the war—including the B-29. The Lancaster's 22,000-pound "Grand Slam" was a weapon for which the bomb bay doors had to be removed and the bomb bay extended fore and aft. Typically, the Lancaster carried bomb loads between 7,000 and 14,000 pounds. It began flying with the RAF in April 1942, had a top speed of 287 miles per hour, and was armed with eight .303-caliber machineguns.

The Soviet air force relied heavily on its **SB-3 bomber;** SB was the abbreviation for *Srednii bombovos,* which translates into "medium bomber."

The Germans used a wide variety of bombers. Among them was the **Dornier Do-17.** A medium bomber developed in 1935, it was armed with six 7.9-mm machineguns and was capable of carrying a 2,200-pound internal bomb load. The **Heinkel He-111** was a twin-engine light bomber that could

Avro Lancaster Mk I

Junkers Ju-87 Stuka

carry a bomb load of 6,000 pounds. The most famous German bombers came from the Junkers factory. The **Junkers 87,** popularly known as the **Stuka,** was a single-engine dive bomber that in the early years of the war was among the most feared weapons in the German arsenal. Although it was an extremely accurate bomber, it was slow and hence vulnerable, as was shown during the Battle of Britain. Its top speed was 232 miles per hour, and it could carry 1,100 pounds of bombs. The **Junkers Ju-88** was one of the most versatile aircraft of the war. It flew at speeds up to 279 miles per hour and could carry a bomb load of 5,000 pounds. It was also used for reconnaissance and as a pilotless missile that was guided to its target by a single-engine fighter temporarily attached to it.

The most prominent Japanese bombers of the war were code-named the Val, the Kate, and the Betty. The **Val** was a single-engine, carrier-based dive bomber that had a top speed of 266 miles per hour, a bomb load of 823 pounds, and an armament of three 7.7-mm machineguns. The **Kate** was a single-engine, low-wing torpedo bomber with a speed of 225 miles per hour. It carried one torpedo and two 7.7-mm machineguns. The **Betty** was the most versatile of Japan's medium bombers. It had a range of 2,262 miles and could carry 2,200 pounds of bombs or torpedoes. Due to ineffective self-sealing fuel tanks, it was nicknamed the **"flying cigar."**

In addition to fighting and bombing, World War II aircraft performed other significant tasks. The **Tiger Moth,** for example, was a single-engine biplane that was widely used as a trainer for the Royal Air Force. The **Fiesler Storch,** or **Storch,** was a German single-engine, high-wing transport and reconnaissance aircraft. Light scouting aircraft used to direct artillery were generally referred to as **grasshoppers.** American reconnaissance and artillery direction missions were often conducted from the **Pipercub Grasshopper,** a single-engine monoplane.

During World War II, transport became an increasingly

Mitsubishi G4M Betty

Gliders: American Waco CG-4A and German DFS-230A

important function of the air forces. Aircraft like the **Junkers Ju-52,** a twin-engine, low-wing model, and the **Douglas C-47,** the twin-engine transport that provided reliable service from 1935 through the Vietnam War, were used extensively to haul troops and supplies. The C-47, which carried 8,000 pounds of cargo or 28 soldiers, was called the **R4D** by the Navy, the **Dakota** by the British, and the **Gooney Bird** by almost everyone.

Gliders, which are fixed-wing aircraft having no power plant, were designed to carry relatively heavy loads to areas that could be practically reached only from the air. The American **Waco CG-4A** could carry 13 soldiers while the German **DFS-230A** carried 10 and the British **Horsa** carried 30. Gliders were normally carried aloft by powered aircraft called **tows.** A Junker Ju-52 could tow up to 10 DFS-230As.

Glider: British Horsa With United States Markings

Although tanks were as much a part of *blitzkrieg* as airplanes, the versatility and roles of tanks were far more limited. The tracked vehicle that provided heavy armor protection and was the principal assault weapon of an army's armored and infantry troops was that army's **main battle tank,** but the actual term was not used until after the war. In addition to a main battle tank, each army of World War II generally employed a number of smaller tanks, which were either medium tanks or light tanks, and special purpose tanks, which were nicknamed **"funnies"** by the British. Among the special purpose tanks were flail tanks, plow tanks, ramp tanks, dual drive tanks, and flamethrower tanks. The **flail tank** was equipped with a series of chains attached to a roller on the front of the tank. When the roller was driven by the tank's engine, the chains beat on the ground ahead of the tank to detonate antitank mines. **Plow tanks** had a blade mounted in front of each tank to dig up mines, but in hard ground the system was ineffective. **Ramp tanks** carried a small bridge or ramp that could be placed over craters or chasms to allow the passage of tanks. One of the British ramp tanks was called the **Armored Ramp Karrier,** or **ARK. Duplex drive,** or **dual drive tanks,** could either operate on land or "swim" by using a propeller attached to the engine. Some tanks lowered curtains from their hulls, thus providing enough water displacement to enable the tanks to float. **Flamethrower tanks,** as well as other special purpose tanks, were generally fashioned from obsolete tanks. Using

pressurized carbon dioxide or nitrogen, the flamethrower tank could project ignited thickened fuel to a distance of about 75 yards.

The first American tank to be introduced in the war was the **M3 medium tank.** Deliveries to the British of these vehicles, with slight modifications of the turret, were begun in 1942. **General Lee** was the British designation given to unmodified M-3s, while the modified versions were dubbed the **General Grant.**[5] Grants had a profound effect on Allied fortunes in the North African desert. The M3 weighed from 30 to 32 tons, and its armament consisted of one 75-mm gun, one 37-mm gun mounted in combination with a .30-caliber Browning machinegun in the turret, and three additional .30-caliber Browning machineguns. **M4 medium tanks,** which came to be known as the **General Sherman,** or simply as **Shermans,** were designed as a direct result of the success of the low velocity 75-mm guns employed by the Germans during the May 1940 *blitzkrieg* in France. Shermans mounted the 75-mm gun and two .30-caliber machineguns, had thicker armor than the Grants and Lees, and weighed over 36 tons. In the final six weeks of the war against Germany, the **M26 tank** was issued to American armored divisions. Originally designated a medium tank, it was changed in 1944 to a heavy tank, and after the war was again called a medium tank. Nicknamed the **General Pershing,** it was the first tank to mount a 90-mm gun. In addition, it had one .30-caliber machinegun in the bow, one .30-caliber machinegun

A "Funny": United States M4 (Sherman) With Flailing Attachment

United States Medium Tanks, M3

United States Medium Tank, M4—General Sherman

United States Medium Tank, M26—General Pershing

British Medium Tank, Mk II—Matilda

mounted coaxially, and one .50-caliber machinegun atop the turret. It could travel 30 miles per hour on roads, and weighed 46 tons.

British tanks were generally too slow and too thinly armored. However, the **Matilda,** which was introduced in 1939, was the most heavily armored tank of its day. With a speed of only 15 miles per hour and a weight of 29.7 tons, the Matilda mounted a 2-pounder main gun and a .303-caliber machinegun. It was succeeded by the **Crusader,** which had a 2-pounder main gun until 1942 and a 6-pounder main gun

afterwards, a speed of 27.5 miles per hour, and a weight of 22.1 tons. The British heavy tank was the **Churchill.** Introduced in 1941, it had a 6-pounder main gun, two 7.92-mm machineguns, a speed of 17 miles per hour, and a weight of 43.1 tons. Throughout the war, the British relied heavily on tanks designed and produced in the United States, especially the **M3/M5 light tank** (called **Stuarts** by the British), the Grants, and the Shermans.[6] Cast armor was first used in tank production by the French in their *Char léger* (light tank) **1935R.** This Renault tank mounted the low velocity 37-mm

British Medium Tank, Mk III—Crusader

British Heavy Tank, Mk IV–Churchill

United States Light Tank, M5A1 (37-mm Gun)–Stuart

French Light Tank, *Char Léger* 1935R

Soviet Medium Tank, T-34

gun retained from World War I. Like British and early American tanks, its armor was too thin and its main gun too light to counter the German *blitzkrieg.*

The Russian **T-26 light tank** was produced from 1932 until 1939. It mounted a machinegun in each of two turrets, a machinegun in one turret and a 37-mm gun in the other turret, or a 45-mm gun and a coaxial machinegun in a single turret. The various models weighed from 7 to 10.3 tons. A convertible wheel/track tank was also produced in the Soviet Union in the 1930s. These **BTs** (*Bystrokhodnic* **tank,** or "fast tank") could travel up to 70 miles per hour on wheels and 40 miles per hour on tracks. They used a **Christie pattern suspension,** which meant that each wheel had independent recoil springs, and large **bogie wheels,** which were the weight-bearing wheels on the inside perimeter of the track that served

to keep the track in place. The BTs were succeeded by the **T-34,** which has been called the "most formidable, and probably the best tank of the war.'" Introduced in 1940, the T-34 mounted a 76.2-mm main gun and two 7.62-mm machineguns. It had a crew of four, a weight of 32 tons, and could travel at a speed of 33 miles per hour. The **T-34/85** was essentially a T-34 with the heavier 85-mm main gun. The 85-mm gun's armor-piercing round could penetrate 4.7 inches of armor at 100 yards at a right angle of impact. The **Klementy Voroshilov (KV) heavy tank** was first introduced in 1939. Weighing 52 tons, early models mounted the 76.2-mm gun, while some later models incorporated the huge 152-mm howitzer into their design. Early models of the KV tank were succeeded by the **Joseph Stalin (JS) tank,** of which three models were produced: the **JS-I, JS-II,** and **JS-III.** The JS-III

German Light Tank, PzKw I Model B

German Medium Tank, PzKw III

had a 122-mm gun, a 7.62-mm machinegun, and a 12.7-mm machinegun. The tank weighed 57 tons, and its top speed was about 20 miles per hour.

The Germans used the term *Panzerkampfwagen* (armored combat vehicle) and numerical suffixes to describe their tanks. The *Panzerkampfwagen I*, or **PzKw I**, was introduced in 1934, weighed 6.4 tons, and was armed with only two 7.9-mm machineguns. The **PzKw II** was introduced in 1938. Weighing 9.8 tons, it was armed with a 20-mm gun and a 7.9-mm coaxial machinegun. Production of the first German medium tank, the **PzKw III**, began in 1937. Early models, continuing the trend toward heavier armor and bigger guns shown in the development of the PzKw I and PzKw II, mounted a 37-mm gun, one 7.92-mm machinegun that was coaxially mounted, and another 7.9-mm machinegun that was mounted in the hull front. In 1940, the **PzKw IIIE** Model appeared with a 50-mm main gun. Shortly after production began on the PzKw III, an even heavier tank with a 75-mm gun was designed. Over 8,000 of these **PzKw IVs** were produced during the war; no German tank was produced in greater numbers. **PzKw V**, better known as the **Panther,** was a heavy tank—although the Germans considered it a medium tank—that had the reputation of being one of the best tanks on any side in the war. It weighed from 45 to 50 tons and had a **torsion bar suspension,** a system that absorbed shock by twisting a transverse rod or bar, which when it untwisted returned the road wheels and hence the track to its normal position. Its main gun was a 75-mm KwK 42, and it mounted

German Heavy Tank, PzKw V–Panther

German Heavy Tank, PzKw VI–Tiger

two 7.9-mm machineguns—one coaxially and one in the hull. The chassis of the Panther was also used to mount an 88-mm gun; such a configuration was known as the *Jagdpanther,* or **Hunting Panther.** It was a highly respected **tank destroyer,** a term that after the war applied to any self-propelled antitank gun. In the **PzKw VI,** or **Tiger,** the 88-mm gun was mounted in the turret, while the two 7.9-mm machineguns characteristic of other German tanks continued to be mounted in the turret and the bow respectively. The Tiger, which was also known as the **Mark VI,** weighed over 62 tons, and its glacis plate was 100mm thick. **Glacis armor** was intended to protect the tank crew and equipment against missiles, just as the glacis of the typical eighteenth century fortification had protected its inhabitants.

Motorized machines dominated the battlefields and the strategy of World War II. Whether a **tank dozer,** which was a tank fitted with a broad blade to fill craters, a **bulldozer,** which was a machine used to fill craters and otherwise alter the earth's surface, or an **armored personnel carrier,** which was a self-powered, armored troop carrier, the war was rarely fought or sustained without the power afforded by internal-combustion engines.

Naval Objects

Throughout World War II, the navies played a far more decisive role in the outcome of the war than they had in any previous major modern war. Moreover, for the first time they were concerned with operations and activities in three arenas: below the sea, upon the sea, and in the air. From beneath the sea, submarines threatened the surface fleet and logistical lifelines. Frogmen and **blockships,** which were ships sunk to

block a channel or a harbor entrance, also influenced the war from beneath the sea. Surface fleets sought to destroy enemy ships, to support amphibious operations, and to project the power of the nation they represented against established objectives. The aircraft carrier became the principal vessel of the fleet, and carrier-based aircraft became a key element in the defeat of the enemy in and near the water.

Ships increased in size, number, and variety. From the **bridge** (the elevated area above a ship's decks from which the ship is controlled) to the **keel** (the longitudinal member that forms the bottom center of the hull) ships were transformed. **Optics,** which is the system of prisms, lenses, and mirrors that facilitate man's ability to see and engage targets and to navigate, improved dramatically. The guns of **main batteries,** which are the groupings of the largest guns on a particular vessel, were increased in caliber and length. The main batteries of most American battleships consisted of nine 16-inch guns. Japanese battleships generally had **14-inch batteries** but some had **18-inch batteries.** The main batteries of destroyers generally consisted of **5-inch naval guns.** The main batteries of World War II warships were enclosed by walls of armor called turrets or **casemates.** The guns protruded from the turret through embrasures like those of eighteenth century fortifications.

World War II torpedoes were devastating weapons. Although generally fired from submarines, they could be launched from surface ships by firing over the **gunwales,** that part of the ship where the hull meets the main deck. To protect a ship from torpedo attack, **torpedo nets** made of steel links were sometimes stretched by booms or attached to floats around a ship, or were extended across a harbor entrance. If a torpedo struck a cargo ship's hull, more than **bilge water,** which was the seepage and condensation that lay in the lowest

Fleet Submarine

part of a ship's hull, would accumulate in the **hold**, that area below decks where cargo was carried.

Throughout the war, submarines were active in the Atlantic, Pacific, and Indian Oceans. The largest submarines in a particular fleet were called **fleet submarines.** During World War II, the United States fleet submarine, which early in the war was a **Tautog** and in later years was a **Gato class submarine,*** was over 300 feet long and could travel at about 20 knots on the surface and 10 knots submerged. At 2.5 knots, it could stay submerged for about 48 hours. It was armed with one or two 5-inch deck guns, a 40-mm gun, a 20-mm gun, and ten torpedo tubes. The Japanese fleet submarines were known as **I-boats.** They were larger and faster than the American boats, but had no radar and could not dive as well. Although their torpedoes were generally superior to Allied torpedoes, I-boats could carry only six to eight torpedoes per vessel.

S-boats, known familiarly as **pig boats,** were the older and smaller American submarines used early in the war. They had five torpedo tubes, a surface speed of 14.5 knots, and a submerged speed of 11 knots. The Japanese attempted to use **midget submarines,** which had a crew of two, carried a single torpedo, and were employed in surprise attacks. Late in the war, the Japanese employed one-man suicide submarines called **Kaiten I.** They weighed about eight tons, had a warhead consisting of 1.7 tons of TNT, and were about 48 feet long. Other Japanese **suicide boats** were called **Shinyo;** they were motor boats designed to ram enemy vessels. Each carried either two tons of TNT in its bow or two depth charges that were strapped across the outer part of the bow.

A **depth charge** is a barrel of explosives that is dropped or catapulted from watercraft for use against underwater targets.

Antisubmarine warfare, or **ASW,** encompasses all of the methods used to locate, track, and destroy enemy underwater vessels. The most significant advances made between the wars in antisubmarine warfare involved the discovery and development of improved detection devices. **Hydrophone listening devices,** which were microphones placed underwater, had been used in ASW in World War I, but they could detect submarines only at relatively close ranges and only when the sub was using its noisy diesel engines. During the 1930s, **echo ranging equipment,** which sent out bursts of energy and could determine the distance, direction, and approximate speed of any object that reflected the energy, was studied in both the United States and Great Britain. When developed, the British echo ranging equipment was called **ASDIC,** an acronym for the name of the committee that headed the project (the Anti-Submarine Detection Investigation Committee). The American detection system was called **sonar,** an acronym for "sound, navigation, and ranging." The energy emitted by sonar was in the form of high frequency sound waves. Once located, submarines became targets for depth charges. These charges were dropped principally from destroyers, destroyer escorts, and **subchasers,** which were fast boats designed for antisubmarine missions. **Depth bombs,** which were lightly cased aerial bombs designed to explode underwater, were also employed against submarines. One of the principal targets on the submarine was the **conning tower,** the command and communications center of the ship. Early in the war, the Germans developed a **snorkel,** which consisted of tubes for air intake and exhaust that allowed the diesel submarine engines to run while the vessel was submerged at periscope depth. The

---

*The term **class** refers to all the vessels of a given design. The class takes the name of the first vessel commissioned of that particular design.

Battleship– USS *Washington*

snorkel complicated the task of ASW fighters because it allowed the submarine to use its powerful engines while submerged—either to charge its batteries for future crises or to travel at speeds up to 14 knots. Submarines had to rely on electrical storage batteries to provide energy to the vessel when it was deeply submerged, however, and on battery power a submarine could only make about 7 knots.

During World War I, submarines had dominated the naval battles. In World War II, however, the **aircraft carrier** took its place as the dominant vessel. The Japanese employed some **seaplane carriers,** which could launch seaplanes, but the aircraft had to land in the water and then be lifted back onto the carrier when the operation was complete. Aircraft carriers had flush, full-length flight decks for the launching and landing of aircraft, and huge elevators to move aircraft from their hangers to the flight deck. The largest and fastest carriers in the American fleet were called the **fleet carriers.** They generally carried a complement of about 85 aircraft. The **escort carrier** was a converted merchant ship that provided air cover for convoys. Because aircraft carriers, which were nicknamed **flattops,** were **thinskinned,** that is, having little armor protection, they generally operated as part of a carrier task force.

Early in the war, the battleship was the major fighting vessel of the high seas fleet, but when the tally at the Battle of Midway was assessed, the aircraft carrier had proven that its

role was dominant. The battleship was a warship that emphasized armor, firepower, and speed, in that order; the battle cruiser emphasized speed first, then firepower, and finally armor. The **pocket battleship** was not a battleship but a heavily-gunned, armored cruiser built by the Germans to evade the prohibitions imposed by the Treaty of Versailles. In 1940, the *Kriegsmarine* re-rated the pocket battleships as heavy cruisers. By World War II, the destroyer had evolved into a versatile escort vessel. Smaller destroyers built after

Destroyer–USS *Willard Keith*

1941, which were designed to protect convoys against U-boats, were called **destroyer escorts.** When destroyers formed a picket line to contribute to the defense of larger vessels, the destroyers were referred to as **screen destroyers.** Smaller than destroyer escorts, the **PT (patrol torpedo) boat** was a small, fast, inexpensive vessel that carried two to four torpedoes. PT boats posed a serious threat to surface vessels of every size. Though limited in range and unable to operate in rough seas, their speed, number, and damage-producing capability contributed immensely to the combat potential of the fleet.

Special purpose naval vessels include **mine layers** and **buoy layers,** which lay sea mines and mark buoys respectively. **Target ships** are generally old or damaged vessels that serve as targets for naval gunfire practice. **Tenders** are repair ships designed to support a specific type of combat vessel.

Battleship	BB	Submarine	SS
Heavy Cruiser	CA	Submarine Tender	AS
Light Cruiser	CL	Mine Layer	CM
Aircraft Carrier	CV	Mine Sweeper	AM
Escort Carrier	CVE	Hospital Ship	AH
Destroyer	DD	Cargo Ship	AK
Destroyer Escort	DE	Oiler	AO
Destroyer Transport	APD	Gasoline Tanker	AOG

Abbreviations for Selected United States Navy Vessels

Logistical requirements for the enlarged navies that fought in the Second World War increased in proportion to the number of ships and sailors involved. Along major shipping routes, outposts that were established in the late nineteenth century as **coaling stations,** places where coal was stockpiled for steam-powered vessels, served as advanced logistical bases. **Anchorages,** areas near bases where large vessels could assemble and drop anchor, were established in appropriate places. Sometimes **bombardons,** cylindrical floats lashed together and anchored to the sea floor, were emplaced to provide a breakwater to protect the anchorage. Where natural harbors were lacking, **artificial harbors,** which consisted of huge barges that could be towed to a spot and sunk to provide a pier for offloading ships, were used.

Ships used in commerce or under contract to haul cargo or passengers were called **merchantmen.** One of the most popular types of merchantmen built in the United States during the war was the **Liberty Ship.** Its design was sufficiently simple to allow it to be readily mass-produced. Japanese transport vessels were called **Marus.** As new destroyers with sophisticated antisubmarine and antiaircraft weapons systems replaced earlier models, the older ships were converted into

troop carriers called **destroyer transports.** Military ships that carried petroleum for replenishment at sea were called **fleet oilers. Tankers** were nonmilitary ships that carried gasoline and other petroleum products. Small fishing and sailing boats that were used to transport supplies along seacoasts were called **luggers.** The long flatboats used in the Orient to move cargo along coasts, or in rivers and harbors, were called **sampans.** They often had a thatched roof over the middle of the vessel, sometimes had a sail, and were generally powered by an oar from the stern. **Junks** were seagoing ships used in Chinese and other waters; they had square sails, a high stern, and usually a flat bottom. Seagoing canoes that had a shaped log attached to prevent the canoe from capsizing were called **outriggers.** They, too, were used extensively in the Pacific for hauling light cargo.

Landing craft were especially designed to move troops, equipment, and supplies during amphibious operations. **Beaching craft** were those boats or vehicles capable of moving onto and over the beach. In the 1920s and 1930s, a New Orleans boat designer, Alexander Higgins, built a 36-foot landing boat that was purchased by the United States Navy and called the **Higgins Boat.** Many subsequent designs for landing craft were based on the Higgins Boat. The ship that carried landing craft to the transport area near the landing beach was called the **mother ship.** From the mother ship, a wide variety of landing craft could be lowered. The **LCM,** or **Landing Craft Mechanized,** was an assault boat capable of beaching and discharging personnel and tanks. Any amphibious vehicle that could move a tank from a larger ship to a beach was called a **tank lighter.** The Marines purchased a number of **LVT(1)s,** or **Landing Vehicle Tracked (1),** which were amphibious personnel carriers that obviated the need for the tank lighter and that were used extensively in the Pacific during World War II and the Vietnam War. The LVT(1) was referred to as an **amphibious tractor** and hence nicknamed the **"amtrac."** It was also called the **Alligator.** The **LVT(A)** was an LVT with a 37-mm gun turret of the Stuart light tank or a 75-mm howitzer turret of the M8 gun motor carriage. The LVT(A) was an **amphibious tank,** and like the LVT(1), it could float and could travel at about seven and a half miles per hour in the water. The **LCVP,** or **Landing Craft Vehicle Personnel,** was a small amphibious assault boat that was capable of beaching. Among the largest of the landing craft was the **LST,** or **Landing Ship Tank.** Nicknamed the **Large, Slow Target,** it was 316 feet long, had a speed of seven knots, and could carry 211 men. In addition to transporting tanks, personnel, and other vehicles, the LST could also carry smaller landing craft. Some small landing craft were armed with rocket-launching batteries and were called **rocket ships.**

Landing Ship Tank (LST)

Defensive Items

During World War II, new weapons necessitated a search for new and better defenses. To protect against air attacks, **barrage balloons,** which were balloons restrained by a ground cable to which nets or wires were attached to impede enemy aircraft, were used. To protect aircraft from the effects of shrapnel, revetments around aircraft parking positions called **blast pens** were built. Tank and other vehicle movement was impeded by the construction of **tank ditches** or, more properly, **antitank ditches.** Tanks and landing craft were sometimes stopped by **tetrahedrons,** which were pyramid-shaped obstacles constructed of structural steel and steel tubing. Infantrymen sought protection from small arms fire in **blockhouses,** which were square concrete shelters often used as command and communication centers. Russian pillboxes or blockhouses were called *tochka,* a term that the Japanese also adopted. Any extensive arrangement of one or more fortified places was called a *Festung* by the Germans. The defensive positions that connected and were diagonal to successive defensive positions parallel to the front were called **switch positions.** Less elaborate field fortifications were called **bunkers,** and if a machinegun was present in a bunker or in a semi-improved firing position, the term **machinegun nest** was applicable. The greatest impediment to tank and vehicular movement was the land mine. **Mine detectors,** which were electrical or magnetic devices used to locate camouflaged or buried mines, were useful, but the time needed to clear an area, especially a large area, often gave the enemy the time that he needed. Also, the Germans employed **wooden box mines,** which could not be detected with a mine detector because they had no metallic parts.

Communications/Electronics

Communications-related items, like items in every other area of technological endeavor, increased dramatically in variety and sophistication both before and during the Second World War. At one extreme, **crystal sets,** which were tiny receivers that had a crystal detector and neither vacuum tubes nor transistors, allowed their users to hear broadcasts of news about the war. Such broadcasts often included code words that were meaningful to the resistance or other clandestine operators. The **crystal** was the key part of the set, for it determined the frequency that was being received. The crystal also determined the frequency in transmitters. At the other extreme, the massive **Chain Home Radar** system provided a means for British pilots and planners to keep informed of German air activities. Built in 1938, the system consisted of 20 radar stations that provided radar cover for the British North Sea and Channel coasts. **Radar,** which is an acronym for "radio detection and ranging," was an **air warning and interceptor system** that projected a radio wave against a reflecting target. The measurement of the wave's travel provided information on the distance, speed, and direction of the target. High power and short wave lengths were needed to extend the range of radar, and this combination was achieved in 1940 when a group of scientists invented the **magnetron,** a cavity resonator that functioned as a transmitting valve at high frequencies. The magnetron also allowed sets of lower power and compact size to be built; hence, aircraft could carry radar to assist in locating targets that might otherwise be obscured by clouds or darkness. The original term for radar—**RDF,** or **Radio Direction Finding**—was used to conceal the principle behind the system's operation. **Radio detection stations,** which used **radio direction finders** to locate a radio transmitter by determining the azimuth of incoming radio waves, were far less sophisticated than radar stations. Nonetheless, throughout the war, radio detection provided valuable **electronic intelligence,** or **ELINT,** a term that refers to intelligence derived from electromagnetic radiation sources. Radar could detect a target without the target's transmitting any radio signal, and thus was the key to an effective **Ground Controlled Intercept,** or **GCI,** system, a system in which a ground radar station first located enemy aircraft and then used signals to direct friendly fighter planes to the enemy craft. In order for GCI to be effective, a means of distinguishing friendly aircraft from enemy aircraft was necessary. The **Identification Friend or Foe,** or **IFF,** system used electronic transmissions to which equipment carried by friendly forces automatically responded by emitting pulses. The nature of the pulses gave the system its nickname, **"pip squeak."** Some detection devices responded to sound rather than radio frequency. These **sound locating sets** could be used to determine the location of gun batteries, generators, or significant vehicle movement.

Electronic countermeasures, or **ECM,** refers to all actions and methods taken to confuse or frustrate enemy ELINT activities. **Radio silence,** which refers to a policy of not transmitting during specified periods of time, is an electronic countermeasure. **Radio simulation,** which refers to the transmission of bogus information, is an **electronic deception measure,** and thus indirectly an electronic countermeasure. Other ECM related activities involve the use of **scrambler radios,** which disarrange transmissions in order to make them unintelligible, the use of **call signs,** which are code words assigned to identify a given user, and the use of codes and ciphers. **Codes** are words, letters, or symbols that have an arbitrarily assigned substitute meaning. **Ciphers** transform a text by substituting letters for other letters or by rearranging the letters of the plain text in a predetermined way.

Disease

In spite of technological and medical discoveries, disease continued to have a profound effect on the fighting strength of the armies of World War II. Particularly rampant in tropical and semi-tropical regions were malaria; **dengue fever,** an acute infectious disease transmitted by mosquitoes; **beriberi,** a disease caused by a lack of or the inability to assimilate vitamin B_1; and **scrub typhus,** or **tsutsugamushi disease,** which, like louse-borne typhus, brought acute fever. **Quinine,** a bitter antimalarial alkaloid, and **atabrine,** another bitter antimalarial drug, were common antidotes, but even when remedies were readily available, disease took a fairly high toll

in casualties. **Sick bays,** the medical rooms on a ship, and army medical facilities were also filled with patients suffering from **war neurosis** and **battle fatigue,** maladies that were as harmful to mind and body as were the bacteriological and viral diseases.

Although World War II technology contributed to greater suffering and destruction, it also brought countermeasures to ease the pain and either prevent or rebuild the destructiveness of total war. The tremendous technological advances made during the war were considerably influenced by the emphasis given to **research and development,** or **R and D,** which is defined as the effort directed toward the increased knowledge of natural phenomena and environment and the application of that knowledge, in support of national goals.

Operations

While the operations of World War II were strongly influenced by civilian leaders, many military officers, in turn, strongly influenced decisions relating to politics and societies. In total war, the distinctions between military, political, and social affairs were often indiscernible, and Clausewitz's dictum that war is a continuation of policy by other means seemed confirmed at the higher levels of authority. **Declarations of war,** which are formal diplomatic statements that announce that a condition of hostility exists between one nation and other specified nations, were issued by most governments. In practice, however, they proved to matter little when mass armies were set in motion. **Suzerainty,** which refers to the control of one nation over the affairs of another nation or area, was determined by force rather than agreement, but the choice to use force remained a political choice. When land areas were **annexed** (placed under the political control of another nation), the choice was political and the means military. Politicians established the broad guidelines, and military planners worked out the details. For example, politicians could decide who would be attacked and generally when, but the choice of **D-Day** or **X-Day,** that is, the actual day that a given operation is to begin, was left to senior military commanders. **Rules of engagement,** or **ROE,** which are the directives that delineate the circumstances under which forces initiate or continue combat engagement, were dictated by political considerations.

The multinational aspect of the war meant that **integrated commands,** commands where appointments to key positions were made without regard to nationality or service, often determined strategic options. Moreover, *modus vivendi,* which refers to workable compromise on issues in dispute without permanent agreement, often prevailed. Commanders, however, were faced by more than political constraints. The problem of limited resources often dictated when, where, and if a *coup de main* (a sudden attack in force) or a *coup de grace* (a decisive finishing act) could be carried out. **Attack on a broad front,** which refers to a concurrent advance of all major units, was workable in theory; but when resources were limited, the advance on a broad front usually proved impractical. **Choke points,** restricted areas through which large numbers of troops or supplies must pass, were often dictated by geography and enemy dispositions. Choke points influenced strategy, and when strategy failed, the options of politicians and soldiers became grim: soldiers often had to **exfiltrate,** or slip out of enemy-controlled areas. Governments had to pay **indemnities,** which were financial penalties imposed upon defeated nations, and defeat brought geographical adjustments. The victors tried to establish a *cordon sanitaire,* a series of buffer states or territories intended to isolate a possible aggressor in a future war.

The methods of the Second World War were as diverse as the possible outcomes were severe. For the United States, World War II was a grand-scale *punitive expedition,* a military operation undertaken to punish a belligerent. It was a series of sudden attacks, or **forays,** intended to inflict the greatest possible amount of damage upon the enemy. **Terror attacks, terror bombing,** and **terror raids,** which suggest that the object of the activity was to strike fear into the hearts and souls of every man, woman, and child possible, were forbidden on occasion, but the totality of the war meant that even these less-than-humane activities were often encouraged.

Tactics

World War II tactics, the ordered arrangements used in fighting or in the movement of combat organizations when fighting may be imminent, were considerably different from the static defense and **wave assaults** that had characterized World War I. **Night attacks,** which require closer coordination, more compact formations, and better morale than daylight attacks, were far more prevalent in World War II than in the Great War. The use of triangular organizations led to the nearly classic solution of almost every problem of applied tactics: **two up and one back.** This tactical application meant that the company with three platoons would typically advance or defend with two platoons on line and one in reserve, the battalion with three rifle companies would advance or defend with two companies on line and one in reserve, and so forth. When terrain or the enemy situation dictated, one of the units on line might be halted and bypassed by adjacent units, with its position on the line assumed by the adjacent units. The bypassed unit then assumed a reserve role, and was said to have been **pinched out.** When units advanced along

narrow corridors within which artillery was forbidden to fire, the corridors were called **no-fire lanes. No-fire lines,** or **artillery control lines,** which were lines short of which artillery could not fire, were other **artillery control measures.**

A common defensive term used in World War II was the **main line of resistance,** or **MLR,** which referred to the line along which an enemy attack would be met by organized resistance that included the coordinated fire of infantry weapons, artillery, and sometimes naval guns. The line of outposts to the front of the MLR was called the **breaker line.** Behind the MLR, units prepared **fall-back positions** when time and terrain permitted. When units were isolated, they often had to form a perimeter or cordon defense around some strongpoint. In World War II, this technique was often referred to as *Igel* **(hedgehog) tactics,** because when a hedgehog, a furry European porcupine, is threatened, it curls itself into a ball, with bristles pointing in all directions. Tanks are not intended to be defensive weapons, but when the situation dictated, they often dug in until their hulls were obscured, which was called **hull defilade,** or until their hulls and turrets were obscured, a condition known as **turret defilade.**

By the 1920s, the ready availability of maps of suitable scale permitted their distribution to the lowest unit levels. Hence, in World War II, graphic techniques that facilitated tactical control of units came into common usage. The term **control measures** refers to the techniques that are represented graphically on maps as a means of assigning activities and responsibilities to operational units. Some common control measures include the designation of zones or sectors, axis of advance, direction of attack or route, lines of departure, control point, and coordinating point. Control measures are displayed on **maps,** which are graphic representations of a portion of the earth's surface drawn to scale. **Overlays,** which are the translucent sheets that graphically delineate the instructions found in operation orders and letters of instruction, are placed on or laid over maps. A **zone,** or sector, is a control measure that is represented by a solid line that runs perpendicularly to the front. It indicates that the unit whose symbol appears at a broken point on the line is responsible for holding or taking the outlined area. The term zone, or **zone of action,** is generally used when a force is advancing; **sector** is generally used when a force is defending. An **axis of advance,** or simply an **axis,** gives greater flexibility to a maneuvering force than zones or sectors because the axis of advance designates only the general direction in which the force must move. An axis of advance is graphically represented by a broad arrow. The **direction of attack,** or **route,** is highly restrictive because it designates the roads or trails that a unit must follow; it is represented by a line with an arrowhead. A **contact point** is a place on the ground where ad-

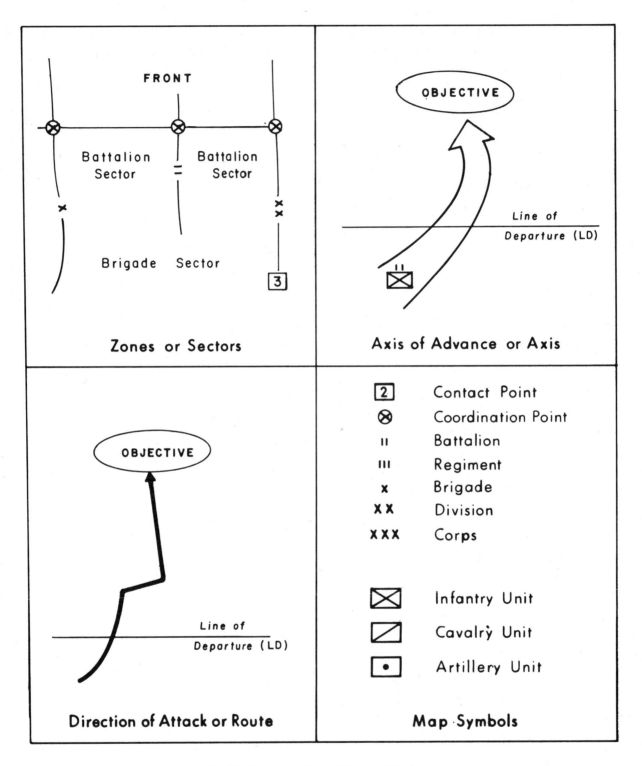

Graphic Representations of Control Measures

jacent units must physically meet; it is represented by a number enclosed in a small square. A **coordinating point** is a place on the ground at the boundary between units where interlocking fires must be planned. A **line of departure,** or

jump-off point, is normally located on or behind that terrain feature closest to the enemy which can be reached without exposure to fire. The line of departure is the boundary that must be crossed at the time designated in an order by the lead element of a unit in order to coordinate the start of a large-scale attack. It is represented by a solid line that is marked "line of departure" or **LD. Phase lines,** which also are represented by solid lines, generally correspond to terrain features that extend across the width of a zone of action. Units can be easily halted at phase lines, and must report the crossing of phase lines.

Operational terms that refer primarily to the enemy include **avenue of approach,** a term that is used in determining the alternative routes that the enemy can use to reach a friendly position; **dispositions,** which refers to the manner in which a unit is deployed; and **deception plan,** which is the statement of the measures to be taken to mislead the enemy into reacting in a manner detrimental to his interests.

Air Operations

Aerial activity was often as important as ground fighting in World War II. Even the **non-rated,** those who were not licensed pilots, quickly learned the importance of **air cover,** the condition in which a ground force had **close air support (CAS)** in the form of fighters available for the strafing and bombing of any enemy forces encountered. When aircraft provide cover for other aircraft or surface forces, they were said to provide an **air umbrella.** Because of the large numbers of aircraft operating in given areas, **air control centers (ACCs)** were established to coordinate and control the allocation and tasking of aircraft. The ACC decided when each aircraft under its control would **sortie,** that is, be dispatched on an operational mission. ACCs also kept track of aircraft flying **ferry routes,** which were routes used to deliver aircraft from factories and bases to operational areas. When carpet bombing was selected as the method of attack, the ACC dispatched all its aircraft to a single area. **Carpet bombing,** or the laying of an **air carpet,** was the dropping of a massive bomb load on an area defined by designated boundaries. The intention was to inflict as much bomb damage on the defined area as possible. Carpet bombing generally preceded an attempted breakthrough of a strongly held enemy position by ground forces, and hence was accomplished in fairly close proximity to friendly frontlines. Most bombing missions were conducted behind the enemy lines in an attempt to interdict the movement of troops and supplies or to strike at strategic targets. **Air raids** were bombing attacks directed against military facilities or cities. An **air strike** was generally a bombing or strafing sortie conducted in support of ground operations. When bombers attempted to fly deep into enemy territory, they often required a **fighter escort,** which meant that fighter aircraft capable of engaging enemy fighters in dogfights accompanied the slower and more vulnerable

bombers. If, as usually occurred, the fighters and bombers took off from different fields, the two groups would meet at a predetermined **rendezvous area.** At a **release point,** units or individual aircraft reverted to the control of their unit or aircraft commander. **Ground fire,** which was antiaircraft fire from surface weapons, always posed a threat to airmen, but except in close-support missions and in the vicinity of protected targets, aircraft could generally avoid surface weapons.

World War II aircraft performed a variety of tasks. Reconnaissance aircraft often took **aerial photographs,** which when scrutinized could reveal valuable intelligence. Reconnaissance aircraft also participated in searches, which were referred to as **single phase searches** when only one pass over a given area was made, and **multiple phase searches** when two or more passes were made. Transport planes were responsible for the **airlifting,** or air movement, of troops and supplies. Bombers were capable of skip bombing, long-range patrol bombing, saturation bombing, precision bombing, and area bombing. **Skip bombing** was a method of aerial bombing in which the bomb was released at such a low altitude that it skipped or glanced along the surface and struck its target above the level of the surface. **Long-range patrol bombing** was a naval air specialty that combined reconnaissance with the bombing of targets of opportunity. **Saturation bombing** was intended to leave nothing in a given area free from destructive effects. **Precision bombing,** which was generally accomplished during hours of good visibility, directed its efforts against targets of comparatively small bulk or area. **Area bombing** was generally done at night or during periods of poor visibility; cities were apt targets for area bombing.

Air Defense

Because of the devastating effect of bombs dropped from high altitude aircraft, **air defense,** which refers to all measures taken to reduce or nullify the effectiveness of hostile air attacks (by both bombers and missiles), became a principal concern in World War II. **Air defense artillery (ADA),** which includes weapons and equipment designed or utilized to combat air targets from the ground, increased in importance, while **civil defense,** which refers to the measures taken by civil authorities to protect and minimize damage to civilians and civilian property during any hostile attack, became a principal concern of local and national governments. **Air raid drills** were conducted to instruct individuals and institutions in the proper actions to take in the case of actual attack. **Alerts,** which were called when attack was imminent, put communities in the best possible posture to survive and minimize the effects of the impending attack.

Airborne Operations

An **airborne operation** involves the movement and delivery of combat forces and their combat support and combat service support elements by air into an objective area for the execution of a tactical or strategic mission. An administrative air movement of personnel, supplies, or equipment is not an airborne operation, even though some of the procedures used in an airborne operation may be applicable. Airborne operations are generally joint operations involving army and air force units. The ability of airborne forces to move rapidly and to land on or near their objective improves the chances of gaining surprise and facilitates the massing of combat power. The presence of airborne forces constitutes a threat that affects the enemy's capabilities by compelling him to deploy his combat power to protect vital installations to his rear as well as in the combat zone. During World War II, in both the European and Pacific theaters, all the major combatant nations conducted airborne operations, achieving varying degrees of success. Perhaps the most successful and decisive operation, if admittedly very expensive in personnel losses, was the German seizure of Crete in 1941. Operations that are intended to prevent enemy incursions that do occur, and to limit the damage caused by natural disaster, are today called **rear area protection,** or **RAP,** operations.

Success in airborne operations depends on the ability of a force to move from the **marshalling area** (where units are organized for air movement) through the **staging area** (where units and equipment make their final preparations for the assault) to the objective—without incurring undue losses. Weather is a key factor in airborne operations, and minimum acceptable conditions are prescribed for each operation. **Administrative drops,** which are generally conducted by paratroopers without any fighting gear, are used for training purposes. The **approach flight** is the final portion of the aircraft's route to the **drop zone,** or **DZ,** the area into which troops and equipment are to parachute. This flight is generally made at very low altitude in order to escape detection by enemy radar. If an aircraft commander prematurely gives the **green light** by signaling for the drop to begin, a **short drop,** or a landing short of the drop zone, occurs. Following the drop, the assault echelons organize as rapidly as possible and rely on aggressive small-unit action to seize initial objectives before the enemy can react effectively. After seizing initial objectives, all units move to occupy the **airhead** line, which insures the continuous airlanding of troops and materiel and provides the maneuver space necessary for projected operations. The **airlanded elements** arrive in gliders when crude fields offer the best landing area, and in transport aircraft when airstrips or airfields are a part of the airhead. Airborne operations generally conclude when the assault echelons are evacuated or when a link-up with conventional ground forces is effected.

Naval Operations

Navies were instrumental in the outcome of many land and air battles in World War II. Especially in the Pacific, joint operations were the rule. From **shipyards** (places where ships were built or repaired) or **drydocks** (storage places where ships could be held out of the water during construction or repair operations) to **scuttling** (the intentional cutting of holes in the bottom of a ship in order to sink it), the life cycles of ships were inextricably linked with the events of war. Ships were an integral part of amphibious operations and **landing operations,** the term used for movements ashore that were unopposed by the enemy. **Shore parties,** which were task organizations responsible for facilitating the movement of men, equipment, and supplies onto and from beaches, undoubtedly preferred to support landing operations rather than amphibious operations. At sea, the navies of World War II continued to use those techniques employed by the first cannon-carrying sailing ships. For instance, **crossing the T,** a technique in which a line of ships passes ahead of and perpendicular to a line of enemy ships, enabled the crossing line of ships to bring more of its guns against the crossed line and thus enjoy the advantages of raking fire. This naval maneuver, of course, did not guarantee victory, just as adherence to any principle of war will not always bring success.

World War II navies had to contend not only with enemy threats from the sea, but also with those from the air. Whenever a threat of engagement from any enemy force was imminent, a condition known as **general quarters** was ordered by the ship's commander. During general quarters, **all hands**—everyone—went to **battle stations,** a term that refers to a sailor's place of duty during naval engagements. For example, during engagements, the engineers on a ship were assigned to duties associated with **damage control.** This term refers to all the measures necessary to preserve or re-establish a ship's watertightness, stability, maneuverability, or firepower; to effect rapid repair of damaged materiel; to extinguish, limit the spread of, and protect personnel and materiel from fire; to limit the spread of, remove the contamination caused by, and provide protection from toxic agents; and to provide for the wounded. When damage control was ineffective, a sailor's only friend was likely to be his **Mae West,** a buoyant lifejacket that caused him to resemble the buxom Hollywood star of the same name.

Intelligence

Like nearly every other aspect of the war, the intelligence operations of World War II were unprecedented in scope, intensity, and sophistication. **Clandestine operations** (any activity that is conducted secretly), **espionage** (spying, especially the spying conducted by agents of one nation against another nation), and **counterespionage** (measures taken to render espionage ineffective) had long been a part of war. In World War II, however, new techniques of discerning and impeding enemy intentions were discovered. **Photo reconnaissance,** the taking of aerial photographs for intelligence purposes, was an effective means of converting **soft information,** or unconfirmed reports, into hard intelligence. Radio intercept proved to be one of the greatest intelligence windfalls of the war when the German cipher system, known to the Allies as **Enigma,** was unraveled by British intelligence. The code name for the project that involved Enigma was **Ultra. Double agents,** who appear to be working in behalf of one side while actually remaining loyal to the opposite side, operated in situations of high risk, as did **stay-behind agents,** individuals who remained in enemy-occupied territory to support intelligence operations. Stay-behind agents were often involved in acts of **subversion** (attempts to undermine the legitimacy of an established military or political authority) and **sabotage** (the malicious damaging of property with the intention of weakening an economic system or a government during a period of national emergency or war). Double agents and stay-behind agents were too often the victims of **summary executions,** the retributive killing of a person without the formality of trial or hearing.

Logistics

The dramatic increase in the scope and intensity of war meant that the systems of supply and services had to increase concomitantly. **Bulk supplies,** which are supplies that are used in very large quantities, were often **bulk loaded,** or stored in such a way that the entire cargo capacity of a vessel was utilized without regard for the order, integrity, or unity of the items involved. **Combat loading,** on the other hand, involved the loading of troops with their essential equipment and initial combat supplies in such a way that rapid debarkation in a desired priority could be achieved. Combat loading often involved **palletizing,** the securement of supplies on a **pallet,** or platform, that could be readily moved by a forklift or crane. **C-rations,** or **combat rations,** the canned meals that sustained many soldiers when kitchens were too remote to be of practical value, were often palletized for rapid delivery to an **advance base,** a place near the front where supplies and services were available to support combat operations. One of the major problems inherent in the complex logistical system of World War II involved **lead time,** the time that passed between the recognition of a materiel need and the fulfillment of that need. Existing bases and lodgments kept the lead time for most items at a minimum. In previous wars, the term **lodgment** had referred to the place in which soldiers were quartered. In World War II, however, the term referred to the base area that resulted when two or more beachheads, or two or more airheads, were consolidated. Even though operational history focuses on the movement of the hard-fighting combat troops, the role and critical importance of logistics should never be overlooked in assessing the reasons for the success or failure of any military force.

Selected Bibliography

The number and variety of books on the era and personalities of World War II are legion. Reference works like the *Rand McNally Encyclopedia of World War II,* edited by John Keegan, abound. Histories, biographies, fiction, and nonfiction fill the shelves of bookstores and libraries, and the number of volumes increases almost daily.

A good account of the events in Europe that preceded the war is found in Raymond Sontag's *A Broken World, 1919-1939.* The best single volume on the Spanish Civil War and its relationship to world and European events is Gabriel Jackson's *The Spanish Republic and the Civil War, 1931-39.* Hitler's role during both the interwar years and his tenure as war leader is well treated in *Hitler: A Study in Tyranny* by Alan Bullock. Hitler's relationship with his generals is the topic of Harold Deutsch's excellent *Hitler and His Generals* and Sir Basil H. Liddell Hart's *The German Generals Talk.* Both reveal candid recollections of the Hitler years based on the authors' talks with prominent German generals immediately after the war. The high commanders on the Allied side and their methods and ambitions are the topic of Herbert Feis' *Churchill, Roosevelt and Stalin: The War They Waged and the Peace They Sought.* The only member of the Big Three to publish his own interpretation and recollection of the war was Churchill. His six-volume masterpiece, entitled *The Second World War* (Volume 1: *The Gathering Storm;* Volume 2: *Their Finest Hour;* Volume 3: *The Grand Alliance;* Volume 4: *The Hinge of Fate;* Volume 5: *Closing the Ring;* Volume 6: *Triumph and Tragedy*), is a monumental history and a testimony to its brilliant statesman-author.

The most complete and detailed version of the war is told in the *U.S. Army in World War II.* This official history, a 78-volume series, is nicknamed the "green books" because of the books' green covers. It consists of 12 subseries, which are nearly all multi-volumed: the War Department, the Army Ground Forces, the Army Service Forces, the Western Hemisphere, the War in the Pacific, European Theater of Operations, Mediterranean Theater of Operations, the Middle East Theater, the China-Burma-India Theater, the Technical Services, Special Studies, and a Pictorial Record.

Less detailed, but still thorough in their coverage, are the two volumes on the war published as the West Point Military History Series by Avery Publishing Group, Inc. *The Second World War: Europe and the Mediterranean* and *The Second World War: Asia and the Pacific.* Still briefer, and containing the excellent maps of Edward J. Krasnoborski, is the *West Point Atlas of American Wars;* most of volume two (1900—1954) of this work is devoted to World War II. An excellent single-volume account of the war that includes a discussion of the impact of war on society as well as brief coverage of the war's operational and political dimensions is Gordon Wright's *The Ordeal of Total War.* A recent and excellent account of America's war against Japan is Ronald Spector's *Eagle Against the Sun.* The story of the war as it involved the United States Navy is found in Samuel E. Morison's officially commissioned and extremely well written *Two Ocean War,* while Morison's multi-volume work *The United States Navy in the Second World War* provides detailed coverage of actions in all theaters of operations. The war in Europe is the topic of Chester Wilmot's *The Struggle for Europe,* and Charles B. MacDonald's

The Mighty Endeavor focuses on American ground forces in the European Theater. Telford Taylor tells the story of Great Britain's battle against Germany during the critical months of 1941 in *The Breaking Wave,* and Russia's fight with Germany is the topic of Albert Seaton's *The Russo-German War, 1941-45.* A Russian perspective of one phase of the war is offered in Vasili Chiukov's *The Battle for Stalingrad.*

The contributions of intelligence to the war effort are covered in Anthony Cave Brown's *Bodyguard of Lies,* which deals extensively with the intelligence preparations for the Normandy invasion; John C. Masterman's *The Double Cross System in the War of 1939 to 1945,* which explores the role and activities of double agents; and Frederick Winterbotham's provocative and sometimes overstated *The Ultra Secret.* The best books on Ultra, however, are the more recent ones: Ronald Lewin's *Ultra Goes to War* and Ralph Bennett's *Ultra in the West: The Normandy Campaign, 1944-1945.*

As a result of the continuing Ultra revelations, Correlli Barnett recently revised his popular work on North Africa, *The Desert Generals.* An exceptionally readable and human trilogy from the pen of journalist-author Cornelius Ryan includes *The Longest Day* (D-Day, June 6, 1944), *A Bridge Too Far* (Operation MARKET-GARDEN), and *The Last Battle* (Berlin 1945). Two new accounts of the events at Normandy are John Keegan's *Six Armies at Normandy* and Carlo D'Este's controversial *Decision in Normandy.* Recent first-hand information on the British airborne landing at Arnhem is found in Major General John Frost's *A Drop Too Many.* Two of the best and most recent works devoted to operations that involved American intelligence failures during the war are Gordon Prange's *At Dawn We Slept: The Untold Story of Pearl Harbor* and Charles MacDonald's *A Time for Trumpets: The Untold Story of the Battle of the Bulge.*

Biographies and first-person accounts by senior military commanders provide another library of war resources. Douglas MacArthur's *Reminiscences* deals with his life from boyhood through retirement, while the middle chapters provide excellent insights into his thoughts on and contributions to the Second World War—at least as he remembered them in his twilight years. The most detailed biography of MacArthur is D. Clayton James' multi-volume *The Years of MacArthur.* Other biographies that focus on the war in the Pacific and Far East include Thomas B. Buell's *The Quiet Warrior: A Biography of Raymond A. Spruance* and Barbara Tuchman's *Stilwell and the American Experience in China 1911-1945.* The United States Army Chief of Staff throughout the war is the subject of Forrest C. Pogue's excellent three-volume biography entitled *George C. Marshall.* The United States Chief of Naval Operations is the subject of Thomas Buell's *Ally of the Trident: A Biography of Fleet Admiral Ernest J. King.* Dwight Eisenhower offers his interpretations of the war in his *Crusade in Europe.* Omar Bradley's *A Soldier's Story,* George Patton's *War as I Knew It,* and Lucien Truscott's *Command Missions, A Personal Story* provide first-person observations on the American participation in the war. Recent and definitive biographies of Patton and Eisenhower are Martin Blumenson's *The Patton Papers* and Stephen Ambrose's *Eisenhower;* both are based on thorough studies of primary sources. General Bradley devoted the last year of his life to a very controversial and sometimes self-centered autobiography entitled *A General's Life.* A resoundingly British view of the war is found in *The Memoirs of Field-Marshal The Viscount Montgomery of Alamein. K.G.* The best British autobiography is First Viscount William Slim's *Defeat into Victory.* German views from the vantage point of high commanders are well presented in Heinz Guderian's *Panzer Leader* and Erich von Manstein's *Lost Victories.*

John S. D. Eisenhower and James Gavin provide valuable insights into the activities and considerations relating to significant aspects of the war in their respective histories: *The Bitter Woods,* which focuses on the Ardennes and the "Bulge," and *On to Berlin,* which treats the events and strategy that preceded the Cold War. The story of Stalin's war with Germany and de Gaulle's fight for France are extremely well told in John Erickson's *The Road to Berlin* and Sir Bernard Ledwidge's *De Gaulle.*

No bibliography of World War II could be complete without some mention of the compelling human stories that made the war both tragic and heroic. Ernie Pyle was one of the best-liked journalists of the war, and his *Brave Men,* in title and content, suggests some of the reasons for his popularity. Life among the troops also inspired one of the wittiest and most humorous books on the war: Bill Mauldin's *Up Front,* which is illustrated with his famous Willie and Joe cartoons. Novels provide insights into war that are beyond the scope of nonfiction works. Ernest Hemingway's *For Whom The Bell Tolls,* which concerns the Spanish Civil

War, and his *Farewell to Arms,* which concerns the war in Italy, are powerful classics that convey the impact of war on man and mankind as no other medium can. Norman Mailer's *The Naked and the Dead,* and Herman Wouk's sequential yet independent works, *The Winds of War* and *War and Remembrance,* deal with the war in a precise and moving style.

Far removed from the novel, but perhaps more significant as they affect the lives of men and the conduct of war, are the doctrinal source materials that guided and directed the American war machine. Official manuals were widely used in the Second World War, and the topics and number of manuals increased dramatically during the war. Even before the American entry into the war, commanders and staff personnel at all echelons were relying on the earliest of the manuals that were newly designated as Field Manuals, or FMs. Those FMs in use early in the war were the *Field Service Regulations, Operations,* first referred to as FM 100-5 in 1939; FM 27-10 *Rules of Land Warfare;* FM 100-10 *Field Service Regulations, Administration;* and FM 100-15 *Field Service Regulations, Larger Units.* As technology and corresponding methodology changed, new manuals appeared. Today, many hundreds of FMs exist; among them is the guide to airborne operations, FM 57-1/AFM 2-51, *U.S. Army/U.S. Air Force Doctrine for Airborne Operations.* Less interesting than novels to most readers, FMs nevertheless remain the lifeblood of well trained modern armies.

Notes

[1]Understandably, much conjecture is involved in arriving at casualty figures for a war as vast and extreme as World War II. Alan Reid in his *A Concise Encyclopedia of the Second World War* (Reading, Berkshire, 1974), p. 213, reports that over 19,000,000 "servicemen died" in the war against Italy and Germany and that nearly 6,000,000 "members of the armed forces died" in the Pacific War. In Appendix B ("Populations, Mobilizations, and Losses of Life in World War II") to Quincy Wright's *A Study of War* (Chicago, 1965), pp. 1541–1543, the number of "soldiers killed or died of wounds" is reported as 16,933,000, but this figure apparently does not include nonhostile deaths, such as those resulting from disease and accident. Reid's 25 million figure, which appears to include death of servicemen from all causes, seems reasonable in light of the Wright appendix.

[2]Francis J. Kelly, *U.S. Army Special Forces, 1961–1971* (Washington D.C., 1973), pp. 3–4.

[3]The early history of the Garand is clouded by some imprecise reporting. Ian V. Hogg and John Weeks in *Military Small Arms of the 20th Century* (London, 1977), p. 183, state that the Garand was "adopted in 1932, began to enter service in 1936, and by 1941 a large part of the American regular army had it as their standard arm." Harold L. Peterson in the *Encyclopedia of Small Arms* (New York, 1964), p. 141, states that "the Garand rifle was adopted by the United States Army on January 9, 1936." Joseph W. Shields, Jr. in *From Flintlock to M1* (New York, 1954), p.

172, also states that the M1 was adopted in 1936, but John Kirk and Robert Young, Jr. in *Great Weapons of World War II* (New York, 1961), p. 312, state that the Garand was "accepted as the basic U.S. infantry weapon in 1940." Barnes in *Weapons of World War II* (New York, 1947), p. 17, writes that in 1939, upon receipt of additional funds, the Ordnance Department "accelerated work on this rifle and made plans for its mass production." Sources agree that the M-1 was in the hands of MacArthur's troops during the fall of the Philippines, and many report that George Patton wrote in 1945 that "the M1 rifle is the greatest battle implement ever devised." See, for example, Barnes, p. 21.

[4]Ray Wagner, *American Combat Planes* (Garden City, NY, 1960), p. 47.

[5]Peter Chamberlain and Chris Ellis, "M3 Medium (Lee/Grant)" in Duncan Crow (ed.), *American Fighting Vehicles of World War II* (Garden City, NY, 1972), pp. 33–52.

[6]Some sources refer to the Stuarts (U.S. designations: M2 light tank, M3 light tank, and M5 light tank) as medium tanks, which confuses them with U.S. M3 medium tanks (the Grants and Lees). See for example John Keegan (ed.), *The Rand McNally Encyclopedia of World War II* (Chicago, 1977), p. 235.

[7]Reid, *A Concise Encyclopedia of the Second World War,* p. 199.

The Age of Small Wars and Nuclear Deterrence

The memory of two great wars in a span of 31 years and the introduction of atomic weapons strongly influenced the generations upon whose shoulders rested the responsibility for establishing post-World-War-II policies and commitments. **Internationalism,** the belief that world peace can be achieved by the friendly association of all nations, helped to create the United Nations, but the conflicting ideologies of the Communist and free worlds influenced world politics and dominated military institutions far more than the internationalistic tendencies. For the Western nations, peace was sought in **collective security,** a policy based on an association whose goal is to maintain peace by using united action to oppose any aggressor who violates the peace.

Major Themes of the Age

The atomic detonations that immediately preceded the end of the Second World War ushered in a new and a potentially self-destructive age in the history of man. The fundamental role of violence in Communist ideology was well understood by many Western leaders for decades before World War II. Chief among the leaders who feared the worst from communism was Sir Winston Churchill. In 1946, he described the boundary that separated Communist territory from Western European territory as the **Iron Curtain.** In spite of the warning of Churchill and others, however, Communist influence spread from the Soviet Union to other areas of the world, and in 1947, President Harry S. Truman announced a policy of **containment,** which sought to prevent the spread of totalitarian influence over free people by either direct or indirect means. Russia's explosion of its first atomic device in 1949, however, changed the balance of power for decades to come. The fear of **atomic Armageddon,** the final and decisive worldwide battle, which according to the book of Revelation

will take place at Armageddon, grew from the potential dangers of nuclear exchange between the superpowers. The threat was intensified by the marriage of nuclear warheads to intercontinental ballistic missiles that are capable of being fired from **hardened silos,** underground storage/launch sites that are virtually indestructible, and missile-firing submarines.

To prevent nations from resorting to the upper level of violence, **deterrence,** which refers to the policy of making the possible costs of aggression greater than the possible gains, became the underlying theme of East-West relations. In the mid-1950s, the United States attempted to insure deterrence by adopting a policy of **massive retaliation,** the threat to use full nuclear force against any aggressor, regardless of whether the aggressor used nuclear or non-nuclear **conventional** warfare. Massive retaliation, however, is itself extreme, and therefore hardly an appropriate response to insurgencies or **wars of national liberation,** the term used to describe efforts by Communist-imbued, and frequently Soviet-sponsored, forces to gain control of vulnerable governments. Massive retaliation was also called **brinkmanship,** because its implementation threatened to bring mankind to the brink of destruction. In the early 1960s, brinkmanship gave way to the principle of **flexible response,** a policy that consisted of a variety of reactions to Communist-inspired activities that threatened the free world. Economic aid to victims, boycott, show of force, military assistance programs, commitment of unconventional forces, commitment of conventional forces, and—ultimately—massive nuclear strikes were all components of the policy of flexible response. The problems of flexible response became apparent in Vietnam, where the policy led to **escalation,** the commitment of an increasing number and variety of forces and weapons in an effort to contain successfully an enemy who was willing himself to increase his commitment. Escalation is often synonymous with **gradualism,** or **graduated response,** which refers to the prac-

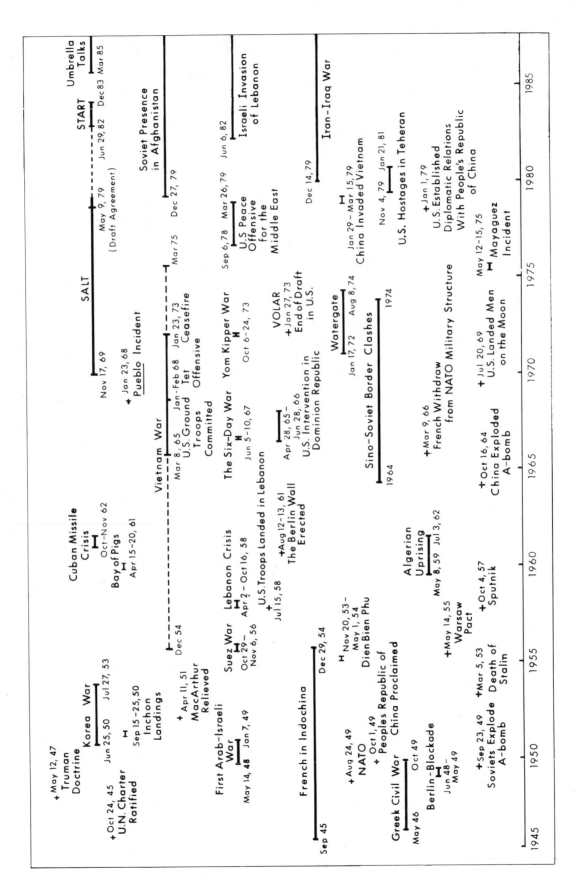

The Age of Small Wars and Nuclear Deterrence

tice of proceeding toward a goal by slight increments of necessary resources rather than by a total commitment of all resources.

That the United States and the Soviet Union recognize the high cost of all-out nuclear war is evidenced by the restraint on the employment of nuclear weapons in the decades after their discovery. This restraint has been reflected in the willingness of the two **superpowers** to engage in **Strategic Arms Limitations Talks (SALT)** in the 1970s, in **Strategic Arms Reduction Talks (START)** in the early 1980s, and in **umbrella talks** that began in 1985 and covered the range of nuclear issues on land and in space. It has also been apparent in **détente,** which refers to the lessening of tensions between the two powers, that has characterized the recent past, and in some mutual triumphs in space. The Soviets were the first to orbit a man-made satellite (named Sputnik); the United States put the first man on the moon. Together the powers have assisted in the conquering of near space. They have agreed in principle to the peaceful uses of space, although since the first experiments in space, the two powers have shown a high degree of competitiveness.

The commitment not to use nuclear weapons is an unwritten and unspoken policy, but an accepted policy, on the part of both Communist and Western nations. Each power has armed itself with atomic weapons, and each reserves the right to use nuclear weapons before the opposing side; to date, however, each has refrained from using them. Although the nuclear warhead has not been used—except as a deterrent—its mere existence has dominated international politics and military planning in the decades since World War II.

Having taken on a new meaning in the nuclear age, cold war now refers specifically to the relationship between the superpowers, the United States and the Soviet Union. Arms control negotiations and **summits,** meetings between the superpower heads of state, have taken place sporadically in the decades since the invention of nuclear arms. Perhaps the existence of those weapons of mass destruction has kept the superpowers from direct military confrontation and spared the world a third general war in this century. Thus, the age of nuclear arms has been an age of nuclear peace.

Other levels of conflict, however, continue unabated. Since the creation of the state of Israel in 1948, the Middle East has repeatedly erupted in the flames of war as Arabs and Israelis clash over rights, religion, and limited resources. In east Asia, mid- to high-intensity conflict came to Korea in the 1950s and to Vietnam in the 1960s, leaving tensions that have led to repeated raids, bloody incursions, and continued strife. The Soviets sent combat forces against the Afghans in December 1979, and in the same month, Iran and Iraq began their prolonged internecine conflict. Africa and Latin America continue to know the pains of armed conflict. Even in Europe,

the heavy hand of the Soviets, felt so sharply in Hungary in 1956, in Czechoslovakia in 1968, and in Poland in 1979, threatens its East European satellites, while terrorist groups, like the Red Army Faction in West Germany and the Red Brigades in Italy, strike too frequently against both military and civilian targets. It is a nuclear age and a nuclear peace, but it is also an age of many small wars. The term **small war,** like "limited war," is relative; it refers to wars in which the major powers do not directly confront each other on the battlefield. Like any war, however, small wars are as deadly serious as any human activity can be.

Participants in the Profession of Arms

At the conclusion of World War II, as at the conclusion of every large-scale, high-intensity war, demobilization of forces was, for many servicemen, the most important military activity in progress. Amidst demobilization, however, **action officers,** who are staff officers assigned to a specific task that generally has a relatively high degree of urgency, grappled with questions of revising organizations and doctrine in light of their wartime experiences. Perhaps because of the urgency associated with their tasks and the apparent need for a bureaucratic shorthand, the "acronymania" that had begun during the war proliferated in the decades after the war. From **SHAPE** (Supreme Headquarters, Allied Powers Europe) and **NATO** (North Atlantic Treaty Organization) to **ROKs** (pronounced "rocks"—troops of the Republic of Korea), **ARVN** (pronounced "ar-vin"—troops of the Army of the Republic of Vietnam), **VC** (Viet Cong, or Vietnamese Communist forces organized and recruited in South Vietnam), and **NVA** (North Vietnamese Army regulars), organizations and individuals became commonly known by their acronymic nicknames.

As in the past, new tasks, new doctrine, and new technology have resulted in new military titles. The helicopter was responsible for the creation of the **air cavalryman,** whose assignment and activities are largely associated with the employment of helicopters for military purposes. Cavalry missions performed in highly mobile ground armored vehicles are the special province of **armored cavalry units. Combat engineers,** a term often used in World War II, are those engineers assigned to combat units, such as infantry divisions. Their principal functions involve work, such as assault bridge building, barrier construction, and mine clearing, that assists the movement of friendly combat units or impedes enemy activities. Combat engineers are distinguished from **construction engineers,** whose principal tasks involve the construction of buildings and fueling facilities, and the construction, improvement, and maintenance of lines of communication in rear areas.

In the duel between the two superpowers, and more generally in the conflicts sponsored by these powers, those who fight on behalf of the Western ideology are known as **free world forces.** Indigenous forces that are staunchly loyal to the Communist ideology or who are full-time Communist soldiers are known as **hard-core.** The term also refers to anyone who is a firm believer in a rigorous lifestyle. When friendly "hard-core" faced the rigors of detention by Chinese Communists, they were subjected to extreme physical and mental torture, sometimes referred to as **brainwashing** because the brain was "cleansed" of rationality and reality and because water was sometimes used as the instrument of torture. During the Vietnam War, too, POWs were cruelly treated, but other age-old methods of gaining military information from enemy soldiers were practiced. Both sides tried to influence individuals aligned with their opponents to **defect,** that is, to leave the forces of one side and join the opposite side. The purposes of the defection programs were to undermine morale, to recover valuable resources (manpower and weapons), and to gain military information. For example, **defectors** were encouraged to leave the VC and NVA during the Vietnam War through the **Open Arms** (*Chieu Hoi*) **Program,** which promised rewards and repatriation to those who would leave the Communist forces.

The end of the United States participation in the Vietnam War brought an end to the American draft system, or **selective service,** a system in which those whose selection was determined by lot had to serve on active duty for two years. The selection process involved the drawing of a chit from a barrel containing a chit for each day of the year; the first "selected" were those who were born on the days drawn first. When the draft ended, the Army relied exclusively on volunteers to fill the **Modern Volunteer Army (MVA)** or **VOLAR** (Volunteer Army). Today's volunteer army provides opportunities for young soldiers to continue their formal education, to learn and practice a wide variety of **military occupational specialties (MOS),** and to serve at posts and installations where the soldier and his **dependents** (the immediate family members supported by the service member) can benefit from the community relationships that have characterized army life for centuries. Ultimately, the questions of who will serve and where are determined by the national policies established by the Government. Since 1812, these questions have been the subject of often intense political debate, and they continue to be a matter of considerable debate today. In an age when the United States is viewed by the rest of the world as the leader of the free world, it is ironic that no other major power (nor most unaligned and Communist nations) relies entirely upon volunteers to fill the ranks of its armed forces. The complex technology of the age requires quality soldiers who have the talent and training necessary to use technology to the fullest extent of its capabilities. Quality leadership and morale are important in every army. In modern volunteer forces, they are essential.

That the United States will continue to use force when sufficiently provoked was confirmed when President Gerald Ford sent American air, sea, and land forces against Cambodian "pirates" who seized the United States merchant ship *Mayaguez* and her crew in May 1975. This action was taken just six months after Congress had overridden a presidential veto of the **war powers resolution,** which limits the President's power to commit forces abroad with Congressional approval by requiring the President to report to Congress within 48 hours after committing forces to a foreign conflict or "substantially" enlarging the number of troops equipped for combat in a foreign country. The resolution was a response to the Vietnam War, which was unpopular within both American society and Congress. The impact of the war powers resolution on the role of force in the age of détente remains unknown, but rulings by the Supreme Court on related issues have already ameliorated its impact.

Principal Organizations

At the conclusion of the Second World War, the military establishment of the United States underwent a major reorganization. The War Department was transformed into the **Department of the Army** and a newly formed coequal, the **Department of the Air Force.** Together with the **Department of the Navy,** these three service departments were subordinate to a newly established Department of Defense. The use of the word "defense" suggested the intended role of the postwar **national military establishment,** but the doctrine of all three services and their pleas for funds to maintain large numbers of troops on active duty and to develop offensively oriented materiel belied the title. The Navy sought bigger and better ships, notably, the **supercarrier** and the more deadly nuclear-powered submarine. The Air Force argued that already proven heavy bombers were needed for the accurate delivery of atomic weapons. Within the Air Force, the **Strategic Air Command,** or **SAC,** was organized to carry out the all-important nuclear delivery mission, and the **Tactical Air Command,** or **TAC,** assumed the ground support missions. Also, the Air Force, rather than the Army, assumed the mission of developing and employing missiles for strategic purposes. The Navy, too, assumed a major role in the strategic missile arena through submarine-launched ballistic missile programs.

The advent of atomic weapons also had a profound effect on operational organizations within the Army, for the threat of nuclear weapons and **tactical nuclear weapons,** which could direct relatively small nuclear warheads against the personnel and equipment of combat and combat supporting units, suggested that smaller, dispersed, and semi-independent units operating on a wide front were needed. In 1956, the **ROCID (Reorganization of the Current Infantry Divison)** plan was implemented to fulfill this goal. The ROCID division was also called the **pentomic division** because five major infantry commands, called **battle groups,** replaced the three infantry regiments of the World War II triangular divisions. In the early 1960s, the need for flexibility to meet the contingencies of war led to another major organizational change. The **ROAD (Reorganization Objectives Army Division)** concept of 1962 established a division base of headquarters and supporting units to which brigades and battalions could be assigned as a given situation dictated. Four standard types of ROAD divisions were created. **Armored divisions** had approximately equal numbers of armored battalions and mechanized infantry battalions. **Mechanized infantry divisions** had more mechanized infantry battalions than armored battalions. **Infantry divisions**—nicknamed **straight leg infantry,** or simply **leg infantry,** because they use their legs rather than fighting vehicles or aircraft as their principal means of moving to and on the battlefield—were made up predominantly of infantry battalions. However, they did contain some armored battalions. The **airborne divisions** were composed of airborne infantry battalions, artillery, and some light armor. A fifth type of division, the **airmobile division,** was organized in 1965. A variety of organizational changes have occurred within the airmobile division since then—even its title was changed to that of the **air assault division**—but its fundamental characteristics remain its lightness and the fact that it is **air transportable,** that is, capable of being rapidly moved by a combination of organic and inorganic aircraft. Helicopters are commonplace in the air assault division, and today all Army ground units receive some support from **helicopter units,** which provide the aircraft, personnel, and facilities to accomplish such missions as reconnaissance, command and control, airlift, and even **firepower,** which refers to the total effect of shells and missiles that can be placed on a target.

On January 11, 1973, another major organizational change within the United States Army was announced by the Secretary of the Army. The **Continental Army Command,** which had been responsible for most training programs, and the **Combat Developments Command,** which concentrated on materiel and technique, were eliminated. They were replaced by **Forces Command (FORSCOM),** which is responsible for TOE units in the United States and for unit training, and by the **Training and Doctrine Command (TRADOC),** which is responsible for the operation of military schools, exclusive of the United States Military Academy and the Army War College, for individual training, and for the promulgation of doctrine.

As the United States became involved in insurgency situations during late 1950s and early 1960s, new nonstandard organizations were created. Military missions expanded into **military advisory groups (MAGs),** and later into **Military Assistance Advisory Groups (MAAGs),** which were multiservice organizations whose principal function was to administer all aspects of military planning and programming that involved resources from the United States.

Special forces units were formed of individuals who were collectively known as the **Green Berets** because of their distinctive headgear. These units, which were given special training in basic and specialized military skills, were organized into small multiple-purpose detachments with the mission to train, organize, supply, direct, and control indigenous forces in guerrilla warfare and counterinsurgency operations. **Studies and Observation Groups (SOGs)** were small detachments organized in Southeast Asia—ostensibly for the purpose of gathering information of a strategic nature. In practice they performed other clandestine missions as well. The war in Southeast Asia also established a need for special engineer construction units. Among these were the **Seabees,** or **C.B.s,** the World War II term for construction battalions, and the Air Force's **Red Horses (Rapid Engineer Deployment Heavy Operation Repair Squadron, Engineering).** Other support functions were performed by **tunnel-rat teams,** which had to search and at times destroy the complex underground mazes that served as Communist havens and headquarters during the Vietnam War, and by **evacuation hospitals,** which were mobile, frequently tent, hospitals used near the front to provide major medical and surgical care and to prepare and sort casualties for further evacuation. The **MASH,** or **mobile army surgical hospital,** was a hospital organization that was prevalent only during the period of the Korean War.

The war in Vietnam led to the creation of several *ad hoc* military organizations. South Vietnam, for example, was divided into four geographical areas, known as **corps areas, corps tactical zones,** and, later, **military regions.** When a large number of American troops operated within a defined area, which often corresponded to a corps area, all such troops became a part of a **field force.** Areas of intense hostile activity within the corps areas were known as **war zones.**

South Vietnamese forces were for a time commanded by a **junta,** a group of generals who possessed political power. The Vietnamese Regular Army was poor by Western standards, and **local units,** which were units recruited, trained, and employed in a relatively small geographical area, were invariably worse than most of the regulars. Each **regional force** was a nationally administered military force assigned to and under the operational control of a province. Forming South Vietnamese political subdivisions, there were 44 provinces. Early in the war, the regional forces were known as the **Civil Guard.** The **popular forces** of the Vietnamese Army constituted a nationally administered military force organized and employed at the village level. These local units consisted of light infantry squads and platoons, whose combat capability was restricted to small but often critical defensive and counterattack operations.

The enemy in Vietnam was highly organized in spite of the relatively primitive nature of his technological base. The areas that the Communists controlled were referred to as **free zones** or **liberated areas.** These Communist-controlled areas were used as bases for the operations of the **field fronts,** which were the NVA units roughly equivalent to the United States divisions, and other **main force units,** the term used in referring to any regular organization of the North Vietnamese Army.

In the 1970s, the American Army reactivated its ranger battalions, and in the 1980s it placed them, along with special forces units, psychological operations battalions, civil affairs battalions, special aviation, and other units, under a **Special Operations Command,** or **SOCOM.** Although the SOCOM units are specifically trained to deal with threats to national security from the low end of the spectrum of conflict—including terrorist threats—they have important roles and missions across the entire spectrum of conflict.

In the mid-1980s, the Army again restructured its divisions in order to exploit the technological advantages of major new items of equipment, such as the M1 Abrams tank, the Bradley Fighting Vehicle, and a multiple-launched rocket system (MLRS), which, although new to the United States Army, had long been a part of Soviet forces. This reorganization was commonly known as **Division '86.** Because Division '86 focused on the heavy divisions designed for mid- to high-intensity conflict, the army subsequently saw a need for light infantry divisions. These divisions, each of which was made up of about 10,000 officers and soldiers, were designed for rapid deployment to regions of likely low-intensity conflict, and hence were intended as a deterrent to conflicts in those regions.

The Technology of the Age

The most militarily significant technological developments in the decades since World War II have occurred in four major areas: electronics, nuclear weapons, missiles, and rotary-wing

aircraft. Yet, because of the level at which some modern insurgencies have been conducted, weapons of great sophistication have been juxtaposed with weapons that rival sticks and stones in their simplicity. **Nuclear weapons,** a term that has replaced "atomic weapons" in the military lexicons and suggests greater levels of destruction, often exist as legitimate weapons of war alongside **punji stakes,** which are sharpened sticks, often covered with dung or other high-bacteria sources, that are placed in shallow holes and camouflaged in an effort to incapacitate the unwary passerby. The nuclear weapons are part of a triad that threatens mass destruction. This triad is known by different acronyms, but regardless of whether it is termed **ABC (Atomic-Biological-Chemical), NBC (Nuclear-Biological-Chemical), CBN (Chemical-Biological-Nuclear),** or **CBR (Chemical-Biological-Radiological),** the threat to mankind is real. These **weapons of mass destruction** have thus far been used only to threaten and coerce, while primitive weapons like the punji stake; the **garrote** (also spelled "garotte" and "garote"), a thin wire stretched between wooden handles and used to strangle; and the **machete,** a large heavy knife commonly used to cut sugar cane and bamboo, have been constantly employed in the hostile environment of insurgency. Technological extremes have never been greater.

Missile Weapons

Small arms have not changed dramatically in the decades succeeding World War II. In fact, during people's wars, the use of small arms of the World War I period has not been uncommon. In the United States Army, the **M-14 rifle** succeeded the M-1 Garand rifle as the principal shoulder arm. The M-14 was developed so that the standard American Army rifle would fire the standard NATO 7.62-mm caliber ammunition. The **M-16 rifle** was developed by Eugene Stoner in the 1950s and was adopted by American airborne and special forces troops in 1963. Because it was smaller and lighter than any existing shoulder arm, and because it fired the lighter 5.56-mm round, the M-16 became the standard rifle of all American forces in Vietnam and of many of the other free world forces fighting there. An unloaded M-16 weighs six and one half pounds and can fire semi- or fully automatically either from a 20-round magazine or from the larger curved **banana magazines.** Today, it is the standard United States rifle. During the Vietnam War, the Soviet **assault rifle AK-47** was the Communist equivalent of the M-16. The AK-47 was developed from the German MP-43 assault rifle and was adopted by some Soviet units in the early 1950s. It subsequently became the standard weapon of all Warsaw Pact nations, and even though it fires a 7.62-mm cartridge, the Soviet

United States Rifle M-14—7.62 -mm

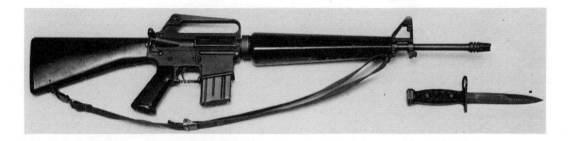

United States Rifle M-16—5.56 -mm

Soviet Kalashnikov Assault Rifle AK-47—7.62 -mm

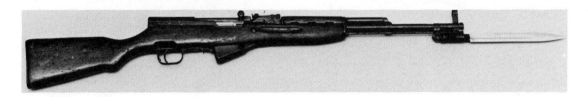

Soviet Simonev SKS—7.62 -mm

7.62-mm rounds are not interchangeable with the NATO 7.62-mm cartridges. The **Simonev SKS** was a 7.62-mm semiautomatic weapon that was introduced to Soviet forces in 1946. Although it was later replaced by the AK-47 assault rifle, it remains in the hands of forces receiving Soviet military aid.

The standard American machinegun of the post-Korean-War period has been the 7.62-mm **M-60.** The design, which dates to the end of World War II, copied the belt-feed system of the German MG-42. Even though the M-60 fires only fully automatically, its low rate of fire of 600 rounds per minute allows skilled users to fire single rounds. it can be mounted on

Soviet SGM Heavy Machinegun—7.62 -mm

a tripod, which weighs 15 pounds, or on its folding and tele-scoping bipod legs. The gun weighs just over 23 pounds, and its 7.62-mm ammunition is fed from a disintegrating link belt. The Soviets also rely on a 7.62-mm machinegun, the **SGM heavy machinegun.** The SGM can be mounted on armored personnel carriers, on a tripod, or on a wheeled carriage. It fires from non-disintegrating metallic belts at a rate of 600 to 700 rounds per minute, weighs just under 30 pounds, and is used by the U.S.S.R. and most satellite nations as a battalion-level machinegun.

M-79 Grenade Launcher—40 -mm

The United States Army adopted the **M-79 grenade launcher** as a light **area-type weapon,** that is, a weapon that projects an exploding shell. The M-79 is effective beyond hand grenade range and short of mortar range against **soft targets** (such as personnel or field fortifications). Effective at ranges up to 400 meters, the M-79 fires a 40-mm high ex-plosive shell. The round weighs nine ounces, and the weapon weighs just over six pounds. Today the capability afforded by the M-79 is available in the **40-mm grenade launcher, M203,** which incorporates the 40-mm system with the M-16 rifle.

In addition to small arms and machineguns, ground soldiers in the mid-1960s were provided with the infrared homing **Redeye guided missile.** A surface-to-air weapon, the Redeye was designed to oppose high-speed close air support fighters. The Redeye is a shoulder arm that weighs 29 pounds, is 4 feet long, and fires a 3-inch diameter missile. The successor to the Redeye was the **Stinger.** Further improved versions are called **Stinger POST** and **Stinger RMP.**

Also in the 1960s, infantry battalions were equipped with a 120-mm recoilless rifle called the **Davy Crocket.** The Davy Crocket was capable of firing a 40-ton TNT-equivalent nuclear warhead to a distance of three miles. To preclude the occurrence of international incidents, both the weapon and its ammunition were closely controlled and, eventually, the rifle was eliminated.

Redeye Guided Missile

M-72A (LAW)–66-mm HEAT Rocket

Antitank weapons used by modern United States and allied infantry units include the **M-72 LAW (Light Antitank Weapon)**, the **Dragon** or **MAW (Medium Antitank Weapon)**, and the **TOW (tube-launched, optically tracked, wire-guided missile)**, which is a heavy antitank weapon. The LAW is packaged with a 66-mm **HEAT (High Explosive Antitank)** rocket in a launcher that is discarded after firing. Its maximum effective range is 200 meters. The Dragon is a man-portable antitank weapon system with an effective range of 1,000 meters. The TOW is a crew-served weapon that can be mounted on vehicles or fired from a ground position. Its maximum effective range is 3,000 meters. Improved versions of the TOW are known as **I-TOW** and **TOW II.**

Conventional field artillery and mortars have changed little since World War II. Improvements have included rocket-assisted rounds to increase the range of existing artillery; a trend toward self-propelled guns; a **beehive round,** which is a flechette-loaded fragmentation shell that was developed as an antipersonnel weapon; and the development of multiple-launch rocket systems. Surface-to-surface missiles, however, changed the United States field artillery inventory dramatically in the decades after World War II. The most popular of these were the Littlejohn, the Sergeant, the Lance, and the Pershing. The **Littlejohn** had a length of 14.4 feet, a firing weight of 780 pounds, and a range of 10 miles. The **Sergeant,** which was deployed in 1961, had a firing weight of 10,100 pounds and a range (with a nuclear or conventional warhead) of from 28 miles to 85 miles. First tested in 1965, the **Lance** had a length of 20 feet, a launching weight of 3,200 pounds, and a range of up to 60 miles with either nuclear or conventional warheads.

In 1976, President Ford secretly authorized the development of a neutron bomb as a warhead for the Lance. The **neutron bomb,** called an "enhanced radiation" warhead, is

M-220 (TOW), Ground-Mounted Missile

essentially a small hydrogen bomb that has only one-tenth of the blast, heat, and fallout produced by hydrogen bombs of comparable yield. Blast and heat damage are confined to a radius of several hundred yards, instead of the radius of several miles that is typical of other hydrogen bombs. The neutron bomb kills by emitting massive numbers of neutrons that destroy the body tissue of its victims. The radiation, however, is short-lived. A blast area can be entered safely within several hours, while it takes months for an area to be free of radiation from an atomic bomb. On April 7, 1978, in preparation for the Strategic Arms Limitation Talks (SALT), President Carter announced that he had decided to defer production of the neutron bomb as a good-will gesture. Consequently, there ensued an often stormy political-military debate concerning the various roles the neutron bomb might play as a strictly tactical weapon, a deterrent to all nuclear war, a "humanitarian" weapon, or a perpetrator of a large-scale nuclear holocaust.

The **Pershing,** which was first fired in 1960, is a two-stage, selective-range ballistic missile that carries a nuclear warhead. With a length of 34.5 feet, a launching weight of 10,000 pounds, and a range of 115 to 460 miles, it remains the largest surface-to-surface missile in the American Army inventory. In December 1983, the first **Pershing II,** or simply **P-II,** rockets were dispatched to West Germany. These were improved versions of the Pershing, but because of their increased range, the concomitant threat to the Soviet

MLRS Launcher

homeland, and the resulting attention brought to the possibility—however remote—of nuclear war, they became the object of massive antinuclear demonstrations in Western Europe, especially in the West German cities of Bonn (the capital), Heidelberg (the headquarters of U.S. Army, Europe), Hamburg, and cities near the deployment sites. Despite these protests, NATO and the West German Government adhered to their decision to deploy the P-IIs. This decision represented a failure for the Soviets, who had tried to divide the West by sponsoring demonstrations and walking out of the START and intermediate-range nuclear forces (INF) talks.

As mentioned earlier in this chapter, the United States Army has developed a rocket system that is comparable to the type the Soviets have used for a number of years. The Multiple Launch Rocket System (MLRS) is organized into batteries, each of which has nine launchers. Each launcher is capable of firing twelve 13-foot long rockets to ranges of over 30 kilometers. One 12-rocket MLRS barrage is considered sufficient to destroy an artillery battery that is in combat firing posture.

Since World War II, the range, accuracy, and lethality of air defense weapons has increased dramatically—largely because the continued advancement in aircraft performance has required more than just better antiaircraft gun systems. For a few years after the war, the term **antiaircraft artillery (AAA)** described the countermeasure to the airborne threat, but when guided missiles were introduced as air defense weapons, the term air defense artillery (ADA) became more prevalent. The first of the guided missile air defense weapons was the **Nike Ajax,** which was initially deployed in the Washington-Baltimore area in 1953. This system made use of improved radar to detect and track hostile aircraft and to direct the Nike Ajax missile to its point of interception. **Nike Hercules**

Pershing II Missile

Chaparral

became operational in 1958 and gradually replaced the Nike Ajax as the free world's major air defense weapon.

Also deployed in 1958, the **Hawk** was designed to counter low- and medium-altitude aircraft. Hawk is sometimes described as an acronym for "Homing All the Way Kills," but whether Hawk refers to the bird or the acronym is a matter of personal preference. In the 1970s, the Hawk was succeeded by an improved Hawk, called **I-Hawk.** Within recent years, emphasis has been placed on air defense of the forward army area and on the protection of airfields. To fill this need, **twin 40s** (two coaxial 40-mm guns) and **quad 50s** (four coax-

ial .50-caliber guns) mounted on tracked or wheeled vehicles are used along with the more sophisticated Chaparrals and Vulcans. The **Chaparral** is a self-propelled tracked cargo vehicle that features launching positions for four surface-to-air missiles. These missiles were adapted from the **Sidewinder,** an infrared homing, air-to-air missile. The **Vulcan** is a very fast firing machinegun of the Gatling type. Because of its short range, the Vulcan is virtually useless against the Soviet Hind helicopter and other close air support systems of potential enemies. As a result, in the mid-1970s, the Army began using the **Sergeant York Division Air Defense (DIVAD) gun,**

Sergeant York Division Air Defense Gun

Patriot Missile

which mounts twin 40-mm cannon, capable of firing 620 shots per minute, on the chassis of old M-48 tanks. The keystone of the theater air defense system is the **Patriot** surface-to-air missile. It can engage multiple targets in an electronic countermeasures environment and has four ready-to-fire missiles per launch unit.

In the 1970s, the most sophisticated of the Army's surface-to-air missile systems was the **Safeguard Antiballistic Missile (ABM)** system. Safeguard employed the **Spartan**—a three-stage solid propellant rocket with a nuclear warhead and a range of about 40 kilometers—as the close-range element of the system. It also included a long-range radar called the **Perimeter Acquisition Radar (PAR),** which was designed to accurately detect and track missiles at ranges of 1,500 to 3,000 kilometers. For political reasons, Safeguard was not deployed. Nevertheless, research and development for defense against intercontinental ballistics missiles (ICBMs) remains an important Army function and is part of the program known as **ballistic missile defense,** or **BMD.** In July 1984, a BMD non-nuclear missile was fired from Kwajalein Island in the western Pacific and successfully intercepted a target ICBM that had been fired from Vandenberg Air Force Base, California. BDM is the core of the **Strategic Defense Initiative**—or, as it is popularly misnamed, the **"Star Wars"** initiative—advocated by President Ronald Reagan in the spring of 1984.

As a continuation of the mission to deliver nuclear warheads via strategic bombers, the United States Air Force has assumed the responsibility for the long-range delivery of nuclear warheads via missiles. The most extensively produced Air Force **intercontinental ballistics missiles (ICBMs),** defined as ballistics missiles capable of traveling over 5,000 miles ground distance, are the Titan and the Minuteman. The **Titan,** a liquid propellant two-stage ICBM, is generally armed with a nuclear warhead. **Titan II** has a length of 103 feet, a diameter of 10 feet, a launching weight of 330,000 pounds, and a range in excess of 6,300 miles. The **Minuteman,** a three-stage solid propellant ICBM, was first fired in 1961. The latest version, the **Minuteman III,** has a length of about 60 feet, a launching weight of 76,000 pounds, and a range in excess of 8,000 miles. The most sophisticated generations of Air Force ICBMs are currently designated the **MX** and **Midgetman.** The **cruise missile,** like the Pershing II, is being deployed to Europe under the NATO decision of December 1979. It is a jet-powered, terrain-following winged vehicle capable of accurately delivering a nuclear warhead by utilizing an approach at extremely low altitudes—too low to be detected by conventional radar. It looks very much like the German V-1 rocket of World War II.

Strategic nuclear weapons gained a new dimension in the mid-1970s, when **multiple, independently-targetable reentry vehicles (MIRVs)** became operational. The **reentry vehicle** is the portion of the missile that can withstand the temperatures produced during re-entry into the earth's atmosphere. The MIRV enjoys two principal advantages. First, the attacker offers several targets to the enemy's ABM capabilities, rather than only one; second, more damage can be done by several small nuclear warheads than by a single warhead of the same total yield. The **Minuteman III** is designed to have three warheads, while the **Poseidon,** the **submarine launched ballistics missile (SLBM)** that succeeded the **Polaris** SLBM, is designed to have ten to fourteen MIRVS.[1]

Bombs and Other Implements of Modern War

Technology continues to produce new and more effective instruments of destruction. Despite these advances, however, weapons do not always function as designed. Any weapon that so malfunctions is called a **dud.** When devices like **blockbusters,** the World War II demolition bombs ranging in weight from two to eleven tons and considered powerful enough to destroy a city block, turned into duds, ingenious and resourceful guerrillas were able to fashion them into **command-detonated mines.** These mines were remotely detonated, generally by an electric charge from a hand-held generator, when personnel targets happened near. **Claymore**

mines, or simply **Claymores,** are antipersonnel mines designed to produce a fan-shaped pattern of fragments. Claymores are either command-detonated or detonated by a trip wire located in the mine's killing zone. Even **butterfly bombs,** which are small fragmentation or antipersonnel bombs with wings that rotate and arm the bomb as it descends, can be rigged as command-detonated mines.

Drones, which are pilotless aircraft; **remotely piloted vehicles (RPVs); laser-guided bombs,** and **television-guided bombs,** commonly known as **Walleyes,** were all used successfully against targets in North Vietnam. The latter two items were called **HOBOS,** or **Homing Bomb Systems.**

One instrument of destruction employed extensively in Vietnam was unusual because its purpose was to destroy plant life rather than animal life or inanimate objects. These **defoliants** were either biological or chemical agents that destroyed leaves or plants so that observation and, hence, security could be enhanced in areas of suspected enemy activity.

Mobility

Like the missile weapons of the post-World-War-II period, mobility devices were improved upon after the war. However, few revolutionary developments occurred. For example, the ramp tanks of the Second World War were succeeded by the **AVLB (armored vehicle launched bridge),** a hydraulically operated system that carried a folded bridge on a tank chassis. Another successor of the ramp tank, the Soviet **ribbon bridge,** is transported on trucks and can be quickly emplaced to facilitate the crossing of water barriers by mechanized or motorized forces. The World War II **Bailey bridge,** a prefabricated set of ten-foot-square panels and

decking that can be constructed on the land adjacent to a gap and pushed into place, remains the principal means of bridging gaps of 130 feet or less. Bulldozers, or simply **dozers,** were used to build roads and airfields in Vietnam, and **Rome plows** (made in Rome, Georgia) were employed to level the jungle either in preparation for construction projects or to permit better observation and, hence, better security.

Fighter aircraft continued to fulfill close air support roles, with jet engines replacing the propeller aircraft that had dominated World War II. During the Korean War, the United States relied heavily on its **F-86 Sabrejet,** a single-seat, single-engine fighter that had first flown in 1947. The first American sweptwing jet fighter to see combat, the Sabre had a speed of 687 miles per hour and a range of about 1,000 miles. It was armed with six .50-caliber machineguns and had attachments for two 1,000-pound bombs. During the Korean War, the F-86 proved to be superior to the Soviet **MiG-15,** which was code-named the **Fagot** by NATO. The MiG-15 had a speed of 668 miles per hour, was armed with one 37-mm cannon and two 23-mm cannons, and had attachments for two 1,000-pound bombs.

During the 1950s, the **century series** of fighter aircraft was developed. The first of the century series was the North American **F-100 Super Sabre.** Flight tested in 1953, it was the first supersonic operational fighter to be used by the United States Air Force. First flown in 1954, the McDonnell **F-101 Voodoo** was a twin-jet, supersonic fighter-interceptor. The Convair **F-102 Delta Dagger** and the further developed **F-106 Delta Dart** were both supersonic, single-seat, all-weather, delta-winged fighters; they were first flown in 1953 and 1956 respectively. The Lockheed **F-104 Starfighter** was a single-seat, single-engine jet fighter with a speed of mach 2.2. First flown in 1954, its armament consisted of Sidewinder missiles, rocket pods, and bombs. The Republic **F-105 Thunderchief**

North American F-86 Sabre

MiG-15 With United States Markings

was a single-seat, single-engine jet fighter-bomber with a top speed of mach 2.1 and a range in excess of 2,000 miles. First flown in 1955, it was extensively used over both South and North Vietnam. The McDonnell **F-4 Phantom** also was widely used in Vietnam. A two-seat, twin-engine jet fighter-bomber, it was first flown in 1958. The **F-111** was designed as a two-seat, twin-jet, all-weather, multipurpose fighter and fighter-bomber with variable sweptwing configurations. First flown in 1964, it had a speed of mach 2.5, a range of 3,800 miles, and could carry up to eight missiles or bombs under its wings. Because its use was confined to low altitude penetration bombing, it was finally redesignated the **FB-111.**

The United States Air Force is currently purchasing **F-15** and **F-16** fighters. Its newest close air support tank killer, the **A-10,** is armed with a 30-mm cannon that uses depleted uranium projectiles to engage and penetrate heavy armor. The Soviet Union is presently producing the **MiG-25,** which, according to Soviet claims, is the fastest aircraft in the world.

The United States Navy today relies upon the **A-6 Intruder,** the **A-7 Corsair,** the F-4 Phantom, and the **F-14 Tomcat,** which is a variable sweptwing, twin-jet fighter-interceptor. The **F-18 Hornet** is being developed as a replacement for the F-4 and a complement to the F-14. Because of the trend toward smaller carriers, the Navy has shown considerable interest in **VTOL (vertical take off and landing)** and **STOL (short take off and landing)** aircraft.

General Dynamics FB-111 With Wings Extended

Convair B-36D

Fighter aircraft in the Vietnam War were often employed from hastily constructed **SATS (Short Airfield for Tactical Support),** which consisted of a runway of World-War-II-vintage **pierced steel plank (PSP),** catapults to assist take offs, and **arresting gear,** that is, steel cables that are hooked by the aircraft and bring it to a stop. **JATO (jet assisted take off)** units were also employed on some aircraft to allow take off from SATS. For instance, JATO was employed by the **C-130 Hercules,** a four-engine turboprop transport capable of carrying 92 men to a range of 4,700 miles. The largest plane ever built is the Lockheed **C-5A Galaxy.** Called a heavy strategic transport (and nicknamed the **Fat Albert**), it was first flown in 1968 and can carry 220,000 pounds of cargo.

Postwar bombers were designed for high-altitude, long-range strategic bombing. The first significant advance over the B-29 was the **B-36,** which was first flown in 1946. The B-36 was the largest and heaviest warplane ever built. It had six pusher (piston) engines, and later models added four jet engines. The aircraft had a crew of 16, a range of 7,500 miles with a 10,000-pound bomb load, and eight remote-controlled turrets mounting two 20-mm cannons each. The B-36 was superseded by the **B-52 Stratofortress,** an eight-jet sweptwing craft first flown in 1952. It has a crew of six, four .50-caliber machineguns in its rear turret, and can carry a bomb load of 60,000 pounds (or a combination of bombs and missiles) to a range of 12,500 miles. In Vietnam, the B-52 was the

Boeing B-52 Stratofortress

CH-21 Shawnee

AH-1 Cobra

CH-47 Chinook

backbone of the Strategic Air Command, and was used extensively for high-altitude, radar-directed night bombing missions, known as **arclights.** Many of the missions of the B-52 have since been taken over by ICBMs.

Problems in the funding of successor bombers to the B-52 were overcome in the first Reagan administration. As a result, the B-52 will be replaced by the **B-1.** In turn, the B-1 will be succeeded by the **Stealth bomber,** which is designed to be virtually undetectable by enemy radar. The Soviet Union uses an advanced long-range bomber, known to NATO as the **Backfire.**

Heavier-than-air craft gained new capabilities with the development of the **helicopter,** an aircraft supported in the air by rotating airfoils rather than fixed wings. Nicknamed the **chopper,** the **whirly-bird,** the **'copter,** the **eggbeater,** and, especially in the Navy, the **helo,** it is an excellent vehicle for short-distance movement of personnel. In addition, it provides an excellent platform for command and control, artillery fire direction, and fire-support weapons. The first widely used troop-carrying helicopter was the twin-rotor **CH-21** (cargo helicopter) **Shawnee,** which was nicknamed the **flying banana** because of its distinctive shape. The **UH-1** (utility helicopter) **Iroquois,** also known as the **Huey,** was the most widely used helicopter of the Vietnam War. Some models could carry as many as 12 soldiers at speeds of 148 miles per hour. Hueys that were unarmed were called **slicks,** while those with external armament consisting of machineguns and cannon were known as **gunships.** In the 1960s, the **AH-1** (attack helicopter) **Cobra** became operational. It can travel at speeds up to 220 miles per hour and employs a variety of weapons systems. The **AH-1Q Cobra TOW** mounts the TOW antitank missile. The popular medium transport helicopter of the Vietnam War was the twin-rotor, turbojet **CH-47 Chinook.** It could carry up to 44 soldiers at a top speed of 172 miles per hour. Also widely used in the 1960s and 1970s were the **OH-6** (observation helicopter) **Cayuse** and the **CH-54 Tarhe,** a heavy cargo helicopter more commonly known as the **Flying Crane.** Its maximum load is 16,000 pounds.

There are two sophisticated helicopters that are coming

UH-60 Blackhawk

into use in the mid-1980s. One is the twin-engine, single-rotor **UH-60 Blackhawk,** formerly called the **Utility Tactical Transport Aviation System (UTTAS),** which is capable of transporting 14 combat-equipped troops. The other is the **AH-64 Apache,** which was formerly called the **advanced attack helicopter (AAH).** It is armed with 2.75-inch rockets, a 30-mm **chain gun** (an advanced machinegun with simpler design than earlier machineguns), and the **Hellfire missile,** a laser-guided missile that is able to direct itself to its target.[2]

AH-64 Apache

British Medium Tank—Centurion

United States Main Battle Tank, M60A1

Soviet Medium Tank, T-62

United States Main Battle Tank—M1 Abrams

Tanks have changed little since the end of World War II, when the British deployed their **Centurion** main battle tank. Armed initially with a 17-pounder main gun, then a 20-pounder main gun, and finally a 105-mm main gun and a .50-caliber machinegun, it remained the standard British tank for 20 years. Although still in service in many countries, the Centurion was replaced by the **Chieftain** as the main battle tank in British service in the late 1960s. The Chieftain carries a 120-mm main gun and two .30-caliber machineguns.

In the United States, the Pershing tank was succeeded by the **M48** in 1952. The M48 had a 90-mm main gun. Various modifications to the M48, especially in the power plant, were made throughout the 1950s. The **M60,** with its 105-mm gun, was the American main battle tank of the 1960s and 1970s. The **M60A2,** adopted in 1968, fired either the **Shillelagh missile** or conventional tank ammunition from its 152-mm gun system. The **M551 Sheridan** also employed the Shillelagh missile. The Sheridan has been known as both an **armored reconnaissance vehicle** and a light tank. It weighed only 15 tons, could be dropped by parachute, and was able to float. Neither the M60A2 nor the Sheridan was particularly successful, however, and both have since been taken out of service. The **M60A3,** which is currently in service, employs a laser range finder.

The Army's main battle tank in the mid-1980s is the **M1 Abrams,** which, like the M60A3, uses a laser range finder. It weighs approximately 60 tons, and its main armament is the 105-mm gun. Ultimately, the German 120-mm smoothbore gun used on **Leopard** and other NATO tanks will be adopted to facilitate ammunition supply and interchangeability with NATO allies.[3] The M1 Abrams is named for General Creighton W. Abrams, a distinguished World War II armor officer who died in 1974 while serving as Army Chief of Staff.

The main battle tank of Warsaw Pact nations is the **T-62 medium tank.** It has a 115-mm main gun, a 7.62-mm coaxial machinegun, and a 12.7-mm antiaircraft machinegun. The Soviets continue to develop new generations of tanks without withdrawing earlier versions from service. The **T-72, T-80,** and others are in various stages of development and fielding.

Half-track armored vehicles, or simply **half-tracks,** are vehicles in which the rear is supported by a complete band track and the front is supported by wheels. Half-tracks were widely used during World War II, especially as personnel carriers. They combined the high speeds needed on improved roads with good cross-country mobility. Post-World-War-II personnel carriers tended to be fully tracked, because fully tracked vehicles could accompany tanks wherever tanks could go. The **armored personnel carrier (APC)** in United States service since 1959 is the **M-113.** Over 30,000 of these vehicles, in several different models, have been produced. The vehicle is designed to carry a driver and a commander as well as 10

M-113 Armored Personnel Carrier

combat-loaded infantrymen. It weighs 10 tons, and has a road speed of 39 miles per hour and a water speed of 3.5 miles per hour. In the 1970s, doctrinal emphasis shifted from armored vehicles for taxiing troops into battle to an **infantry combat vehicle (ICV),** which fights as part of the mechanized infantry squad throughout its operations. The **infantry fighting vehicle (IFV)** of the 1980s carries a nine-man infantry squad, has firing ports that allow the squad to fire from the vehicle, and mounts the TOW antitank weapon and the rapid-firing 25-mm cannon with a 7.62-mm coaxial machinegun. When used by infantry, the vehicle is called the **M2 Bradley fighting vehicle,** while the cavalry version is called the **M3 Bradley fighting vehicle.** Both are named for General of the Army Omar Bradley.

M2 Bradley Fighting Vehicle

Aircraft Carrier and Destroyer

Naval Objects

Since World War II, the Navy has been transformed by technological developments. The aircraft carrier emerged from the war as the dominant fighting vessel in the fleet, but in the nuclear age, any one of a number of ships could become the dominant warship of the future. Nuclear power was introduced as the propelling force on a number of vessels, and additionally, the nuclear warhead capable of being launched from under the sea has made the **ballistic missile submarine (SSBN)** a candidate for the next dominant warship. A few oil-powered attack submarines (SSs) and over 80 nuclear-powered attack submarines (SSNs) make up the balance of the American underwater fleet. The battleship is making a comeback, too, as World War II battleships are refitted—in spite of the fact that when the USS *New Jersey* was refitted during the Vietnam War, she was ill-suited for the assigned missions.

The world's largest warships are the nuclear-powered carriers (CVNs) of the United States Navy.⁴ There is no typical carrier, but the USS *Nimitz*, commissioned in 1975, is a logical standard for comparison. The *Nimitz* is 1,092 feet long and 252 feet in extreme beam; displaces 91,400 tons; draws 38 feet, has two nuclear reactors (with a fuel life of 15 years) whose energy produces 280,000 horsepower to drive four propellers; and has a top speed that, although classified, is known to be in excess of 30 knots. Carrying an air group, which consists of 100 attack and reconnaissance planes and helicopters, the *Nimitz* accommodates about 570 officers and 5,720 enlisted men. She is not only a floating airfield but a floating city as well.

Today's cruisers are all missile ships. Some are oil burning **(CGs)**, while others have nuclear power plants **(CGNs)**. They are armed with **Standard** and **Harpoon missiles** for use against air or surface targets, and with **antisubmarine rockets (ASROC)** for use against submarines. Destroyers continue to serve as screening vessels. Some are conventionally armed (the DDs), while others have launchers for firing surface-to-air missiles (the DDGs). The **Spruance (DD963) destroyers** form the Navy's newest class. They have gas turbine-drive engineering plants that represent yet another change in ship propulsion. During World War II and much more recently, cruisers and destroyers were "all-gun" ships. Today, all cruisers and most destroyers have missiles as their main weapons.

The American destroyer escorts (DEs) of World War II were officially designated **frigates (FF)** in 1975. They are primarily involved in antisubmarine operations, although some, designated **FFGs,** carry missiles. Small combatants, mine warfare ships, and amphibious warfare ships make up the remainder of the fleet's fighting force. Some experimental work is being done with **nondisplacement type hulls,** such as **hydrofoils** and **surface effect systems,** for there are many advantages to be gained from traveling just over the water rather than in it.

The Soviet Navy, which has emerged as a first-rate sea force, commands an extensive submarine fleet. In light of this, antisubmarine warfare also promises to be an area of high interest.

Communications/Electronics

Advances in the world of electronics have had a significant impact on military technology in general and on communications in particular. Radios have been transistorized, and television has been used extensively in security and monitoring roles. **Electronic sensing and surveillance equipment** can detect sounds and movement that the human senses cannot. Even **man sniffers,** or **people sniffers,** were used in Vietnam. Mounted in aircraft, they could detect the presence of chemicals found in human urine, and thus assisted in the location of large concentrations of enemy troops. Computers, too, not only enhance man's ability to perform the complex administrative and logistical bookkeeping necessary in modern war, but also increase the efficiency of such combat related functions as fire direction, missile guidance, intelligence production, and airspace management. Command and control is also facilitated by computers, television, and other electronic means. In an age when television dominates people's lives from childhood on, it is not surprising that electronics dominates armies.

Operations

The end of the Second World War, like the end of other major wars, brought rapid American **demobilization,** that is, the discharging from service of personnel and the mothballing of equipment. The **mobilization base,** those resources available for remobilization, remained, but the level of readiness for any level of war was seriously reduced. The small wars characteristic of the decades after demobilization have been deceptive, for even though the level of violence has been low, the need for resources has been immense.

The **enclave theory** sought to establish areas free of insurgent activities that would serve as bases from which operations could be launched to extend the perimeter of the enclave. This theory required soldiers not only for the offensive operations but also for **base security,** that is, the protection of the heart of the enclave. Activities within the enclave, such as **base development,** which is the improvement or expansion of the resources and facilities in a base to better support military operations, also consumed manpower and equipment. The French referred to the enclave theory as the *tache d'huile* (spot of oil), because, as a drop of oil spread on water, so would an enclave spread until the insurgents in many small wars were eliminated. **Strategic hamlets,** the communities created and organized by American civil affairs advisers to repulse any insurgency influence, became many drops of oil that were to spread and join in an effort to end the war in Vietnam. Like the troops that attempted to secure the many drops, however, the oil became **overextended,** too thin to be of practical value.

Tactics

When combat units leave relatively secure base areas, they rely on **fire tactics,** that is, the use of firepower, rather than on **shock tactics,** the use of shock action, or the close hand-to-hand fighting that characterized earlier wars. In Vietnam, fire tactics were practiced during what were known as **search and destroy, cordon and search,** and **search and clear** operations, terms that imply what was all too commonly known—that the enemy was hard to find. **Point units,** the vanguard of forces advancing in an insurgent environment, sought the enemy. Even when enemy presence was only suspected, **fire missions,** or calls for supporting fires, were transmitted via radio to **fire support bases,** which were areas, often carved out of the jungle, where artillery and mortars were placed to support the ground forces. At the fire support bases, the **Fire Direction Center (FDC)** computed the range and direction to the suspected target and passed this information to the gun crews. When several fire support bases could reach the target area, guns from these bases shot **overlapping fires.** Occasionally, the enemy was tenacious. At these times, **relief columns,** reserves that could advance by road or trail, sometimes brought additional support. Often, **eagle flights,** which were reaction forces flying in helicopters and awaiting word that an enemy force had been located, brought the added firepower.

Because fire tactics are less effective at night than during periods of good visibility, free world forces in Vietnam generally prepared defensive positions at night. The laager defense was recalled from earlier wars. The "**Custer ring,**" an allusion to Custer's final "circling of the wagons" on the banks of the Little Big Horn in 1876, was another term that described the circular defensive positions commonly used. Interlocking fires of automatic weapons were planned, and the **final protective fires,*** which were the preplanned fires to be used in case of an all-out attack, were a part of each night's defensive plan. The **killing zones,** areas in which firepower was concentrated, were nearly impossible to cross successfully. The effects of fire tactics were generally apparent in the next day's **body count,** the reported number of enemy killed, and a term that unfortunately received too much attention as a measure of American success. Because of the terrain in Vietnam, political and military encouragement to succeed, and the enemy's extraordinary efforts to recover their dead, inaccurate body counts were all too common.

Airmobile Operations

Airmobile operations involve the helicopter transport and airlanding of troops, equipment, and supplies in the conduct

---

*The technique of devising final protective fires dates back to World War I.

of ground combat missions.⁵ Because the airlanded force includes few antitank weapons, airmobile objectives should be on terrain that is unsuitable for deployment of mechanized forces. **Airmobile assaults** allow a commander to threaten areas in the enemy rear, overcome distances quickly, overfly barriers, bypass enemy defenses, extend the area of operations, have greater flexibility in the deployment of reserves, and reduce the vulnerability of friendly forces to nuclear attack by enabling the forces to be dispersed and concentrated quickly when contact is imminent or established. Airmobile forces can operate from **heliports,** which are improved helicopter landing areas that generally have maintenance and other support activities. More often, however, airmobile forces are picked up at a **landing zone (LZ),** which can be any relatively flat area that is free of obstructions. A few hours of work with demolitions, chain saws, and machetes are usually sufficient for a small unit to clear a landing zone in even the thickest jungle. **Air mobility** is a tremendous asset in modern war, but without adequate air cover or air superiority, airmobile forces become lucrative and virtually defenseless targets.

Electronic Warfare Operations

In modern wars of all types, the commander must consider the electromagnetic environment as a battlefield extension where an invisible but very real struggle known as **electronic warfare** takes place.⁶ Electronic warfare (EW) is made up of electronic combat and defensive EW. **Electronic combat** consists of **electronic warfare support measures (ESM),** such as signals intercept and direction finding to provide target acquisition data, and **electronic countermeasures (ECM),** which is the jamming and deception of an enemy's command, control, intelligence, and weapons systems using electronic emitters. **Defensive EW,** also known as **electronic counter-countermeasures (ECCM),** refers to the electronic tactics used to protect friendly emitters from the enemy's jamming and target acquisition efforts.

Modern armies are capable of selectively depriving adversaries of control of the electromagnetic environment. In addition, certain situations may incidentally or unintentionally deprive forces of control. For example, the **electromagnetic pulse (EMP)** that emanates from a nuclear detonation can damage radios and other electronic equipment, and hence seriously interfere with command and control communications and target acquisition systems. The successful commander will find alternate means of communication and apply countermeasures to insure the accomplishment of the mission assigned to him.

Logistics

The proper securing of the logistical base is as important in small wars as it is in other forms of war. Large stockpiles of equipment, supplies, ammunition, and **POL (petroleum, oil, and lubricants)** are prime targets for raids and sabotage. **Vehicle and tank parks,** where large numbers of vehicles are concentrated for shipment, storage, and repair, are also vulnerable targets. In an insurgent environment especially, forces cannot be certain that even main supply routes (MSRs) are clear of mines and free of ambuscades on a day-to-day basis. The logistical problems of the often unseen insurgent, infiltrator, or terrorist are also great. He must often rely on manpower and local sources to provide the materials necessary to maintain his tools of war. He lives among the people and draws his logistical strength from the people. Massive stockpiling is rare for insurgents, infiltrators, and terrorists, but ubiquitous **caches,** or hiding places, often can conceal significant amounts of war materiel. In the Vietnam War, the North Vietnamese and their allies often relied on **sanctuary areas,** areas that were inviolable to free world forces because of political agreements and other political pressures, as bases for supplies. The complexity of today's logistical systems demands that all routes and bases of supply be adequately protected from all threats.

Bibliography

Single volumes that deal with the wars of the post-World-War-II decades are difficult to find because of the diversity and open-endedness of the period. However, some general works on the period, especially the early part, do exist. Two such works are Bernard Brodie's *Strategy in the Missile Age* and his critical *War and Politics*. Herman Kahn's *On Thermonuclear War* grapples with some of the intense questions related to the use of nuclear weapons, while Robert E. Osgood's *Limited War: The Challenge to American Strategy* calls for a limited war strategy as a solution to the problems posed by atomic arsenals.

A concise one-volume study of the Korean War is David Rees' *Korea: The Limited War*. The Chinese intervention in Korea and its political and military background are the focus of Allan Whiting's *China Crosses the Yalu*. An interesting and readable first-person account of the war is found in Matthew B. Ridgeway's *The Korean War*. The official volumes on the Korean War, which appear under the general rubric of The United States Army in the Korean War, include Colonel James F. Schnabel's *Policy and Direction: The First Year;* Roy E. Appleman's *South to the Naktong, North to the Yalu,* which gives a detailed account of operations from June to November 1950; and Walter G. Hermes' *Truce Tent and Fighting Front,* which covers the last two years of the war. *Selected Readings in Warfare Since 1945,* which is part of the West Point Military History Series, contains brief but accurate accounts of the Korean War, the rise of Red China, and several of the Arab-Israeli wars.

The insurgent wars of the post-World-War-II period are dealt with in a variety of scholarly and popular works. The French experiences from the late 1940s through the early 1960s are the topic of Peter Paret's *French Revolutionary Warfare from Indochina to Algeria*. Bernard Fall produced many books on Indochina and Vietnam, including *Street Without Joy* (the story of the French-Indochina War, 1946-1954), *Hell in a Very Small Place* (the tale of Dien Bien Phu), and *Two Vietnams: A Political and Military Analysis*. A critical and provocative work on the American involvement in Vietnam is found in Frances Fitzgerald's *Fire in the Lake: The Vietnamese and Americans in Vietnam*. The background, the battles, and the effects of the 1968 Tet offensive are well reported in Don Oberdorfer's *Tet!* A view of the war from the side of the insurgent is presented in Douglas Pike's *War, Peace and the Viet Cong*. Views of American soldiers include Dave R. Palmer's *The Summons of the Trumpet: U.S.-Vietnam in Perspective* and William C. Westmoreland's *A Soldier Speaks*. An excellent and comprehensive account of America's war in Vietnam is found in Stanley Karnow's *Vietnam: A History*. John Clark Pratt's *Vietnam Voice: Perspectives on the War Years, 1941-1982* tells the story of the war through news releases, poems, GI graffiti, speeches, diaries, songs, and official reports. The United States political side of the Vietnam War can be gleaned from Arthur Schlesinger's *A Thousand Days: John F. Kennedy in the White House,* and the war is the major issue in Herbert Schandler's *The Unmaking of a President: Lyndon Johnson and Vietnam*. A highly favorable view of America's conduct of the war is found in Guenther Lewy's *America in Vietnam*.

The Army Center of Military History, which prepared the official histories of World War II and the United States Army in the Korean War, has also published a series of accounts on various aspects of the Vietnam War. The authors of these volumes held positions of responsibility in the conflict. Included in the series are *Airmobility, 1961-1971,* by Lieutenant General John J. Tolson; *Allied Participation in Vietnam,* by Lieutenant General Stanley Robert Lawson and Brigadier General James Lawton Collins, Jr.; *Base Development in South Vietnam,* by Lieutenant General Carroll H. Dunn; *Cedar Falls-Junction City: A Turning Point,* by Major General Bernard W. Rogers; *Command and Control, 1950-1969,* by Major General George S. Eckhardt; *Communications-Electronics 1962-1970,* by Major General Thomas M. Rienzi; *The Development and Training of the South Vietnamese Army, 1950-1972,* by Brigadier General James Lawton Collins, Jr.; *Division-Level Communications,* by Major General Charles R. Myer; *Field Artillery, 1954-1973,* by Major General David Ewing Ott; *Financial Management of the Vietnam Conflict, 1962-1972,* by Major General Leonard B. Taylor; *Law at War: Vietnam, 1964-1973,* by Major General George S. Prugh; *Logistic Support,* by Lieutenant General Joseph M. Heiser, Jr.; *Medical Support of the U.S. Army in Vietnam, 1965-1970,* by Major General Spurgeon Neel; *Mounted Combat in Vietnam,* by Lieutenant General Donn A. Starry; *Riverine Operations, 1966-1969,* by Major General William B. Fulton; *The Role of Military Intelligence,* by Major General Joseph A. McChristian; *Sharpening the Combat Edge: The Use of Analysis to Reinforce Military Judgment,* by Lieutenant General Julian J. Ewell and Major General Ira A. Hunt, Jr.; *Tactical and Materiel Innovations,* by Lieutenant General John H. Hay, Jr.; *U.S. Army Engineers, 1965-1970,* by Major General Robert R. Ploger; *U.S. Army Special Forces, 1961-1971,* by Colonel Francis J. Kelly; and *The War in the Northern Provinces, 1966-1968,* by Lieutenant General Willard Pearson. The origins of America's involvement in Vietnam are discussed in Ronald H. Spector's *Advice and Support: The Early Years, 1941-1960,* produced by the Center of Military History in 1983. The fall of South Vietnam is presented in Arnold Isaac's *Without Honor: Defeat in Vietnam and Cambodia*.

The background and fighting of the Arab-Israeli wars are addressed in *The Israeli-Arab Reader,* edited by Walter Laquer, while first-person insights into the 1956 conflict can be found in Moshe Dayan's *Diary of the Sinai Campaign*. The 1967 conflict is well treated Chaim Herzog's *War of Atonement*. The 1982 Israeli invasion of Lebanon is covered in Jonathan C. Randal's eyewitness account, *Going All the Way: Christian Warlords, Israeli Adventurers, and the War in Lebanon*. Other recent small wars are covered in a collection of articles edited by Howard J. Wiarda, *Rift and Revolutions: The United States in Central America;* Max Hastings and Simon Jenkins' *The Battle for the Falklands;* and Thomas T. Hammond's *Red Flag over Afghanistan*.

Good books on the arms race include David Holloway's *The Soviet Union and the Arms Race;* the Harvard Nuclear Study Group's *Living with Nuclear Weapons;* Michael Sheehan's *The Arms Race;* and David N. Schwartz's *NATO's Nuclear Dilemmas*.

Doctrine concerning air mobility operations is found in FM 90-4, *Airmobile Operations*. General policies and guidance for today's Army are found in FM 100-1, *The Army*.

Although much of the material concerning special operations is classified, a great deal can be learned from Tony Geraghty's *Who Dares Wins: the Story of the Special Air Service, 1950–1980.* "Who dares wins" is the motto of the British Special Air Service, which is roughly equivalent to the elite elements of United States special operations forces.

Comprehensive and current information on the latest developments in the world of military weapons and equipment can be found in the first-rate yearbooks known simply as Jane's. The Jane's enterprise was founded in 1909 by Fred T. Jane, and today's volumes include the following titles: *Armour and Artillery; Military Vehicles and Ground Support Equipment, Weapons Systems;* and *All the World's Aircraft.*

Notes

[1]Bernard and Fawn Brodie, *From Crossbow to H-Bomb,* revised edition (Bloomington, IA, 1973), p. 296.

[2]William D. Siuru, Jr. (U.S.A.F.), "New Army Helicopters," Department of Engineering, United States Military Academy, West Point, New York.

[3]William D. Siuru, Jr., "Army's Swift New Battle Tank," *Popular Science,* CCXIV (February 1979), 97.

[4]The following discussion is based on *The Bluejacket's Manual,* 20th revised edition by Bill Wedertz (Annapolis, 1978), Chapter 3.

[5]United States Army, FM 90-1, *Employment of Army Aviation Units in a High Threat Environment,* September 1976.

[6]United States Army, FM 100-5, *Operations,* August 1982, pp. 4–5.

Toward A More Perfect Understanding

10

The future belongs to those who dare. And daring requires courage born of loyalty, and wisdom born of experience. Courage and wisdom cannot be acquired from books or professors—no matter how brilliant and erudite they might be. Loyalty and experience, however, can be acquired through understanding and education, and history can provide countless examples that contribute to the further appreciation of both. Upon these foundations, the serious student can build a lifetime of service that will in some way—big or small—contribute to mankind's quest for a better world.

Major Themes of the Future

Although historians neither claim nor seek to deal with the future, the assiduous study of the past does provide insights into future events. One general truth that emerges from this insight is that the collective body of man's knowledge will continue to grow. This growth has been likened to a circle of light, representing the known, in a sea of darkness, representing the unknown. As the circle expands, we become more aware of the sea of darkness surrounding us. Thus, the more we learn, the more we come to recognize our own ignorance. For some, the circle of knowledge evokes pessimism; for others, the circle of knowledge gives encouragement, because each time the frontier of darkness is pushed back, we improve the quality of our 'ives and the lives of others. Especially for the soldier, to cease to strive is to cease to grow and, ultimately, to cease to be effective. The consequences of ineffectiveness are tragic for soldiers, for they often involve the suffering of man and the fate of nations.

A second theme of the future involves the roles that the profession of arms is likely to play in the complex societies characteristic of the twentieth century. Man's recognition of the role of force, as either a deterrent to hot war or an active participant in hot war, will insure that better societies will exist in the twenty-first century.

> . . . the sanction of force lies at the root of social organizations. A careful reading of history reveals that physical force, or the threat of it, has been applied to the resolution of social and political problems since man formed the first primitive tribal group. That force, orchestrated as the situation demands, has continued to persist—to assure order; to combat enemies abroad or suppress revolts at home; to uphold what is right, hopefully not to undermine it.[1]

Similar roles will surely persist in the foreseeable future.

A third theme of the future is that technology will become more sophisticated and that more diligent study and training will be required of soldiers who hope to perform successfully on the battlefield of the future. The cost of technology will continue to increase, too, hence, the competition for resources will increase. Greater value must be derived from the resources available, and greater enthusiasm, efficiency, and energy must be injected into those activities that matter most.

A final theme involves the importance of preparedness, a word that has become synonymous with such luminaries from the American military past as John C. Calhoun, Emory Upton, Elihu Root, and Leonard Wood. Modern military forces must be prepared to fight intensely, anywhere, on short notice. Such a state of preparedness requires that soldier-leaders be attuned to political developments and, more importantly, that they and all elements within their realm of responsibility be physically and faithfully ready to act. They and their subordinates must especially be prepared as individuals and units, physically and spiritually, to perform in a manner that would make their forebears and progeny proud.

The Threads of Continuity Revisited

Man's understanding of the future is bound to the past. Because the past is a blend of tradition and change, a review of the threads of continuity that were explained at the start of this volume can aptly summarize where the institutions of war have been and, if not suggest where the institutions of war are headed, at least give evidence that tradition and change will continue to characterize man's advance through time. The specifics of these changes and traditions will depend on both the resources available to and the decisions made by the nation's future civilian and military leaders.

Military Professionalism

Since attitude distinguishes the professional from those who are not professionals, it is difficult to definitively assess the military professionalism of a given age. Nevertheless, certain major themes within the area of military professionalism can be discerned. The early Greek city-states generally relied on citizen-soldiers to fight their wars. Because the Army was composed of part-time soldiers, its commitment to better the conduct of war was limited. The Army of Alexander the Great, on the other hand, was composed of many long-term soldiers who demonstrated a high degree of expertise in military engineering, siege warfare, and battle in the open field. Hannibal's Carthaginian Army was made up of volunteers, local recruits, and mercenaries. Usually, the mercenaries were responsible for performing the more complex tasks, such as cavalry reconnaissance, fighting, and engineering. The early Roman armies were composed of citizen-soldiers, but as the Empire grew, the tradition of a long-standing army, typified by the centurions, grew, also. Foreign soldiers, often lacking motivation and dedication, permeated the ranks of the Roman Army as the Empire receded before the onslaught of horse-borne barbarians. Poorly organized and armed, the barbarians operated in relatively fragmented groups and achieved success in proportion to the brute force they could wield.

In the Middle Ages, military professionalism was a part of the feudal relationship that dominated societies. Landowners exacted military service from their serfs and trained their most likely warriors in the requirements of individual combat. For the most part, armies remained undisciplined hordes, although individual warriors might display some professionalism. The Swiss brought professionalism to a higher plane when their military leaders began to train their citizen-soldiers in the use of the pike and halberd and the value of the compact, disciplined formation known as the Swiss square.

Beginning in the fourteenth century, firearms necessitated new formations and increased training; military professionalism increased concomitantly. In the seventeenth and eighteenth centuries, aristocrats led the armies; they were tutored by students of war and understood the seriousness of war. The most successful were those who exacted high standards of discipline and performance from their soldiers. During the reign of Louis XIV in France, modern military professionalism came of age. Standing armies, uniformed, trained, equipped, and paid by the resources of the state, were established, and long service provided the opportunity for continued study and improvement of the military institutions of the day. During the American and French revolutions, two significant factors spurred the growth of military professionalism. First, the citizen-soldiers, motivated by patriotism, replaced the woeful "dregs of society" that had made up the armies of the seventeenth and eighteenth centuries. Second, leaders were selected increasingly on the basis of merit and education rather than birth. Students of war wrote books on the theory of war—books that were read, studied, and reflected upon by soldiers who had to lead large units in combat. Technical schools were founded in the decades before and just after Napoleon's rise to power, and staff colleges like the *Kriegsakademie* in Berlin became committed to the diligent study of the conduct of war.

Throughout the nineteenth century, soldier-leaders recognized the increasing importance of formal study and learning to the profession of arms. Schools were established, both to prepare soldiers for commission and to provide advanced military training. In the United States, the Civil War provided an impetus to professionalism by discrediting the concept of amateurism as it had been applied on the battlefields of that war. Army leaders, in particular, subscribed to the ideas associated with scientific management, successfully arguing that their fundamental function was to inspire and lead men, especially under trying conditions. When total war descended on the major powers of the world, first in 1914 and again in 1939, mass armies were filled by conscripts and hastily trained leaders. Those whose formative years had been in association with military institutions became the backbone of the enlarged armies. Modern technology, tactics, strategy, and statesmen demand that the modern soldier-leader be dedicated to the profession of which he is a part and capable of faithfully and effectively executing the complex and difficult missions with which he is charged.

Tactics

Tactics refers to the ordered arrangements of military forces that are used when engagement by rival forces is imminent or

		EXTERNAL FACTORS		INTERNAL FACTORS		
		Technology	Political Social Economic	Generalship Theory & Doctrine Logistics & Admin	Strategy & Tactics	Military Professionalism
800 B.C.	Punic Wars / Persian Wars	Galleys / Bow / Spear / Sword	City-states	Hannibal / Caesar / Alexander	Phalanx / Legions / Wars of Conquest / Age of Cavalry	Hoplites / Citizen soldiers / Knights
A.D. 1470	Crusades / 100 Years War	Stirrup / Gunpowder	Dark Ages / Religious Domination / Growth of cities / States / Enlightenment / Nations / Money Economy	Machiavelli / Coup d'oeil	Tercio / Total War / Swedish Brigade / Napoleonic Tactics / Limited Warfare	Mercenaries / Conscription / Nation-in-Arms / Guerrillas / Military Academies / Aristocratic Officers
1789	30 Years War / Seven Years War / American Revolution	Field Artillery / Bayonet / Flintlock / Sailing Ships	*French Revolution / Industrial Revolution	Gustavus / Marlborough / Frederick / Jomini / Clausewitz	Corps / Napoleon / Wellington / Battles of Annihilation	Journals / Staff Colleges
1815	American Civil War / Napoleonic Wars	Percussion locks / Iron warships	Liberalism	Mahan / General Staffs / Railroads / Moltke / Lee / Grant	Grant / Lee / Moltke	War Colleges
1865	World War I / Wars of German Unification	Breechloaders / Rifling / Radio	Imperialism / Alliances / Arms Races / Socialism	Joffre / Haig / H-L / Foch / Pershing	Schlieffen / Total War / Trench Warfare	Branch Schools / Conscription
1918	World War II	Smokeless powder / Machine guns / Submarine / Tanks / Airplanes	Pacifism / Depression / Appeasement / Communism	Douhet / Field Manuals / Mao / Stalin / Giap	Hitler / Marshall / Eisenhower / MacArthur / Blitzkrieg / Total War	Conscription
1945	Korean War / Vietnam War / Arab-Israeli Conflicts	Aircraft Carriers / Electronic Warfare / ICBMs / Helicopters	United Nations / Cold War / Détente / Atomic Energy		Airland Battle / Low-Intensity Conflict / Air Assault	Volar / Guerrillas
present						

The Threads of Continuity Revisited

underway. In ancient Greece, tactics was evidenced by the solid masses of soldiers formed in phalanxes. Alexander, following the lead of Epaminondas, used groups of phalanxes to provide greater strength through flexibility. Roman legions generally used more numerous, more flexible, and smaller formations than their Greek predecessors, but the clash of arms continued to occur between tightly closed formations of spear-throwing and sword-wielding infantry. During the Dark Ages, tactics was characterized by disorder, the melee, and by individual combat. Disciplined formations returned when the Swiss recognized the advantage of the closed phalanx in combating mounted warriors.

Firearms changed tactics, too, for when firepower replaced shock action as the dominant force on the battlefield, linear formations were found to be more conducive to the massing of fire than close, compact, deep formations. During this era of warfare, intelligence and security were enhanced by the use of small detachments of skirmishers operating in front of the main body. In the eighteenth century, tactical formations were still linear, and long hours of drill were an indispensable prelude to success in battle. The mass armies of the Napoleonic period, however, found that the time required to learn the proper execution of linear tactics was often limited and that linear formations could be successfully defeated by columns, which were more easily trained. Both columns and lines were employed by the French in the attack. In the British Army, however, the line prevailed and generally brought success. Technology altered tactics in the nineteenth century, for when rifles replaced muskets, breech-loaders replaced muzzle-loaders, and repeating weapons replaced single-shot weapons, compact formations served better as targets than as viable fighting formations. Formations thus became more dispersed, and cover became essential to survival.

The American Civil War, in which the attacking force suffered inordinate casualties in the face of an enemy who fought from well-prepared defensive positions, was a harbinger of the tactical problems that were to confront the military forces of the World War I participants.

During World War I, firepower was devastating, and cover was so necessary to survival that the fighting front had to disappear into an underground world. The Germans attempted to break the stalemate of the trenches by employing small, mobile infiltration teams supported by short, intense, and accurately registered artillery fire. The Allies turned to technology to break the stalemate, and mobile protection in the form of the tank ultimately restored mobility to tactics. In World War II, mobile firepower provided by tanks, airplanes, and warships dominated the fighting and dictated the tactics. Today, dispersion seems desirable in light of the threat of nuclear weapons, but mass is still needed to exploit the opportunities provided by highly mobile firepower-

producing arms. In modern small wars, stealth and surprise are the principal components of tactical success. Small, well-trained units capable of functioning independently must be prevalent in modern armies—if those armies are to succeed on either the conventional or nuclear battlefield.

Like army tactics, naval tactics have developed largely in response to technological developments. The warships of the ancient powers were propelled by oars, and ramming and boarding were the prevalent forms of naval combat. In the sixteenth and seventeenth centuries, warships became wind-powered rather than oar-powered. The age of sail began, and since cannon became effective instruments of destruction in this same period, naval tactics became "artillery duels" rather than infantry clashes. In the late nineteenth century, the age of sail was supplanted by the age of steam. Before the age of steam, navies and armies had operated relatively independently, but in the twentieth century, amphibious landings brought naval and army tactics into a close relationship: the navy provided transport and gunfire support to landing forces. At sea, the great iron warships, developed in the naval races of the early twentieth century, fought only one major battle—the Battle of Jutland—during World War I. Like the merchant ships that had moved the lifeblood of empires, the ponderous warships were vulnerable to the silent death that submarines capably delivered. The military value of large capital ships was further diminished by the development of the bomber, and aircraft carriers assumed a dominant role in naval tactics at the Battle of Midway during World War II. Since World War II, the cost of sophisticated capital ships has further diminished their role in modern navies, and the missile-carrying submarine now competes with the large aircraft carrier as the most highly prized vessel in the fleets of the great powers. With its nuclear-powered ships and nuclear-armed warheads, the navy, more than the army, has felt the impact of the nuclear age.

Strategy

Strategy refers to the organized way in which a unit commander or any leader combines and orchestrates the resources at his disposal to attain a given goal. A great variety of strategies have been demonstrated through the ages. The strategy of the ancient Greek city-states was a defensive strategy, for they rarely fought except to protect their homeland, however poorly defined the boundaries of their homeland may have been. Alexander practiced an offensive strategy, or a strategy of conquests. Hannibal, too, attempted a strategy of conquest, but the Roman strategy of exhaustion thwarted his efforts. When political entities began to coalesce in the period known as the Renaissance, the need for sound defensive arrangements against states that wished to grow through military conquest was recognized. Religious differences became causes of war in the early seventeenth century, and religious fervor often dictated that a strategy of annihilation be pursued. In the wake of the Thirty Years' War, reason linked with linear tactics and the high cost of battle brought strategies of maneuver into prominence. Revolutionary fervor and the Industrial Revolution of the late eighteenth century and the early nineteenth century returned strategies of annihilation to the forefront. The increased lethality of weapons in the late nineteenth century suggested that the only sane strategy was that of the indirect approach, but theorists overlooked the practical in favor of the moral, and evangelically encouraged the preparation of a strategy of the offensive. It failed miserably in the face of the mass fire weapons of World War I. Limited funds and the destructiveness of modern total war seemed to dictate that a defensive strategy be the highest priority of post-World-War-I military forces. The strategy of *blitzkrieg,* however, was an offensive strategy used with great success by Nazi Germany during the early years of World War II; the Western Allies adopted a strategy of unconditional surrender in their struggle to defeat their totalitarian adversaries in the war. Since World War II, the emerging nations of the world have adopted a strategy of survival because of encroachments by the great powers. Today, no military force can rest idly, for the strategy of world domination gained by violent means remains a fundamental tenet of Communist ideology.

Logistics and Administration

The complexity and magnitude of logistics and administration have increased markedly during the centuries of recorded history. In ancient times, armies could carry all the necessities of war with them on campaign or procure them in the area of operations. Men and horses, women and weapons, food and forage, could all be readily obtained wherever armies traveled. Weapons could be made and repaired, recovered and reused in the course of a long campaign. Families accompanied the warriors, and relatively few records and requisitions were required. During the Dark Ages, feudal lords outfitted their peasant-warriors at the start of each campaign, while knights generally outfitted themselves. Supplies and services were obtained from local resources; rape and pillage were the most common means of procurement. As weapons became more sophisticated, as rulers recognized the evils of rape and the destructiveness of pillage, and, most notably, as gunpowder became more prevalent, armies began to transport certain necessities from places of manufacture to the forces on campaign. Baggage trains accompanied armies,

while supply trains brought stores of weapons, gunpowder, and other artifacts of war to the front via road or water. By the eighteenth century, extensive systems of depots and magazines were established along the principal invasion routes in and near the frontiers of the great powers and their allies. Secure lines of communication were essential to the survival of an army in the field. Although Napoleon often relied on local procurement to sustain his armies, he was never totally free of his line of communication to Paris and to key industrial areas in France. Logistics in the mid-nineteenth century was profoundly altered by the construction of railroads, which could transport greater quantities of men and supplies with greater speed and reliability than was possible with animal-drawn wagons. Though vulnerable to interdiction, the railroads became the heart and soul of the army supply system. Steam-powered ships, too, brought increased reliability and capacities to transoceanic lines of communication, and the Industrial Revolution insured that the quantities of goods produced kept pace with needs of the modern mass armies. Early in the twentieth century, motor transport and cargo aircraft with capacities in the range of tens of thousands of pounds—and, later, hundreds of thousands of pounds—further increased the logistical capabilities of modern armies. Improved communications have facilitated the administration of military forces, and, today, computers assist immeasurably in the efficient administration of manpower and the masses of supplies that are required to fuel it.

Military Theory and Doctrine

Ideas about war—including both those that do not meet with official acceptance and those that are duly sanctioned—have in all probability existed since the first warrior took the field. Until the invention of the printing press, however, doctrine was seldom written, and official sanction occurred more through usage than edict. Many ancient scholars reflected on the meaning, the instruments, and the conduct of war, and a considerable wealth of knowledge about war survives in the writings of Xenophon, Thucydides, and Vegetius. In the Dark Ages, many ideas concerning war were lost, and many lessons had to be relearned by subsequent generations. Machiavelli was among the first of the Renaissance writers to organize his extensive thoughts on war for the benefit of subsequent generations. He well understood the role of force in the history of man. Early in the seventeenth century, Hugo Grotius recognized the need for a code of laws to govern the conduct of war between nations. War, which had been intensified by fierce religious ardor, as evidenced in the Thirty Years' War, became influenced by the rationalism that prevailed during the scientific enlightenment of the sixteenth century. Throughout the seventeenth and eighteenth centuries, military writers dealt with such topics as the evolutions of individuals and units on the battlefield, the construction of camps and fortifications, and the geometric and mathematical formulas that might assist commanders in the conduct of military operations. The Age of Reason was felt in the realm of military theory.

The writers of the post-Napoleonic period, Jomini being chief among them, were greatly influenced by the classical thought of the eighteenth century. Consequently, they sought the secrets or principles of success, clearly demonstrated, they maintained, by the campaigns of Napoleon, Frederick the Great, and earlier successful commanders. Paralleling Jomini's work in terms of its time of preparation, but coming later in terms of its influence, was Clausewitz's *On War*. It sought to unravel the complexity of war, to determine war's true nature, to discover war's relationship with politics, and to better understand the role of moral forces in war. To this day, no book on war approaches the impact or the breadth and depth of thought and analysis of Clausewitz's masterpiece. The nineteenth century was the age in which military schools were established for the promulgation and study of doctrine. In Berlin, the excellence of the famous *Kriegsakademie* was evidenced by the success of the Prussian armies in the Wars of German Unification. A companion institution, the German General Staff, also contributed immensely to the German success, which was achieved by a combination of sound doctrine and the doctrinal unity that was a product of the well-established school and staff.

The mass armies of modern war necessitated new guidelines for the conduct of war, while new technology demanded handbooks and instructions on the function and employment of modern instruments of war. The result has been a proliferation of official field manuals, regulations, guides, handbooks, and other publications relating to military theory and doctrine.

Generalship

Generalship is the employment of the qualities and attributes necessary to command major units. It involves concepts of professionalism, tactics, strategy, logistics and administration, and theory and doctrine. It also involves personal courage, exemplary action, and dedication.

Epaminondas relied on strategy, and especially upon tactics, to gain his victory over the Spartans at the Battle of Leuctra. Alexander the Great was an able administrator, strategist, and tactician, and one whose example of physical courage and success merited the respect of his own age and of subsequent generations. Hannibal, too, was an able and exemplary tactician, strategist, and leader, but ultimately he failed because his strategic goal of destroying the Roman confederation eluded him. Belisarius and Genghis Khan were courageous leaders whose strategy and tactics brought repeated success on long campaigns.

Charlemagne, who was only semi-literate, relied on charisma and courage to inspire the rabble provided by his vassals to overcome the threats to his domain. Gustavus Adolphus worked ceaselessly to improve all facets of the Swedish Army and the entire Swedish society. His successful leadership of the Protestant forces during the Thirty Years' War was a tribute to his ability to inspire loyalty and his tactical and strategic genius. His theory became doctrine as he schooled his cavalry in shock tactics and his infantry in evolutions facilitated by his creation of the Swedish brigade.

Frederick the Great was a master tactician and a master strategist. He knew how to maximize the use of the limited resources at his disposal. He was a thinker, a writer, and an able administrator, and his logistical system of depots and magazines served his army well. In the American Revolutionary War, George Washington and Benedict Arnold provided vivid contrasts in generalship. The former was solid rock; the latter was boisterous, with a character like shifting sand.

Tactically, strategically, administratively, and logistically, Napoleon repeatedly showed the brilliance of genius. But ultimately, at least in a military sense, he failed miserably. Abdication, escape, and exile are not the laurels of the complete general.

The American Civil War was rife with examples of generalship, good and bad. Lee, brilliant at Chancellorsville but besmirched by repeated failure at Gettysburg less than two months later, and Grant, whose tarnished character and indifference to theory were forgiven in light of his tactical and strategic success, are two examples.

Increasingly, the twentieth century has produced generals whose strengths lie in the realm of strategy and administration. One of the first of the great generals who achieved success largely as a result of his strategic and administrative talents was Helmuth von Moltke (1800–1891). His reputation was principally based on the leadership he demonstrated as Chief of the Prussian General Staff (after 1871, the German General Staff) rather than his ability as a field commander. Even though strategic and administrative talents have become more important qualities of generalship in the twentieth century than they were in earlier periods, many generals have established themselves as a result of personal courage, combat leadership, and other qualities more characteristic of previous eras. George Patton was such an individual, but no general combined the old virtues with the new more effectively than Douglas MacArthur.

Today, tactics has become a product of textbooks, while logistics has become the province of civilians and computers. Courage is demonstrated in the midst of tough decision making, but rarely in the midst of physical threat. The swashbuckling and impulsive traits of generals of centuries long past have given way to the collective wisdom of complex bureaucracy, but courage and genius remain essential traits of any individual who strives to be the best in the realm of generalship.

Political Factors

The ideas and actions of governments or organized groups have changed dramatically in the millennia of recorded history. The Greek city-states, groups with extremely local interests, fought long and well against would-be conquerors. Groups with fervent religious and commercial interests, like the ancient Romans and Carthaginians, became involved in intense wars. Weak governments have been the victims of both external and internal wars. Authoritarian governments have known both brilliant success and abject failure in war. In modern times, governments determine the participants, the means, and the objectives of war. Political considerations dictate the course of wars.

Social Factors

Social factors are closely related to political factors. For example, when formal political organizations do not exist, societies govern themselves, and social and political factors coincide. Societies can be democratic, autocratic, homogeneous, or stratified, and military organizations are affected profoundly by the type of society they represent. A democratic society tends to require military service of all its members. Autocratic societies tend to draw military forces from certain classes within the society or from groups outside the society. Few societies are homogeneous, most of them reflecting some degree of stratification. Stratification, which results from economic, religious, political, racial, or intellectual differences, restricts entry into certain levels of military rank.

Social factors, in conjunction with political factors, often determine the source of military manpower. Citizen-soldiers have repeatedly responded to the call to arms when danger is near—especially in democratic societies. Mercenaries dominated the ranks in early modern autocratic societies. Volunteers have swelled the ranks in democratic societies on the eve of many major modern wars, but when popular sentiment proved to be mercurial, conscription was necessary to provide the manpower needed to sustain the fighting forces until a political settlement could be achieved. Armies have been instruments of social change, and armies have been instruments of individual social mobility. Ultimately, armies reflect the good and the bad that exist in the societies they represent. However, in spite of the fact that an army reflects the society it serves, today's soldiers must recognize that they "work in a social environment which is at best indifferent and at worst hostile to their activities."[2]

Economic Factors

The activities and ideas that affect the resources of a government or a society are called economic factors. Like political and social factors, economic factors have a decided impact on the nature and conduct of war. Control of precious resources, such as gold, bauxite, rubber, and oil, can become the objects of aggression. Wealth can buy mercenaries and equip armies. Defense budgets can contribute to economic growth or, unwittingly, contribute to depression and double-digit inflation. The availability of vital resources determines the intensity, duration, and often the outcome of wars at all levels. At the same time, it is well recognized that military forces are becoming "so expensive to maintain that either the economy has to be distorted to provide anything of reasonable size or the forces must be pared down in quantity to the extreme limit."[3] Economic factors will surely influence the size, shape, and materiel of future armed forces.

Technology

Technology is the use of knowledge to create or improve practical objects or methods. It affects tactics, strategy, logistics, administration, theory, and doctrine. Technology differs from other factors that affect war because it continuously and progressively builds upon all known previous technology. The javelin was an improvement over the spear, the percussion lock was an improvement over the flintlock, the rifle was more effective than the musket, the breech-loader was better than the muzzle-loader, smokeless powder is better than black powder, and iron warships are better than wooden ships. Whether all technological development is good, however, remains a topic worthy of debate. To list definitively the most significant technological developments in the history of war would be difficult at best, but to suggest and ponder some important developments can be a profitable exercise. In the age of antiquity, the discovery of fire, torsion, and iron ranks high. The saddle and stirrup markedly improved the effectiveness of cavalry in the Dark Ages. Gunpowder and resultant firearms corresponded with man's emergence from the Dark Ages. Flintlocks and bayonets succeeded matchlocks and pikes in the seventeenth century. Technologically, the eighteenth century was relatively dormant—ironic, because it was an intellectually active age, the Age of Enlightenment. The Industrial Revolution profoundly affected warfare in the late eighteenth and nineteenth centuries. Interchangeable parts, mass production, steam engines, breech loading, rifling, smokeless powder, repeating small arms, telegraphy, iron warships, and recoil systems each had an impact on the conduct of war in the nineteenth century. Radio, internal-combustion engines, heavier-than-air craft, radar, television, jet-propelled aircraft, intercontinental ballistics missiles, transistors, helicopters, atomic energy, and computers have been among the most militarily significant technological developments in the present century.

The threads of continuity are a means to an end—the end being a better understanding of war and a greater appreciation of mankind through the study of the past.

The Future of War

The debate between pacifists and staunch advocates of military preparedness will undoubtedly continue. In this regard, it is particularly appropriate that words spoken over one hundred years ago should be relied upon when discussing the future possibilities, requirements, and study of war:

> War is itself a school of the sternest and most thorough character; it teaches not only the theory of the military art, but its practice also. The army, which is often engaged in war, which frequently puts in practice the theory of the art, preserves the knowledge of the theory and instructs the recruits which from time to time swell its ranks, at least in practice. And, after all, the utility of the theory is only that the practice may be perfect. . . .

> . . . The art of commanding men is not the art of giving orders, and inflicting punishment if they be not obeyed. It is the art of obtaining from men a willing obedience; the art of drawing forth their utmost efforts; the art of inspiring them with such sentiments of duty and devotion that no obstacle can daunt them, nor danger can appall. The most important qualities for the man who seeks to command men are not intellectual qualities; they are moral; it is a firm, honest, unselfish, and just character that commands the respect of men, attracts their confidence and secures their obedience. . . .

> . . . We know that the whole material universe, from the atoms to the stars, is in motion; that in it motion is life and rest is death. So it is in the realm of mind; the mind must move, and if it does not go forward it will go back.

> And, moreover, the military art is a constantly changing and constantly progressing art—never before so rapidly changing and progressing as now, and even if you could stand still, your profession in a few short years would leave you far behind. . . .

> Do not think that I am holding up to you the duty of leading a hard, cold, and cheerless life. Pleasures and enjoyments within the bounds of reason are as necessary as labor to man's best development, both of body and of mind. But I beg you to believe that I speak the opinion and tell you of the experience of all thoughtful men . . . when I say that a life of honest and honorable effort is the only happy life. The profession of arms has been the object, sometimes, of the admiration, and, sometimes, of the execrations of mankind. By turns it has deserved them both. Which it should receive depends upon the manner and the spirit in which, and the objects for which it is pursued.[4]

The object of war must be to preserve, not destroy, what is good and worthy. Moreover, at no time should man seek to fight the last war, for the last war will but complete the cycle that began in Eden.

Participants in Future Wars

The past is prologue, and greater understanding of the motivations and thoughts of past participants in the profession of arms can serve to suggest better courses of action for future participants. The following points are those of a retired soldier and gentleman, who served long and faithfully:

First: May we Army folk never lose our capacity to enjoy, our ability to put on a cheerful countenance in the face of whatever aggravation or adversity comes our way. To do otherwise is to imperil ourselves, and add nothing worthy to the Army which we serve. . . . Wordsworth's "Happy Warrior," I am sure, is more than a mere poetic cliche. . . .

Second: . . . Ritual self-approval leads to complacency and can end in disaster . . . or a reasonable facsimile thereof. A feeling of self-worth, tempered by the habit of salty self-criticism, is a balance hard to come by, but crucially important to . . . continuing health.

Third and Last: This point is somewhat akin to my second, in that it has something to do not only with attitudes, but words and their meanings. I allude to the words *professional* and *professionalism,* words in our daily lexicon, which have passed from mere currency and have reached the *incantatory* stage; debased coin I'm afraid, spent with more prodigality than appreciation for their full meaning. The danger of using *professional* and *professionalism* in the incantatory fashion is that we incline to limit their meaning to a kind of superficial finesse in the manipulation of things and people, to the reaching of tidy solutions, to the unconscious neglect of the often awkward human ingredient in our affairs, to the canonization of the "system" as a self-justified entity with a rationale for being all its own. The "system" is a construct. . . . And all too often we lose sight of the important last words of the dictionary definition of the word *professional* to the effect ". . . members of a profession are committed to a kind of work which has for its *prime purpose the rendering of public service.*" These qualifying words, it seems to me, are crucial to the real meaning of *professionalism* and elevate the humane significance of our calling to its proper place.

Now, it is generally accepted that *amateur* and *amateurism* are the reverse of the former two words; and too often the word *amateur* carries with it the odor of *dilettantism*—a kind of self-indulgent dabbling in things in which one is not formally schooled and not paid for. The other implication is that the *amateur* can produce a variety of mischiefs. In ascribing this pejorative flavor to the word *amateur* we overlook its origins; it comes, of course, from the Latin TO LOVE, and the *amateur* is one who ". . . has a marked fondness, a liking, a taste for something." Many of my dearest colleagues—and I'm sure my experience is not unique in this—have been amateurs, especially during the war. They brought to the battlefield not only a high level of technical competence but, and this is more important—a *marked fondness* for caring about their soldiers and leading them in a good cause; a sparkling enthusiasm which marks the man who does things for the joy of doing them, and a belief in what he is doing.

What I'm driving at is this: let's beware of intoning the word *professionalism* in a ritual way, unless it carries with it the inseparable purpose of rendering public service. That's the crux. Technical proficiency without this purpose is a truncated thing and can lead by itself to grief as far as the humane aspect of our calling is concerned. And let's temper our *professionalism* with those admirable qualities of *amateurism* which express themselves in enthusiastic keenness; a joy in the job.[5]

The future will be better if General Sutherland's words are heeded.

Soldiers' lives are arduous, but while understanding the seriousness and urgency of their assigned tasks, they can and must seek enjoyment and, through enjoyment, convey the enthusiasm that contributes to the successful and faithful accomplishment of any task. Ideally, they will be professional amateurs in the best sense of both words.

Notes

[1]Colonel Thomas E. Griess, "The Profession of Arms," an address to members of the Class of 1982, United States Military Academy, West Point, New York. First Revision, September 5, 1978, p. 1.

[2]Michael Howard, "Military Science in an Age of Peace," *Journal of the Royal United Services Institute for Defence Studies,* CXIX (March 1974), 4.

[3]*Ibid.,* p. 8.

[4]Excerpts from Major General A.H. Terry's address to the United States Military Academy's graduating class of 1883 as quoted in Hugh T. Reed, *Elements of Military Science and Tactics,* 7th edition (Chicago, 1889), pp. 541–545.

[5]Brigadier General (Ret.) Edwin V. Sutherland, "Thoughts of an Academic Board Member on Leaving West Point," *Assembly,* XXXVI (September, 1977), 31.

Index

Page numbers for definitions are in boldface.

1st Special Service Force, **129**
2-I-C, **77**
2.36-inch rocket launcher, **132**
3-inch Ordnance Gun, **85**
3-pounder, **51**
3-pounder, British artillery, **66**
3.5-inch rocket launcher, **132**
4-pounder, French artillery, **66**
4.2-inch mortar, **133**
4.5-inch multiple rocket launcher, **134**
5-inch naval guns, **153**
6-inch howitzer, French, **66**
6-pounder, British artillery, **66**
6-pounder, French artillery, **66**
7.92-mm parabellum machinegun, **106**
8-pounder, French artillery, **66**
9-mm machine pistol, **106**, 107
9-pounder, British artillery, **66**
12-pounder, British artillery, **66**
12-pounder, French artillery, **66**
12-pounder, Napoleon, **85**
14-inch batteries, **153**
18-inch batteries, **153**
18-pounder, British artillery, **66**
24-pounder, British artillery, **66**
40-mm grenade launcher, M203, **174**
57-mm recoilless rifle, **132**
60-mm mortar, **133**
"75," **107**
75-mm rapid-fire field gun, **107**
75-mm recoilless rifle, **132**
81-mm mortar, **133**
82-mm rocket, **134**
98K, **103**
105-mm howitzer, **132**
132-mm rocket, **134**
150-mm howitzer, **132**
155-mm gun, **132**
155-mm howitzer, **132**
320-mm railroad artillery, **107**
320-mm spigot mortar, **133**
600-mm self-propelled howitzer, **132**
800-mm railroad gun, **132**

A-6 Intruder, **180**
A-7 Corsair, **180**
A-10, **180**
A-20, **143**
AA guns, **106**
AAA (antiaircraft artillery), **176**
AAH (advanced attack helicopter), **183**
Abatis, **88**
ABC (atomic-biological-chemical), **172**
ABM (antiballistic missile), **178**
A-bomb, **135**
Abrams, General Creighton W., 185
Absolute ceiling, **137**
Absolute war, **2**

ACCs (air control centers), **161**
Ace, **97–98**
Action, **12**
Action officers, **169**
Active defense, **23**
Active duty, **95**
Ad hoc, 129
ADA (air defense artillery), **161**, 176–178
Adjutant, **78**
Adjutant general, **78**
Administration, **7**, 197, 198, 199
Administrative drops, **162**
Admiral, **78**
Advance base, **163**
Advance guard, **16**, 68–69
Advance to contact, **16**
Advisers, **70**
Advisory assistance, 69, 70
AEF (American Expeditionary Force),
  95, **100**, 102
Aerial observers, **97**
Aerial photographs, **161**, 163
Aerial reconnaissance, **117**
Aero companies, **101**
Aero squadrons, **101**
Aft, **53**
Age of Enlightenment. *See* Age of Reason.
Age of Napoleon, 59–70, 102, 108, 196
Age of oars, 37, 196
Age of Reason, 41–58, 197–198, 200
Age of sail, 196
Age of small wars and nuclear deterrence,
  167–189, 196
Age of steam 196, 198
Aggrandizement, **123**
AH-1 Cobra, 182, **183**
AH-1Q Cobra TOW, **183**
AH-64 Apache, **183**
Aides-de-camp (aides), **61**, 63
Air assault division, **171**
Air carpet, **161**
Air cavalryman, **169**
Air control centers (ACCs), **161**
Air cover, **161**
Air defense, **161**
Air defense artillery (ADA), **161**, 176–178
Air force, **101**
Air group, **129**
Air guns, 35
Air line of communication (ALOC), **12**
Air marshals, **125**
Air mobility, **189**
Air operations, 161
Air raid drills, **161**
Air raids, **161**
Air service, **101**, 127
Air strike, **161**
Air superiority, **117**

Air transport fleet, **129**
Air transportable, **171**
Air umbrella, **161**
Air warning and interceptor system, **157**
Airacobra, **138**
Airbase, **137**
Airborne, **125**
Airborne division, **171**
Airborne operations, **162**
Airborne soldiers, **125**
Aircraft, 97–98, 109–110, 111, 171, 172, 176,
  177, 179–183, 188, 197. *See also* Airplanes;
  Dirigibles; Helicopters.
Aircraft carrier, **155**, 187, 196
Airdrome, **137**
Airfield, **136**
Airfoils, **134**
Airhead line, **162**
AirLand Battle, **6**
Airlanded elements, **162**
Airlanding division, **128**
Airlifting, **161**
Airmen, **125**
Airmobile assaults, **189**
Airmobile division, **171**
Airmobile operations, **188–189**
Airplanes, 97, **109–110**, 111, 135, 136–146,
  179–182, 196
Airport, **136**
Airstrip, **137**
Albacore, 141, **144**
Alerts, **161**
Alexander the Great, 6, 7, 27, 29, 32, 33, 36,
  38, 39, 194, 196, 198
All hands, **162**
All-around defense, **24**
Alliances, **93**
Allies, the, 100, 101, 107, 108, 109, 116, 135,
  140, 143, 163, 196, 197. *See also*
  Triple Entente.
Alligator, **156**
ALOC (air line of communication), **12**
Amateur, 201
Amateurism, 201
Ambulance, **109**
Ambulance companies, **101**
Ambulance service, **101**
Ambuscade, **16**
Ambush, **16**
American Army, 77, 78, 126. *See also*
  United States Army.
American Civil War, 12, 49, 73–91, 93, 94,
  95, 98, 99, 101, 102, 109, 115, 194,
  196, 198
American Expeditionary Force (AEF), 95,
  100, 102
American Indians, 98, 102
American Navy, 77. *See also* United States
  Navy.

American Revolution, 17, 39, 49, 65, 73, 101, 194, 198
American Waco CG-4A, **146**
Ammonal, **108**
Ammunition and other implements of destruction, **52–53**, 56, 66, 86, 89, 108–109, 118, 119, 130–131, 134–135. *See also* Bombs and other implements of modern war.
Ammunition parks, **135**
Amphibious assault, **118**
Amphibious landing, **118**, 196
Amphibious operations, 114, **118–119**, 120, 121
Amphibious tank, **156**
Amphibious task force, **118**
Amphibious tractor, **156**
Amphibious truck, **136**
Amtrac, **156**
Anchorages, **156**
Ancient Greece, 29, 31, 32, 33, 34, 37, 38, 194, 196, 199
Ancient Rome. *See* Romans.
Ancients, 29, 31, 32, 33, 34, 35, 36, 37, 38, 39, 194, 196, 197, 198, 199
"Angel of Mons," **100**
Annexed, **158**
Antiaircraft artillery (AAA), **176**
Antiaircraft weapons, **106–107**, 176–178, 18
Antiboat guns, **132**
Antimilitarism, **123**
Anti-Submarine Detection Investigation Committee (ASDIC), **154**
Antisubmarine rockets (ASROC), **187**
Antisubmarine warfare (ASW), **154**
Antitank ditches, **157**
Antitank weapons, **132**, 175, 180, 183, 186
ANZAC (Australian-New Zealand Army Corps), **98**
APC (armored personnel carrier), **185–186**
Appeasement, **123**
Applied tactics, **5**
Approach, **54**
Approach flight, **162**
Approach march, **23**
Arbalest, **43**
Arbalester, **43**
Archers, 32, **35**
Archies, **106**
Arclights, **182**
Area army, **128**
Area bombing, **161**
Area defense, **24**
Area of operations, **12**
Area-type weapon, **174**
ARK (Armored Ramp Karrier), **147**
Armada, **33**
Armies, **128**
Armies of occupation, **88**
Armistice, **68**
Armor (body), 31, 34, 38, 63, 114
Armor piercing bombs, **135**
Armor piercing rounds, **135**
Armored cars, **113**
Armored cavalry units, **169**
Armored cruisers, **114**
Armored divisions, **171**
Armored fighting vehicles, 32, 169. *See also* Tanks.
Armored gunboats, **87**
Armored personnel carrier (APC), **153**, **185–186**
Armored Ramp Karrier (ARK), **147**
Armored reconnaissance vehicle, **185**
Armored vehicle launched bridge (AVLB), **179**
Armories, **81**
Arms race, 95
Army, **11**
  activities of, 12–13
  anatomy of, 11–12
  environment of, 12

purpose of, 13
Army, The, 9
Army Air Forces, **127**
Army Field Forces, **127**
Army groups, **101**
Army of Northern Virginia, 80
Army of Tennessee, 80
Army of the Potomac, 80
Army Service Forces, **127**
Army surgical hospitals, **171**
Army's Distinguished Service Cross, 127
Arnold, Benedict, 198
Arquebus, 47, 48
Arquebusiers, 43
Array, **33**
Arresting gear, **181**
"Arrival," **118**
Arrow, **35**
Arrowhead, **35**
Arsenals, **81**
Artificial harbors, **156**
Artificiers, 63
Artillery, **50**, 63, 66, 97, 107–108, 115, 117, 175
Artillery control lines, **159**
Artillery control measures, **159**
Artillery train, **53**
Artillery units, 63, 64, 117
ARVN (troops of the Army of the Republic of Vietnam), **169**
ASDIC (Anti-Submarine Detection Investigation Committee), **154**
Askaris, **98**
Asphyxiating, **109**
ASROC (antisubmarine rockets), **187**
Assailable flank, **19**
Assault battalions, **102**
Assault fire, **117**
Assault guns, **132**
Assault position, **68**
Assault rifle AK-47, **172–173**
Assembly areas, **117**
Assigned, **128**
ASW (antisubmarine warfare), **154**, 155
Atabrine, **158**
Atomic Armageddon, **167**
Atomic bomb, **135**
Atomic weapons, **135**. *See also* Nuclear weapons.
Attached, **128**
Attack from march column, **68**
Attack of an organized position, **17**
Attack on a broad front, **159**
Attack position, **68**
Attaque brusque, **125**
Australian Army, 77
Australian soldiers, 101
Australian-New Zealand Army Corps (ANZAC), **98**
Austrian Army, 98, 101, 115
Autocratic society, 199
Automatic, **104**
Automatic weapons, **104–106**
Auxilia, **31**
Available rate of supply, **121**
Avenger TBF, **144**
Avenue of approach, **161**
Aviation units, 33, 172
Aviators, 97
AVLB (armored vehicle launched bridge), **179**
Avro Lancaster Mk I, **144**
Awards and decorations, **99**
Ax, **35**
Axe, **35**
Axis, **159**
Axis of advance, **159**

B-1, **182**
B-17, **143**
B-24, **143**
B-25, **143**

B-29, **143**, 181
B-36, **181**
B-52 Stratofortress, **181–182**
Back (of bow), **35**
Backfire, **182**
Backward planning, **119**
Bacon, Roger, 41
Bacteriological weapons, **135**
Baggage train, **53**
Bailey bridge, **179**
Ballista, **37**
Ballistic missile, **134**
Ballistic missile defense (BMD), **178**
Ballistic missile submarine (SSBN), **187**
Balloon bomb, **135**
Balloons, 66, 67, 109
Banana magazines, **173**
Band, **33**
Bangalore torpedoes, **109**
Banner, **79**
BAR (Browning automatic rifle 1918A1), **130**
Barbary Coast pirates, 67
Barges, **114**
Barracks, **39**
Barrage (artillery term), **55**, 118
Barrage (naval term), **113**
Barrage balloons, **157**
Barshot, 53, 86
Base, **3**
Base development, **188**
Base hospital, **101**
Base of operations, **12**
Base of supply, **12**
Base security, **188**
Basic load, **83**
Bastion, **54**
Battalion, **46**
Battalion of the train, **64**
Battery, **46**
Battle (military confrontation), **12**
Battle (military organization), 33, **45**
Battle cruisers, **114**
Battle fatigue, **158**
Battle groups, **171**
Battle in the open field, **23**
Battle of annihilation, **6**
Battle stations, **162**
Battleax, **35**
Battle-bowler, **114**
Battles of decision, 93
Battleships, 67, 187
Battlework, **87**
Bayonet, 47, 49, 56, 64, 81, 102, 130, 200
Bayonet strength, **98**
Bayonne, France, 47
Bazooka, **132**
Beachhead, **119**
Beaching craft, **156**
"Beans and bullets," **119**
Beaufighter, **141**
Beehive round, **175**
BEF (British Expeditionary Force), 98, **100**
Belisarius, 198
Bell, **138**
Belligerent, **1**
Belly (of bow), **35**
Beriberi, **158**
Besiege, **12**
Betty, **145**
"Big Bertha," **108**
Bilge water, **153**
Bill (pole arm), **46**, 47
Binoculars, **116**
Biological weapons, **135**
Biplane, **109**, 137, 138
Bismark, Otto von, 95
Bivouac, **39**
"Black Jack," 95, **102**
Blakely gun, **85**
Blakely rifle, **85**
Blank, **86**
Blast pens, **157**

Blighty, **98**
Blighty wounds, **98**
Blitzkrieg, **125**, 147, 151, 197
Blockade, **57**
Blockade runner, **57**
Blockbusters, **178**
Blockhouses, **157**
Blocking force, **17**
Blocking positions, **68**
Blockships, **153**
Blowguns, 35
"Blowtorch," **109**
Bludgeoning weapons, 33, 34, 35
Blue Max, **99**
Blue-water strategy, **116**
BMD (ballistic missile defense), **178**
Boat lanes, **119**
"*Boche*," **98**
Body count, **188**
Body snatchers, **98**
Boeing, **138**
Boeing B-17F Flying Fortress, 142
Boeing B-29 Superfortress, 143
Boer War, 100
Bofors, **132**
Bofors Armament Works, 132
Bogie wheels, **151**
Bolsheviks, **102**
Bolt (of crossbow), 35
Bolt (of gun), **83**
Bolton Paul Defiant, **143**
Bomb, **86**
Bombard, **50**, 51
Bombardment, **86**
Bombardons, **156**
Bombers, **109**, 110–111, **137**, 180, 181, 196
Bombproof, **88**
Bombs and other implements of
    modern war, 178–179
Bonaparte, Napoleon, 6, 7, 8, 21, 59, 60, 61
    62, 64, 65, 66, 67, 70, 73–91, 107, 194,
    197, 198
Booby traps, **109**
Booms, **87**
Bore (of gun), **51**, 52
"Bought the farm," **98**
Boulette, 81
Bourbons, 41
Bow (of ship), **53**
Bow (weapon), 34, 35
Bowsprit, **53**
Bowstring, 35
Box defense, **24**
Boxcars, **113**
Bradley, Omar, General of the Army, 186
Bradley fighting vehicle, 172, **186**
Brainwashing, **170**
Branch schools, **127**
Brass hat, **97**
Brattices, **38**
Breach, **87**
Break contact, **25**
Breaker line, **159**
Breakout, **68**
Breakthrough, **68**
Breastwork, **54**
Breech, **48**, 51, 85
Breechblocks, **51**
Breechloader, **50**, 196, 199
Bren Gun-Mk I, **130**, 131
Brevetting, 77, 95
Brewster Buffalo, **141**
Bridge (of a ship), **153**
Bridgehead, **27**
Bridgehead line, **27**
Brigade, **46**
Brigade group, **129**
Brigadier, **77**
Brigadier general, **77**
Brigantines, **67**
Brigs, **67**
Brinkmanship, **167**

Britain, Battle of, 139, 141, 145
British Army, 66, 77, 96, 97, 98, 99, 100,
    101, 102, 196
British Bomber Wing, **101**
British Bren Gun, **130**, 131
British Expeditionary Force (BEF), 98, **100**
British Horsa, **146**
British Navy, 77, 95, 98, 101
British Royal Air Force, 125
British 13-pounder gun, **106–107**
British 37-mm direct fire cannon, **108**
British 60-pounder, **107**
Broadax, **35**
Broadside, **56–57**
Broadsides, **57**
Broadsword, **35**
Brooke, John, 85
Brooke gun, **85**
Brown, Jacob, 77
Brown Bess, **49**
Browning, John, 130
Browning automatic rifle, 131
Browning automatic rifle 1918A1 (BAR), **130**
Browning machinegun, Model 1917, **104**, 105
Browning .303, **132**, 140, 141, 143
Browning .50 caliber M-2, **132**
Bruderkrieg, **93**
BTs (*Bystrokhodnic* tank), **151**
Bucellarii, **32**
Buffer territories, **116–117**
Bulk loaded, **163**
Bulk supplies, **163**
Bull Run, Battle of, 80
Bulldozer, **153**, 179
Bullet, **81**, 86, 89
Bulwarks, **87**
Bunkers, **157**
Buoy layers, **156**
Bureau, **80**
Bureau chief, **80**
Bureau of Military Justice, 80
Bureaucracy, **199**
Burial teams, **102**
Burnside, Ambrose, 80
Bursting rounds, **135**
Bushido, **127**
Bushwhacker, **89**
Butterfly bombs, **179**
Buzz bomb, **134**
Byrnie, **38**
Bystrokhodnic tank (BTs), **151**
Byzantine Empire, 29
Byzantines, 29, 33, 34, 35, 38, 56

C-5A Galaxy, **181**
C-130 Hercules, **181**
Caballero, **31**
Cabinet, 6, **100**
Caches, **189**
Cadres, **99**
Caesar, Julius, 6, 7, 29
Caissons, **66**
Calhoun, John C., 193
Caliber, **51**
Call signs, **158**
Camouflage, **114–115**, 117
Camp, **39**, 198
Camp followers, **56**
Campaign, **12**
Campaign of 1704, 8
Campaign strategy, **6**
Campaigning season, **39**
Canister (ammunition), **53**, 66, 86
Canister (of gas mask), **109**
Canna, 50
Cannae, Battle of, 17, 19, 20, 29
Cannon, **50**, 51, 52, 53, 86
Cannonade, **86**
Cantonment, **39**
Canute, King, 35
Cap (percussion), **81**
Capital ships, **67**, 196

Capitalism, **93**
Captain, **44**
Captain (naval), **78**
Caput, 44
Carabiniers, **63**
Carabiniers u cheval, **63**
Caracole, **55**
Caravel, **53**
Carbine, 63, **65**
Carpet bombing, **161**
Carrack, **53**
Carre, 35
Carreau, 35
Carrier group, **129**
Carrier task force, **129**
Carro-ballista, 37
Carry a flank, **23**
Carter, Jimmy, 176
Carthaginians, 6, 29, 34, 35, 38, 194, 199
Cartridges, **66**
CAS (close air support), **161**
Case shot, **53**, 86
Casemates, **115**
Casemates (naval), **153**
Castled, **53**
Castles, **38**
Castrametation, **39**
Casualties, **79**
Cat, **38**
Catalina, **143**
Catalina PBY, **143**
Catapults, **36–37**
Cavalry, 32–33, 34, 55–56, 63, 64, 98, 101,
    102, 169, 186, 194. *See also* Cavalrymen.
Cavalry pickets, **70**
Cavalrymen, 31, **32**, 34, 35, 38, 63, 64,
    65, 98
    air, 169
    heavy, **32**, 63
    light, 63, 64
    See also cavalry.
CBN (chemical-biological-nuclear), **172**
CBR (chemical-biological-radiological), **172**
C.B.s, **171**
CC (combat command), **129**
Cease-fire, **68**
Center, **11**
Central position, **21**
Central Powers, 95, 116
Centurion (Roman officer), **32**
Centurion (tank), **184**, **185**
Century series, **179**
CGNs, **187**
CGs, **187**
Chain gun, **183**
Chain Home Radar, **157**
Chain of command, **77**
Chain shot, **53**, 86
Chamber, **49**
Chancellorsville, 198
Change front, **20**
Chaparral, **177**
Char d'assaut, 113
Char léger (light tank) 1935 R, **149**, 150
Charge, **47**
Chariots, 34, 37
Charlemagne, 29, 31, 33, 38, 198
Charleville musket, **64–65**
Chassepot, **82**
Chasseurs à cheval, 63
Chasseurs à pied, 63
Chassis, **109**
Chauchat, **106**
Chemical warfare, 109, 179
Chemical-biological-nuclear (CBN), **172**
Chemical-biological-radiological (CBR), **172**
Cheval-de-frise, **88**
Chevalier, 31
Chevaux-de-frise, **88**
Chevaux légers, 63
Chevaux 8-Hommes 40, 113
Chief engineer, **78**

Chief of artillery, **78**
Chief of staff, **44**
Chief of Staff of the Army, 77
Chief of the German General Staff, **44**
Chief of the Imperial General Staff
   (CIGS), **97**
Chief of the Prussian General Staff, 198
Chief petty officer (CPO), **125**
Chieftain, **185**
Chieu Hoi, **170**
"Chinese attack," 118
Chivalry, 31
Choke points, **159**
Chopper, **182**
Christie pattern suspension, **151**
Churchill (tank), **149**, 150
Churchill, Sir Winston, 6, 21, 113, 167
CH-21 Shawnee, 182, **183**
CH-47 Chinook, 182, **183**
CH-54 Tarhe, **183**
CID (Committee of Imperial Defense), **97**
CIGS (Chief of the Imperial General
   Staff), **97**
CINC (commander-in-chief), **125**
Ciphers, **158**
Citadels, **87**
Citizen militia, 31
Citizen-soldiers, 31, **194**
Civil defense, **161**
Civil Guard, 172
Civil war, **3.** *See also* American Civil War
Clandestine operations, **163**
Class (age group), 61
Class (of ships), **154**
Classes of supply, **121**
Clausewitz, Carl von, 2, 3, 75, 158, 198
Claymore mines, **178–179**
Claymores, **178–179**
Clearing operations, **118**
Close air support (CAS), **161**
Close hauled, **56**
Close-order tactics, **68**, 196
Club, 34, 65
Coal boxes, **108**
Coaling stations, **156**
Coalition cabinet, **100**
Coalition war, **116**
Coast artillery, 63, 66
Coast watchers, **127**
Coastal defense, **68**
Coastal division, **129**
Coaxial machineguns, **132**
Cock (of gun), 48
Codes, **158**
Cohortal legion, 33, 38
Cohorts, 33
Cold Spring, NY, 85
Cold war, 1, 169
Collective security, **167**
Collier, 114
Colonel, **44**
Colonel general, **97**
Colonello, 44
Colonialism, 93, 95
Colored regiments, **101–102**
Colors, 79
Colt Company, 83
Colt revolver pistol, **83**
Combat air group, **129**
Combat arms, **11**
Combat bombload, **137**
Combat command (CC), **129**
Combat Developments Command, **171**
Combat engineers, **169**
Combat loading, **163**
Combat outpost, **23**
Combat patrol, **89**
Combat power, **79**
Combat rations, **163**
Combat service support units, 11, 12
Combat strength, **79**
Combat support units, 11, 12

Combat team, **64**
Combat units, **11**
Combined arms, **46**
Combined arms concept, **46**
Combined command, **61**
Comitatenses, **32**
Command and control, **88–89**
Command channels, 77
Command post, **64**
Command post exercises (CPXs), **99**
Command relationships, **128**
Command-detonated mines, **178**–179
Commander, **78**
Commander-in-Chief, 77, **125**
Commanding General of the Armies of the
   United States, 77
Commanding General of the Army, 77
Commando, **125**
Commerce raiding, **67**
Commissary, **89**, 90
Commissary officer, **78**
Commission, 44
Commissioned officer, **44**
Committee of Imperial Defense (CID), **97**
Commodore, **78**
Communication trench, **87**
Communications, **67**, 68, 88, 116, 157–158,
   187–188, 197
Communications zone, **128**
Communism, **93**
Communist Manifesto, 93
Commutations, **75**
Compact formations, 196
Companions, **32**
Company, **32**
Company grade officer, **78**
Company officer, **78**
Composite battery, **46**
Composite bow, **35**
Composite regiments, **80**
Compulsory service, **102**
Computers, 188, 197
Concealment, **55**
Concentration, **55**
Concentration camps, **79**
Concentration off the battlefield, **21**, 22
Concentration on the battlefield, **21**, 22
Concentric advance, 67, **68**
Concussion bombs, **108**
Condottieri, **43**
Confederacy, the, **88**
Confederate Army, 77, 79, 81, 88, 89
Confederate Chief of Ordnance and
   Hydrography, 85
Congress, 77, 78, 129
Congress of Vienna, **73**
Congreve, Sir William, 66
Conning tower, **154**
Conscription, 61, 102, 194, 199
Conservatism, **73**
Consolidated, 118, 138
Consolidated B-24H Liberator, 142
Constabulary troops, **98**
Constantine, 31
Constantinople, 29, 41
Construction engineers, **169–170**
Contact point, **159**
Containment, **167**
Continental Army, **39**
Continental Army Command, **171**
Continental strategy, **116**
Continuity, threads of, 3–8, 194, 195, 200
   external, 7–8, 199–200
   internal, 4–7, 194, 196–199
Continuous front, **117**
Contraband, **75**
Control measures, **159**
Control officer, **125**
Control ships, **119**
Conventional warfare, **167**
Convoy, **64**
Coordinated attack, 17

Coordinating point, **160**
Coordinating staff, **78**
Copernicus, 41
'Copter, **182**
Cordite, **108**
Cordon and search operations, **188**
Cordon defense, **24**
Cordon sanitaire, **159**
Corduroyed, **86**
"Corkscrews," **135**
Coronel, 44
Corporal, **44**
Corps, 61, **64**, 80, 101
Corps area, **128**
Corps areas (in Vietnam), **171**
Corps groups, **101**
Corps of Engineers, 11, 80
Corps system, 61, **64**
Corps tactical zones, **171**
Corsairs, **67**
Corvettes, **54**, 67
Corvus, **38**
Cossacks, **63**
Cotton clads, **87**
Council of war, **44**
Counterattack, **17**
Counterbattery fire, **86**
Counterespionage, **163**
Counterinsurgency, **2**
Counterintelligence, **89**
Countermanded, **89**
Countermarch, **56**
Countermining, **88**
Counteroffensive, **17**
Counterscarp, **54**
Coup, **2**
Coup de grace, **159**
Coup de main, **159**
Coup d' état, **2**
Coup d'oeil, 70
Couriers, **88**
Courts of inquiry, **79**
Courts-martial, **44**
Cover (protection), **55**, 196
Cover (security mission), **68–69**
Covered way, **54**
Covering force, **25**
Covering force area, **23**
Covering post, **25**
Cowling, **137**
Cowpens, **17**
Coxswain, **126**
CPO (chief petty officer), **125**
CPXs (command post exercises), **99**
C-rations, **163**
Creeping barrage, **55**
Crew-served weapons, 36–37
Crimean War, **75**
Critical terrain, **67–68**
Croix de guerre, **99**
Crossbow, **35**, 36
Crossfire, **55**
Crossing areas, **26**
Crossing force, **26**
Crossing front, **26**
Crossing sites, **26**
Crossing the T, **162**
Cruise missile, **178**
Cruisers, 67, **114**, 187
Cruising speed, **137**
Crusader (tank), **149**
Crusaders, 32, 38
Crusades, 31, 38
Cryptographer, **98**
Crystal, **157**
Crystal sets, **157**
Cuirass, **38**, **43**, 54, 63
Cuirassier, **43**, 63
Culverin, **52**
Curtain, **38**
Curtis Hawk 75A, **139**
Curtis P-40, 138, **139**

Custer, George, 188
"Custer ring," **188**
Cutting weapons, 33, 34–35
CVNs **187**
Cyclist battalions, **102**
Cylindrical-conoidal bullet, **83**
Cynicism, **123**

Da Vinci, Leonardo, 41
Daggers, 34, 102
Dahlgren, John A., 85
Dahlgren gun, 85, 86
Daimyo, **125**
"Daisy cutters," 108–109
Dakota, **146**
Damage control, **162**
Dardanelles Committee, **100**
Dark Ages. *See* Middle Ages.
Darts, **35**
Darwin, Charles, 93–94
Darwinism, 93–94
Dauntless, **144**
Davy Crocket (rifle), **174**
Day bombers, **137**
Day fighter, **137**
DD963 destroyers, **187**
D-Day, **158**
DDGs, **187**
DDs, **187**
De Cordoba, Gonzalo, 45
Deactivation, **117**
Death march, **129**
Deception plan, **161**
Deck officer, **126**
Deckhand, **126**
Declarations of war, **158**
Decoration, **99**
Decoy ships, **114**
Deep order, **68**
Defeat in detail, 18, 21, 68
Defect, **170**
Defectors, **170**
Defense
  forms of 23–24
Defense in depth, **24**
Defense sector areas, **23**
Defensive EW, **189**
Defensive items, 33, 34, 38, 54–55, 87–88,
  114–116, 157
Defensive operation, 12, 21–24
Defensive weapons, 104, 106, 108
Defensive-offensive, **17,** 20
Defoliants, **179**
"Degommered," **100**
Delay, **25**
Delbruck, Hans, 6
Deliberate attack, 16, **17,** 70
Deliberate river crossing, 26, 27
Deliberate withdrawal, **25**
Demi-brigade, 63–64
Demilune, **88**
Demobilization, 169, **188**
Democracy, 73–75
Democratic society, **199**
Demonstration, **17**
Dengue fever, **158**
Department, **80**
Department of Defense, 61, 170
Department of the Air Force, **170**
Department of the Army, **170**
Department of the Cumberland, 80
Department of the Navy, **170**
Department of the Ohio, 80
Department of the Tennessee, 80
Department of War, **61**
Dependents, **170**
Deploy, **68**
Deployed for combat, **16**
Deployed in depth, 23–24
Depot, **56,** 197
Depth bombs, **154**
Depth charge, **154**

Deputy commander, **77**
DEs, **187**
Desert Mounted Corps (DMC), **101**
Desertion, **79**
Destroyer escorts, **156**
Destroyer transports, **156**
Destroyers, **114,** 187
Detachment, **45**
Détente, **169**
Deterrence, **167**
"Deuce and a half," **136**
Developing nations, **69**
Developing the situation, **70**
Dewoitine D. 520, **143**
DFS-230A, **146**
Dictionary of United States Army Terms, 39
Diesel engine, **135**
"Diggers," **98**
Dilettantism, **201**
Diplomacy by force, **1**
Direct fire, **55**
Direct fire weapons, 107–108
Direct order, **88**
Direct pressure force, **17**
Direct support, **55**
Direction of attack, **159**
Directive, **89**
Directory, **59**
Dirigibles, **109**
Disappearing cupolas, 115–116
Disarmament, **123**
Disease, 99, 116, 135, 158
Disengage, **25**
Dispatches, **89**
Dispositions, **161**
Distinctive insignia, **79**
Distinguished Flying Cross, **127**
Distinguished Service Cross (DSC), **99**
Distinguished Service Medal (DSM), **99**
Distinguished Unit Citation, **127**
Distribution of forces, **16**
DIVAD (Sergeant York Division
  Air Defense) gun, 177–178
Dive bombers, **137**
Diversion, **17**
Division, **46**
Division '86, **172**
Division area defense, **24**
Division mobile defense, **24**
Division river crossing, **26**
DMC (Desert Mounted Corps), **101**
DMZs (demilitarized zones), **68**
Doctrine, 7, 13, 198, 199
Dogfights, **109**
Dornier Do-17, **144**
Double agents, **163**
Double envelopment, **19,** 20, 21
Double lines of operations, **68**
"Doughboy," 98–99
Douglas Corporation, **143**
Douglas C-47, **146**
Douglas SBD Dauntless, **144**
Dozers, **179**
Draft, **61,** 102, 170
Dragon (medium antitank weapon), **175**
Dragon (musket), **43**
Dragoon (musket), **43**
Dragoon musket, **65**
Dragoons (mounted infantrymen), 43, **63,** 65
Drawdowns, **117**
Dreadnoughts, **114**
Dreyse needle gun, **81,** 82, 103
Drones, **179**
Drop tanks, **137**
Drop zone, **162**
Drum fires, **118**
Drunge, **33**
Drydocks, **162**
DSC (Distinguished Service Cross), **99**
DSM (Distinguished Service Medal), **99**
Dual drive tanks, **147**
Dual Monarchy, **95**

Duce, **125**
"Duck," **136**
Duckboards, **115**
Dud, **178**
Dugouts, **115**
DUK W, **136**
Dumps, **121**
Duplex drive, **147**
Duty, **61**
DZ, **162**

Eagle flights, **188**
Earthwork, **54**
Eastern Front, 101, 113
Eastern Roman Empire. *See* Byzantines.
Eastern strategy, **116**
ECCM (electronic counter-counter
  measures), **189**
Echelon, **19**
Echo ranging equipment, **154**
ECM (electronic countermeasures), **158,** 189
Economic factors, **8,** 199
Economy of force (principle of war), **10**
Economy-of-force measure, **22**
"Eeps," **100**
"Eggbeater," **182**
Eisenhower, Dwight D., 6
Elastic defense, **117**
Electric telegraph, **88**
Electronic combat, **189**
Electronic communications, **116**
Electronic counter-counter measures
  (ECCM), **189**
Electronic countermeasures (ECM), **158,** 189
Electronic deception measure, **158**
Electronic intelligence (ELINT), **158**
Electronic sensing and surveillance
  equipment, **188**
Electronic warfare, **189**
Electronic warfare support measures
  (ESM), **189**
Electronics, 116, 157–158, 172, 187–188, 189
Elephant (assault gun), **132**
Elevating screw, **66**
ELINT (electronic intelligence), **158**
Embrasure, **54**
EMP (electromagnetic pulse), **189**
Emplacement, **115,** 116
Encampment, **39**
Encirclement, **21**
Encircling force, **17**
Enclave theory, **188**
Encounter battle, **16**
Encryption systems, **116**
End run, **20**
Endurance, **137**
Enfilade fire, **55**
Engaged, **25**
Engagement, **12**
Engagement on two fronts, **68**
Engels, Friedrich, 93
Engineers, 64, 65
Engineers, Corps of, 11, 80
Engines of war, 36–37
England. *See* British.
Enigma, **163**
Enlightenment, **41**
Enlistment, **75**
Enola Gay, **143**
Ensign, **78**
Entrenchments, **87**
Envelopment, **19**
Epaminondas, 18–19, 196, 198
Ersatz reserve, **102**
Ersatz units, **102**
Escalation, 167–168
Escort carrier, **155**
Escort fighters, **137**
Escuadron, **46**
ESM (electronic warfare support
  measures), **189**
Espionage, **163**

Espontoon, **46**
Esprit de corps, **79**
Estimate (financial), **69**
Estimate of the situation, **69**
Evacuation hospitals, **171**
Evacuations, **117**
Evolution, **56**
EW, **189**
Exchanged, **79**
Exec, **77**
Executive, **77**
Executive officer, **77**
Exfiltrate, **159**
Expedition, **3**
Expeditionary force, **64**
Expert (marksmanship rating), **89**
Exploding shell, **86**
Exploitation, **17**
Exploration, **3**
Explosive shell, **66.** *See also* High
  explosive shell.
Exterior lines, **21**

F-4 Phantom, **180**
F4U Corsair, **141**
F6F Hellcat, **141**
F-14 Tomcat, **180**
F-15, **180**
F-16, **180**
F-18 Hornet, **180**
F-86 Sabrejet, **179**
F-100 Super Sabre, **179**
F-101 Voodoo, **179**
F-102 Delta Dagger, **179**
F-104 Starfighter, **179**
F-105 Thunderchief, **179–180**
F-106 Delta Dart, **179**
F-111, **180**
Fabian strategy, **6**
Fabian way of war, **6**
Fabius, **6**
Fagot, **179**
Fall-back positions, **159**
Faschamp, **104**
Fascines, **115**
Fascists, **123**
"Fat Albert," **181**
Fat Man, **135**
FB-111, **180**
FDC (fire direction center), **188**
Feigned withdrawal, **20**
Feint, **17, 23**
Feldgrau, **115**
Feldkappe, **115**
Feld-Marschal, **44**
Females (tanks), **113**
Ferdinand assault gun, **132**
Ferry routes, **161**
Festung, **157**
Feudalism, **29, 31–32, 33, 194, 197**
Feuerwalz, **118**
Fez, **115**
FFGs, **187**
FFs, **187**
Field armies, **101**
Field artillery, **53, 55, 65, 66, 67**
Field commander, **77**
Field division, **128**
Field duty, **12**
Field force, **171**
Field fortification, **54**
Field fronts, **172**
Field glasses, **116**
Field grade officer, **77**
Field gun, **53, 65, 67, 107, 113**
Field hospital, **101**
Field marshal, **44**
Field of fire, **55**
Field officer, **77**
Field radio, **116**
Field sanitation, **116**
Field telegraph, **88**

Field work, **54**
Fiesler Storch, **145**
Fighter aircraft, **109–111, 179–180**
Fighter escort, **161**
Fighter-bombers, **137**
Fighters, **137**
File, **11**
Final protective fires, **188**
Finance Corps, **11**
Finnish Army, **132**
Fire and maneuver, **55**
Fire and movement, **55**
Fire bomb, **135**
Fire dance, **118**
Fire direction center (FDC), **118, 188**
Fire fight, **23**
Fire missions, **188**
Fire rafts, **87**
Fire step, **115**
Fire support, **55**
Fire support areas, **119**
Fire support bases, **188**
Fire tactics, **188**
Fireballs, **37**
Firepower, **171, 196**
Firing pin, **82**
First contingent of untrained *Landsturm,* **102**
First duty sergeant, **125**
First line troops, **98**
First quartermaster general, **44**
First sergeant, **125**
First World War. *See* World War I.
Fix the enemy, **19**
Fixed ammunition, **66**
Flag, **79**
Flag of truce, **79**
Flag officer, **78**
Flagships, **67**
Flail tank, **147**
Flak, **107**
Flame projector, **109**
Flame thrower, **109**
Flamenwerfer, **109**
"Flames," **101**
Flamethrower tanks, **147**
Flank, **11**
Flanking movement, **19**
Flanking position, **19**
Flanking towers, **38**
Flares, **116**
Flatboats, **67, 87**
Flatcars, **113**
Flattops, **155**
Fleet, **33**
Fleet admiral, **125**
Fleet carriers, **155**
Fleet oilers, **156**
Fleet submarines, **154**
Flexible defense system, **117**
Flexible response, **167**
Flight officer, **125**
Flights (naval organization of aircraft), **33**
Flights (on bolts), **35**
Flintlock, **48, 199, 200**
Flintlock musket, **49, 64**
Float planes, **137**
Flogging, **79**
FLOT (forward line of troops), **23**
Flotilla, **33**
Flugabwehrkanone, **107**
"Flying banana," **183**
Flying boats, **137**
Flying bomb, **134**
Flying Cigar, **145**
"Flying Circus," **101**
Flying column, **129**
"Flying Crane," **183**
Flying Fortress, **143**
Focke-Wulfe Fw-190, **140**
Focke-Wulfe Fw-190A, **141**
Fodder, **90**
Foederati, **32**

Fokker D-VII, **110**
Fokker DR-1, **110, 111**
Foot artillery, **63, 65**
Foot soldiers. *See* Infantrymen.
Forage, **90**
Forage caps, **115**
Foraging, **56**
Forays, **159**
Force, role of, **193, 197**
Forced march, **38–39**
Forces Command (FORSCOM), **171**
Ford, Gerald, **170, 175**
Fords, **86**
Fore, **53**
Formal siege approach, **54**
FORSCOM (Forces Command), **171**
Fort, **39**
Fort Monroe, VA, **78**
Fortification (s), **33, 36, 37, 38, 39, 63, 64, 115–116, 198**
Fortress, **39, 63**
Forty and Eights, **113**
Forty-five pistol, **106**
Forward defense area, **23**
Forward line of troops (FLOT), **23**
Forward observer, **97**
Forward supply bases, **89**
Foundry, **51**
Four-wheel drive, **135**
Fowling pieces, **65**
Foxholes, **88**
Fragmentation bomb, **135**
Franc tireurs, **98**
France
  Army of Italy, **59**
  Napoleonic wars, **62**
  See also French.
Francisca, **35**
Franco-Prussian War, **82, 85, 95, 98, 127**
Franks, **34, 35, 38**
Fraternization, **79**
Frederick the Great, **6, 7, 8, 18, 19, 21, 99, 198**
Frederick William III, **127**
Fredericksburg, Battle of, **89**
Free world forces, **170**
Free zones, **172**
Freikorps, **102**
French, William Henry, **89**
French Army, **61, 63–64, 65, 66, 67, 68, 96, 97, 98, 99, 100, 117**
French Empire, **61, 73**
French light machinegun, Model 1915, **106**
French Revolution, **59, 61, 73, 194**
French 8mm rifle M1886, **103**
French 75-mm gun, **107**
Frequency, **116**
Friction primer, **85–86**
Frigate, **54, 67, 187**
"Fritz," **98**
Frizzen, **48**
Frocked coat, **97**
"Frog," **98**
"Frog eater," **98**
Frogman, **126**
Front, **11, 101**
Frontal attack, **18**
Frontier, **75**
Frontier companies, **102**
Frontlets, **38**
Frontline, **23, 39, 99**
Frontline ditch, **115**
Führer, **125**
Full general, **77**
Fully automatic weapon, **104**
Fulminate of mercury, **81**
"Funnies," **147**
Furlough, **79**
Fuse, **86**
Fusilade, **55**
Fusiliers, **63**
Fuze, **86**

G-1, **127**
G-2, **127**
G-3, **127**
G-4, **127**
Galileo, 41
Galleass, **53**
Galleon, **53**
Galley, 37-38
Gambeson, **38**
Garand, John, 130
Garote; garotte. *See* Garrote.
Garrison, **79**
Garrison artillery, 63, 66
Garrison duty, **12**
Garrison troops, **12**
Garrote (garote; garotte), **172**
Gas, **109**
Gas masks, **109**
Gasoline-powered truck, **109**
Gatling, Dr. R.J., **104**
Gatling guns, **104**, 177
Gato class submarine, **154**
Gauge, **51**
Gauntlets, **38**
GCI, **158**
Gendarmerie, **63**
Gendarmerie d'élite, **63**
Gendarmes, **63**
General, **44**
General Grant (tank), **147**
General headquarters (GHQ), **100**
General headquarters (Napoleonic era), **63**
General in chief, **61**
General Lee (tank), **147**
General of brigade, **61**
General of division, **61**
General of the army, **125**
General order, **89**
General outpost, **23**
General Pershing (tank), **147**, 148
General quarters, **162**
General Sherman (tank), **147**, 148
General staff, **63**
General staff, German, **198**
General staff officers, **63**
General support, **55**
General war, **2**
Generaloberst, **125**
"*Generaloberst der Waffen SS,*" 125
Generalquartermeister, **56**
Generalship, 7, 29, 198-199
Genghis Khan, 198
Genocide, **123**
Gens d' armes, **43**, 63
Geographic organizations, **80**
Geographical crest, **24**
German Army, 96, 97, 98, 99, 100, 101, 102,
     103, 106, 108, 114, 126, 127, 128, 129, 196
German DFS-230A, **146**
German Empire, 95, 99
German General Staff, **198**
German Mauser, Model 1871, **103**
German Mauser, Model 1898, **103**
German Maxim, Model 08-15, **104**, 105
German MG-42 machine gun, 131, 173
German MP, **106**, 107, 172
German Republic, 123
German Unification, Wars of, **93**
German V-1 rocket, 134, 178
German V-2 ballistic missile, 134
German 5.9-inch howitzer, 108
German 9-mm parabellum submachine gun
     MP 18, **106**, 107
German 15-inch gun, **107**
German 88, **132**
German 88-mm flat trajectory gun, **108**
German 88-mm siege howitzer, **108**
German 420-mm siege howitzer, **108**
Gettysburg, 198
GHQ (general headquarters), **100**
GHQ (Grosses Haupt-Quartier), **100**
G.I., **126**
Glacis, **54**

Glacis armor, **153**
Gladiator, **141**
Gladius, **35**
Glide bomb, **135**
Glider troops, **125**
Gliders, **146**

Goedendag, **46**
"Gone west," **98**
Gonne, **48**
"Gooney Bird," **146**
Gorge, **54**
Gotha, **110**, 111
GQG (Grand Quartier Général), **100**
Gradualism, 167-168
Graduated response, **167**
Grand admiral, **125**
Grand blessés, **98**
Grand Cross, **127**
Grand division, **80**
Grand Fleet, **113**
Grand Quartier Général (GQG), **100**
"Grand Slam," 144
Grand strategy, **6**
Grand tactics, **5**
Grant, Ulysses S., 77, 80, 198
Grape (ammunition), 53, 86
Grapeshot, 53, 59, 66
Grasshoppers, **145**
Great Britain. *See* British.
Great Captains, **61**
Great Depression, **123**
Great War. *See* World War I.
Greaves, **38**
Greek city-states. *See* Ancient Greece.
Green Berets, **171**
Green light, **162**
Grenade launchers, **174**
Grenades. *See* Grenade launchers;
     Hand grenade; Tear gas grenades.
Grenadier, **43**, 63, 64
Greys (British regiment), 101
Gribeauval, General Jean, 66
Griess, Thomas E., quoted, 193
Grossadmiralen, **125**
Grosses Haupt-Quartier (GHQ), **100**
Grotius, Hugo, 197
Ground Controlled Intercept, **158**
Ground fire, **161**
Ground reconnaissance, **117**
Group captain, **117**
Growth. *See* Industrialization and growth.
Grumman, 138, 141, 144
G-sections, **127**
Guards units, **129**
Guerrileros, **63**
Guides, **114**
Guidon, **79**
Gun, 48, **51**
Gun motor carriage M7, **132**
Gunboats, **67**
Guncotton, **108**
Gunne, **48**
Gunpowder and gunpowder weapons, 29, 31,
     33, 35, 38, 41-46, **47**, 48-58, 64, 102-108,
     114, 197, 199-200
Gunships, **183**
Gunwales, **153**
Gurevich, Mikhail, 143
Gustav, **132**
Gustavus Adolphus, 6, 7, 46, 198

Halberd, **46**, 47, 194
Halbert, **46**
Half armor, **54**
Half-rations, **90**
Half-track armored vehicles, **185**
Half-tracks, **185**
Hall rifle, **81**
Halleck, Henry, 77
Halyard, **79**
Hammer, **81**, 86

Hammerhead, **117**
Hand cannon, **47**
Hand grenade, **49**, 108
Hand spike, **66**
Hand weapons, 34, 35, 36
Handle (of bow), **35**
Handley-Page, **110**, 111
"Hanging on the wire," **98**
Hannibal, 6, 19, 29, 34, 194, 196, 198
"Hans Wurtz," **98**
Hansa-Brandenburg, 110
"Happy Warrior," 200
Hara-kiri, **127**
Hard intelligence, **89**
Hard-core, **170**
Hardened silos, **167**
Hardtack, **90**
Hari-kari, **127**
Harpoon missiles, **187**
Hasta, **34**
Hastati, **32**
Hasty attack, **16**, 70
Hasty river crossing, **27**
Hauberk, **38**
Haversacks, **90**
Havoc, **143**
Hawk, **177**
Hawk 75A, **139**
Hawker Hurricane, **141**
Head logs, **88**
Headquarters company, **101**
HEAT (high explosive antitank) rocket, **175**
Heaume, **38**
Heavier-than-air craft, 102, 109, 182. *See
     also* Airplanes; Helicopters.
Heavy bombers, **137**
Heavy cavalry, 32, 63
Heavy infantry, 32, 34
Heavy tank, **113**
Heavy weapons company, **129**
Heer, **128**
"Heine," **98**
Helicopter units, **171**
Helicopters, 169, 171, **182-183**, 188, 189
Heliographs, **88**
Heliports, **189**
Hellfire missile, **183**
Helm, **38**
Helmet, 38, **114**
Helo, **182**
Henry rifle, 83, 84, 85
Higgins, Alexander, 156
Higgins Boat, **156**
High command, 67-68, **97**, 100
High explosive antitank (HEAT) rocket, **175**
High explosive shell, 108-109. *See also*
     Explosive shell.
High Seas Fleet, **113**
Hilt (part of sword), **34**
Hindenburg Line, **116**
Hiroshima, Japan, 135, 143
Hiroshima Little Boy, 136
Hispano-Suiza cannon, 143
Hitler, Adolf, 125, 127
HMS, **114**
HMS *Dreadnought,* 114, 115
HOBOS (homing bomb systems), **179**
Hofkriegsrat, **61**
Hohenzollerns, 41
Hold (of a ship), **154**
Holding attack, **19**
"Holy Joe," **98**
Home front, **117**
Home Guard, **102**
Home island defense force, **102**
Homing bomb systems (HOBOS), **179**
*Hommes-*Forty, **113**
Homogeneous society, **199**
Honveds, **98**
Hooker, Joseph, 80
Hoplites, **32**
Horsa, **146**

Horse artillery, **53**, 63
Hospital organization, the, **101**
Hospital wagons, **66**
Hostile (injury), **79**
Hot pursuit, **118**
Hot shot, **53**, 66
Hot war, **1**, 193
housecarls, **31**
Howitzer, **52**, 53, 108
Howth rifle, **103**
Huey, **183**
Hull defilade, **159**
"Hun," **98**
Hundred Days, **65**
Hundred Years' War, 41
Hunting Panther, **153**
Hurricane barrages, **118**
Hussar, **43**, 63
Hydraulic buffer systems, **107**
Hydrofoils, **187**
Hydrophone listening devices, **154**
Hypaspists, **32**

I-boats, **154**
ICBMs (intercontinental ballistics missiles),
    167, **178**, 182
ICV (infantry combat vehicle), **186**
Idealism, **123**
Identification Friend or Foe (IFF), **158**
IFV (infantry fighting vehicle), **186**
Igel (hedgehog) tactics, **159**
I-Hawk, **177**
Illumination devices, **135**
Impact fuze, **108**–109
Imperial Guard, **64**
Imperialism, **93**
Implied missions, **69**
In contact, **25**
"In the clear," **116**
Incantatory stage, 201
Incarcerated, **79**
Incendiary projectiles, **37**
"Incoming," **118**
Indemnities, **159**
Independent Air Force, **101**
Indian Scouts, **129**
Indian Wars, 75
Indirect approach, **21**
Industrial Revolution, **64**, 75, 81, 93, 197,200
Industrialization and growth, 75
Infantry, 33, 34, 56, 63, 64, 65, 98, 101, 102,
    106, 115, 117, 118, 169, 171, 175, 186,
    196. *See also* Infantrymen.
Infantry combat vehicle (ICV), **186**
Infantry divisions, **171**
Infantry fighting vehicle (IFV), **186**
Infantry group, **129**
Infantrymen, **32**, 33, 63, 64, 67, 98, 114
    heavy, **32**, 34
    light, **32**, 34, 63, 171
    line, **63**
    See also Infantry.
Infiltration, **89**
Inspector general, **44**
Insurgency, **2**, 69. *See also* Small wars.
Insurrection, **3**
Integrated commands, **158**
Intelligence, 69–70, 89, 117, 163
Intelligence officer, **78**
Intelligence operations, 69–70
Intelligence services, **70**
Intelligence source, **89**
Intendant, **63**
Inter-Allied Supreme War Council, **100**
Interceptors, **137**
Intercontinental ballistics missiles (ICBMs),
    167, **178**, 182
Interdict, **57**
Interior lines, **21**, 22
Interlocking fire, **55**
Internal security, **88**
Internal warfare, **2**

Internal-combustion engines, 102, 109, 110
International war, **1**
Internationalism, **167**
Intervention, **3**
Intrenchments, **87**
Invasion, **1**
Invested, **87**
Investment divisions, **101**
Iowa Temperance Regiment, 80
Irish Rebellion, 103
Iron Cross, **99**, 127
Iron Curtain, **167**
Iron ramrod, **49**
Iron ration, **121**
Iron warships, 102, 114, 196
Ironclad boats, **87**
Ironclads, **87**
Irregulars, **43**
Isolationism, **123**
Istrebital, 143
I-TOW, **175**
"Ivan," **98**

Jack, **79**
"Jack Johnson," **107**
Jackson, Andrew, 73
Jagdpanther, **153**
Jäger, **43**
Japanese Army, 127, 129
Japanese Type II light machine gun, **130**, 132
Japanese Zero, 138
Japanese 50-mm grenade launcher, **133**
JATO (jet assisted take off), **181**
Javelins, **34**, 35
Jeep, **136**
Jerkin, **38**
"Jerry," **98**
Jet, **137**
Jet assisted take off (JATO), **181**
"Johnny," **98**
Johnny Rebs, 75
Joint campaigns, **118**
Joint force, **127**
Joint operations, **118**
Jomini, Antoine-Henri, 8, 75, 198
Joseph Stalin (JS) tank, **151**
Josephine (field gun), **107**
JS-I, **151**
JS-II, **151**
JS-III, **151**
Juggernaut, **125**
Jump-off point, **160**
Junior officer, **78**
Junkers Ju-52, **146**
Junkers Ju-87 Stuka, 145
Junkers Ju-88, 145
Junkers 87, 145
Junks, **156**
Junta, **171**
Justinian, 29, 32
Jutland, Battle of, 196

Kaiser Bill, **98**
Kaiten I, **154**
Kamikaze, **127**
KAR-98K rifle, **130**
Kate, **145**
Katyusha rocket launcher, **134**
Keel, **153**
Keep, **38**
Kepi, **115**
Key, Francis Scott, 66
Key terrian, 67–**68**
KIA (killed in action), **79**
Killing zones, **188**
Kills, **98**
Kipling, Rudyard, 98
Kite balloons, **109**
Kitty hawks, **139**
Klementy Voroshilov (KV) heavy tank, **151**
Knee mortar, **133**

Knights, **31**–32, 35, 37, 38, 197
Knight's Cross, **127**
Knight's sword, **35**
Knives, 102
Knots, **114**
Kohlenhasten, **108**
Kommando, **128**
Koniggrätz, 21
Konteradmiral, **125**
Korean War, 2, 21, 104, 130, 169, 171,
    173, 179
Krag-Jorgenson, Model 1892, **103**
Krieg, **128**
Kriegsakademie, 194, 198
Kriegsmarine, **128**, 155

Laager defense, **24**
Labor battalion, **129**
Labor camp, **129**
"Ladies from Hell," **101**
Ladle, **66**
Laid on a target, **118**
Lancaster, **144**
Lance, **34**, 63, 64, 102
Lance (missile), **175**
Lance corporal, **125**
Lancers, **34**, 63, 98
Land cruiser, **113**
Land destroyer, **113**
Land mines, **135**
Land ship, **113**
Landing area, **119**
Landing craft, **156**
Landing craft, infantry (LCI), **114**
Landing Craft Mechanized (LCM), **156**
Landing Craft Vehicle Personnel
    (LCVP), **156**
Landing operations, **162**
Landing Ship Tank (LST), **156**, 157
Landing Vehicle Tracked (1), **156**
Landing zone (LZ), **189**
Landser, **126**
Landsknechte, **43**
Landsturm, **102**
Landwehr, **102**
L'anglais, **98**
Lanyard, **86**
"Large, Slow Target," **156**
Laser-guided bombs, **179**
Last hope, **121**
Latécoère, **144**
Lateral communications, 67, **68**
LAW (Light Antitank Weapon), **175**
Lawrence patent primer, **83**
Lay siege, **12**
LCI (landing craft, infantry), **114**
LCM (Landing Craft Mechanized), **156**
LCVP (Landing Craft Vehicle Personnel),
    156
LD (line of departure), **161**
Lead elements, **23**
Lead time, **163**
Leather gun, **52**
Leave, **79**
Leave of absence, **79**
Lebel mam'se-le, **103**
Lee, Robert E., 198
Lee Enfield .303, **130**
Leeward, **56**
Leg infantry, **171**
Legio, **33**
Legion of Merit, **127**
Legionnaires, **33**, 34
Legions (*legio*), **33**, 196
Leopard, **185**
Lepanto, Battle of, 53
Lethal gas, **109**
Letter of instruction, **88**–89
Leuctra, Battle of, 19, 198
Leuthen, Battle of, 19
Levée en masse, **61**, 64

Level bombers, **137**
Levens projector, **109**
Lewis, Isaac N., 106
Lewis machinegun, **106**
Liaison officers, **63**
Liber, 75
Liberalism, 73–75, 93
Liberated areas, **172**
Liberation, 3
Liberator, **143**
Liberty Ship, **156**
Liddell Hart, Basil H., 21
Lieu, 44
Lieutenant, **44**, 78
Lieutenant colonel, **44**
Lieutenant commander, **78**
Lieutenant general, **44**, 61 (Napoleonic era)
Lift (a siege), **13**
Light Antitank Weapon (LAW), **175**
Light assault guns, **106**
Light bombers, **137**
Light cavalry, 63, 64
Light cruisers, **114**
Light infantry, 32, 34, **63**, 171
Light machinegun M1919A6, **130**
Light machineguns, **106**
Light tank, **113**, 185
Lighters, **87**
Lighter-than-air craft, 66, 109
Lightning war, **125**
Limber, **53**, 66, 67
Limitanei, **31**
Limited war, **1–2**, 169
Limited warfare, **59**
"*Limoged*," 100
Lincoln, Abraham, 83
Line ahead, **56**
Line bombers, **137**
Line effectives, **98**
Line infantry, **63**
Line of battle, **56**
Line of circumvallation, **54**
Line of command, **78**
Line of communication (LOC), **12**, 68, 197
Line of contravallation, **54**
Line of departure (LD), **160**
Line of operations, 67, **68**
Line of retreat, **24–25**
Line officer, **77–78**
Line troops, **63**
Linear formations, 68, **196**
Linear tactics, **56**, 68, 196, 197
Lines of contact, **39**
Lines of engagement, **21**
Lines of investment, **87**
Lines of supply, **68**
Linkup, **68**
Linstock, **66**
Listening posts, **23**
Litter bearers, **98**
Little Big Horn, **188**
"Little Boy," **135**
Littlejohn, **175**
Loaded, **83**
LOC (line of communication), **12**
Local units, **171–172**
Locks (of guns), **47**
Lodgment, **163**
Logistic trains, **64**
Logistical tail, **56**
Logistics, **6**, 56, 89–90, 119–121, 163, 197–198, 199
Logistikos, **56**
Loiter time, **137**
Long ships, **53**
"Long Tom," **107**
Longbow, **47**
Long-range patrol bombing, **161**
Long-range reconnaissance patrol (LRRP), **64**
Loophole, **54**
Lorica, **38**

Lorry, **109**
Louis, Prince, 8, 9
Louis XIV, King of France, 194
Low-intensity conflict. *See* Small wars.
LRRP (Long-range reconnaissance patrol), **64**
LST (Landing Ship Tank), **156**, 157
Luft, **128**
Luftwaffe, **128**, 141
Luger automatic pistol, **106**
Luggers, **156**
Lunette, **54**
LVT (1), **156**
LVT (A), **156**
Lyddite, **108**
LZ (landing zone), **189**

M1 Abrams tank, 172, **185**
M-1 carbine, **130**
M-1 Garand .30 caliber, 104, **130**, 172
M2 Bradley fighting vehicle, **186**
M3 Bradley fighting vehicle, **186**
M3 medium tank, **147**, 148
M3/M5 light tank, **149**
M4 medium tank, **147**
M-14 rifle, **172**, 173
M-16 rifle, **172**, 173, 174
M26 tank, **147**
M-48 tanks, 178, **185**
M-60 (machinegun), **173–174**
M60 (tank), 184, **185**
M60A2, **185**
M60A3, **185**
M-72 LAW (Light Antitank Weapon), **175**
M-79 grenade launcher, **174**
M-113, **185–186**
M551 Sheridan, **185**
MAAGs (Military Assistance Advisory Groups), **171**
MacArthur, Douglas, 198
Mace, **34**
Macedonians, 32, 33, 34
Machete, **172**
Machiavelli, Niccolo, 197
Machine pistols, **106**
Machinegun company, **101**
Machinegun nest, **157**
Machineguns, 104–107, 173–174, 177, 183, 185, 186
Machines (siegecraft), **36–37**
Macht, **128**
Mae West (lifejacket), **162**
Magazine (ammunition and arms storage place), **56**, 197
Magazine (section of firearm), **103**
Magazine rifle, **103**, 104
Maginot Line, 33
Magnetron, **157**
MAGs (military advisory groups), **171**
Magyars, **98**
Mail, **38**
Main attack, **15**
Main batteries, **153**
Main battle area, **23**
Main battle fleets, **113**
Main battle tank, **147**
Main body, **16**
Main force units, **172**
Main line of defense, **23**
Main line of resistance (MLR), **159**
Main supply base, **89**
Main supply routes (MSRs), **121**, 189
Maison, **61**, 63
Major, **77**
Major general, **77**
Malaria, **116**, 158
Males (tanks), **113**
Man of war, **53**
Man sniffers, **188**
Man-at-arms, **31**, 63
Maneuver, **12**, 18–21, 93
Maneuver (principle of war), 10

Maneuvers, **127**
Manifest Destiny, **75**
Maniples, **33**
Manipular legion, 33, 34
Map exercises, **99**
Map reconnaissance, **117**
Maps, **159**
March column, **23**
Maréchal, **44**
Maréchal de camp, **44**
Maréchal des logis, **44**, 56
Marine, **128**
Marines, the, 141, 156
Marius, Gaius, 33
Mark IV, 112, **113**
Mark V Heavy Tank, **113**
Mark VI, **153**
Mark A tank, 112, **113**
Marksman, **89**
Marksmanship, **89**
Marlborough, the Duke of, 8, 9
Marshal, **44**, 61, 63
Marshalling area, **162**
Marsilly carriage, 86
Martin 167, **144**
Martinet, **44**
Martinet, Jean, 44
Marus, **156**
Marx, Karl, 93
Marxism, 93, 95
Maryland I, **144**
Maschinen-pistole, **106**
Maschinen-pistole 18 (MP-18), **106**, 107
Maschinen-pistole 38 (MP-38), **130**
MASH (mobile army surgical hospital), **171**
Masked, **118**
Mass (principle of war), 10
Massacre, 3
Massive retaliation, **167**
Match, 47, 49
Matchlock, 47, 48, 200
Matchlock musket, **48–49**
Matilda, **149**
Maurice of Nassau, 46
Mauser, Peter Paul, 103
Mauser Model 71/84, **103**
Mausers, 103
MAW (Medium Antitank Weapon), **175**
Maxim, Hiram, 104
Maxim heavy machinegun, Model 1893, **104**, 105
Mayaguez, 170
Mechanization, **123**
Mechanized infantry divisions, **171**
Mechanized unit, **128**
Medal of Honor, 99
Medical Department, 80
Medical Service Corps, 11
Medieval period. *See* Middle Ages.
Medium Antitank Weapon (MAW), **175**
Medium bombers, **137**
Medium tank, **113**
Meeting engagement, **16**
Melee, **33**, 196
Mercenaries, **31**, 61, 194, 199
Merchantmen, **156**
Mercury fulminate, **81**
Messerschmitt Bf-109, **139**, 140
Messerschmitt Bf-110, **140**
Messerschmitt Me-262 Schwalbe, 141
Messerschmitt 262, **140**
MG-42, **130**
Michelangelo, 41
Middle Ages, 13, 29, 31–32, 33, 34, 35, 36, 37, 38, 41, 43, 53, 56, 61, 63, 66, 79, 194, 196, 197, 199, 200
Middle East, 102, 169
Middle Military Division, 80
Midget submarines, **154**
Midgetman, **178**
Midway, Battle of, 144, 155, 196
MiGs, **143**

MiG-15, **179**, 180
MiG-25, **180**
Mikoyan, Artem, 143
Militarism, 95
Military academy, **78**
Military Assistance Advisory Groups
    (MAAGs), **171**
Military Balloon Corps, 80
Military cabinets, **100**
Military crest, **24**
Military division, **80**
Military Division of the Mississippi, 80
Military Information Bureau, 80
Military installations, **128**
Military mission, **100**, 171
Military occupational specialties (MOS), **170**
Military organization, rudiments of, 11–13
Military Police Corps, 11
Military professionalism. *See*
    Professionalism, military.
Military regions, **171**
Military strategy, **6**
Military Telegraph Corps, 80
Military theory and doctrine. *See*
    Theory and doctrine, military.
Militia, **31**, 102
Mills, Sir William, 108
Mills bomb, **108**
Mine detectors, **157**
Mine layers, **156**
Mine thrower, **108**
Mineral jelly, **108**
Mines, **87**, 108, 109, 113–114, 178–179, 189
Mine-sweepers, **113–114**
Minewerfer, **108**
Minié, Claude, 82–83
Minié ball, **82–83**, 85
Mining (with breach), **87**
Minister of War, **61**
Minor tactics, **5**
Minuteman, **178**
Minuteman III, **178**
MIRVs (multiple, independently-targetable
    reentry vehicles), **178**
Missile weapons, **33**, 34, 35–37, 47–52,
    64–66, 67, 81–86, 103–108, 130–134, 170,
    172–178, 183, 185, 187, 196
Mission-type orders, **69**
Mitrailleuse, **104**
Mitsubishi A6M Type Zero, 138
Mitsubishi G4M Betty, 145
Mitsubishi Type O, **138**
Mixed brigade, **129**
Mixed force, **129**
Mixed order, **68**
MLR (main line of resistance) **159**
MLRS (multiple-launched rocket system),
    172, 175, 176
Moat, **38**
Mobile army surgical hospital (MASH), **171**
Mobile defense, **24**
Mobile reserve forces, **16**
Mobile strike force, **24**
Mobility and mobility devices, 32, 34, 37, 53,
    63, 66, 70, 86–87, 109–113, 118, 135–153,
    179–186
Mobilization base, **188**
Mobilize, **12**
Model 38 6.5-mm rifle, **130**
Model 1795 Musket, **65**
Modern Volunteer Army (MVA), **170**
Modus vivendi, **158**
Mole, **38**
Monoplane, **137**, 141, 143, 145
Montigny, 104
Mopping up, **118**
Morale, 78–**79**, 159
Morane-Saulnier M.S. 406, **143**
Morning star, **34**
Morse, Samuel F.B., 88
Morse Code, **88**
Mortars, **52**, 54, 86, 108, 115, 133, 175

MOS (military occupational specialties), **170**
Mothballed (equipment), **68**
Mother ship, **156**
Motor boats, **114**
Motorized unit, **128**
Mountain regiment, **129**
Mountain unit, **129**
Mounted, **16**
Mounted regiments, **80**
Mounted riflemen regiments, **80**
Movement to contact, **16**
MP-18 (*Maschinen-pistole* 18), **106**, 107
MP-38 (*Maschinen-pistole* 38), **130**
MP-38/40, **130**
MP-40, **130**
MSRs (main supply routes), **121**, 189
"Muddling through," **100**
Muleskinner, **129**
Muleskinner units, **129**
Muleteer, **129**
Multiple, independently-targetable reentry
    vehicle (MIRVs), **178**
Multiple lines of operations, **68**
Multiple phase searches, **161**
Multiple-launched rocket system (MLRs),
    172, 175, 176
Munitions, **63**
Musket, **48**, 49, 56, 196, 199
Musketeers, **43**
Musketoon, **65**
Mussolini, Benito, 125
Mustard gas, **109**
Mutiny, **100**
Mutual support, **11**
MVA (Modern Volunteer Army), **170**
MX, **178**

Nagasaki, Japan, 135
Nagasaki Fat Man, 136
Nambu, General, 130
Napalm, **135**
Napoleon. *See* Bonaparte, Napoleon.
Napoleon III, 85
Napoleon Gun Howitzer, 84, **85**
Napoleonic era. *See* Age of Napoleon.
Narrow gauge (gage) system, **113**
Nashville, Battle of, 19
Nation, **61**, 63
Nation in arms, **61**
National armies, **61**
National Guard (World War I), **102**
National Guard (World War II), **129**
National military establishment, **170**
National Socialist Workers' Party, 123
National strategy, **6**
Nationalism, **73**, 93, 95
NATO (North Atlantic Treaty Organization),
    169, 176, 178, 179, 182, 185
  standard ammunition, 172, 173
Naval aviation, **137**
Naval division, **33**
Naval objects, 33–34, 37–38, 53–54, 67, 87,
    113–115, 153–156, 187. *See also*
    Seapower and sails.
Naval operations, **162**
Naval organizations, 33
Naval power, **56**
Naval race, **95**
Naval squadron, 33
Naval tactics, **196**
Navy Cross, **127**
Nazi, **123**, 129
NBC (nuclear-biological-chemical), **172**
NCO (non commissioned officer), **44**
Near East, 38
Needle gun, **81**
Neutrality, **88**
Neutron bomb, **175–176**
New World, 41, 47
Newton, 41
Night attacks, **159**
Night bombers, **137**

Nike Ajax, **176**
Nike Hercules, **176–177**
Nimitz, Chester W., 6
Nisson huts, **115**
Nitroglycerine, **108**
No man's land, 98, **115**, 117
Nocks (of bow and arrow), 35
No-fire lanes, **159**
No-fire lines, **159**
Non commissioned officer (NCO), **44**
Nondisplacement type hulls, **187**
Nonhostile (injury), **79**
Non-rated, **161**
Norden bombsight, **143**
North American P-51 Mustang, 139
North Atlantic Treaty Organization.
    See NATO.
North Vietnamese Army regulars (NVA),
    169, 170, 171
"Notbehelf," **100**
Nuclear deterrence, age of, 167–189, 196
Nuclear war, **2**
  general, **2**
  limited, **2**
  strategic, **2**
Nuclear warheads. *See* Nuclear weapons.
Nuclear weapons. 167, 170, 171, **172**, 174,
    175–176, 178, 187, 196
Nuclear-biological-chemical (NBC), **172**
Nuclear-powered vessels, 170, 187, 196
Numeri, **32**
Numidian cavalrymen, 34, 35
NVA (North Vietnamese Army regulars),
    169, 170, 171

Oak leaf cluster, **127**
Ober, **128**
Ober Kommando-Wehrmacht, **128**
Oberstehheeres Leitung (OHL), **100**
Objective, **68**
Objective (principle of war), **10**
Objective point, **68**
Obligation, **61**
Oblique fire, **55**
Oblique order, **18**, 19
Observation posts, **70**
Obstacles, **88**
Offensive (principle of war), **10**
Offensive operation, **12**, 16–21
Offensive à Poutrance, **116**
Offensive items, 33–38, 108, 109, 170
Offensive operation, 70
Offensive weapons. *See* Offensive items.
Officer, **44**
Officer candidate schools, **99**
Officer of the day, **127**
Officer of the deck, **127**
Officer training schools, **99**
OHL (Oberstehheeres Leitung), **100**
OH-6 Cayuse, **183**
OKW, **128**
"Old Contemptibles," **98**
On the leeward gage, **56**
On the weather gage, **56**
On War, **2**, 198
Onager, 36, **37**
One and one-half ton truck, **136**
Open Arms Program, **170**
Open battle, **23**
Open warfare, **117**
Open-order tactics, **68**
Operation order, **15**
Operational control, **128**
Operational level of war, **5**
Operational organizations, 80
Operations, 38–39, 55–57, 67–70, 88–90,
    116–121, 158–163, 188–189
Operations officer, **78**
Operations security (OPSEC), **11**
Opolchenie, **102**
OPSEC (operations security), **11**
Optics, **153**

Ordnance, 134
Ordnance Department, 85
Ordnance stores, 89
Ordre mince, 68
Ordre mixte, 68, 69
Ordre profond, 68
Organic, 128
Organic equipment, 128
Organic weapons, 128
Organized reserve, 129
Origin of Species, 93
Ottoman Empire, 43
Outer line of defenses, 23
Outposts, 23
Outriggers, 156
Over the top, 117
Overextended, 188
Overlapping fires, 188
Overlays, 159

P-II, 176
P-26, 138
P-35, 138
P-39, 138
P-40s, 139
P-47 Thunderbolt, 139
P-51 Mustang, 139
P-70s, 143
Pacification, 3
Pacifists, 123, 200
Pack animals, 86
Paddle wheels, 87
Paddle-wheelers, 87
Palatians, 32
Palisades, 38
Pallet, 163
Palletizing, 163
Pan (of early small arms), 48
Pandours, 43
Panje, 135
Panoply, 38
Panther, 152–153
Panzer, 113, 125
Panzer division, 129
Panzer forces, 125
Panzer grenadiers, 129
Panzerfaust, 132
Panzerkampfwagen, 152
Panzerkampfwagen I, 152
Paper cartridge, 81
PAR (Perimeter Acquisition Radar), 178
Parabellum pistol Model 1900, 106
Parachute troops, 125
Parados, 115
Parallel, 54
Paramilitaries, 63
Parapet, 54
Paratroopers, 125
Parent organization, 128
Paris gun, 107, 108
Parliament, 6
Paroled, 79
Parrott, Robert P., 85
Parrott guns, 85
Parties, 102
Partisan war, 69. *See also* Small wars.
Passage of lines, 23
Passing box, 66
Pathfinders, 125
Patriot, 61
Patriot (missile), 178
Patrol, 89
Patrol torpedo (PT) boat, 156
Patton, George, 198
PBY, 138, 143
P.C. (poste de command), 64
Peace, 1, 2, 3
Peaceful coexistence, 1
Peloton, 46
Peltasts, 32
Pendant, 79
Penetration, 18

Penetration of opportunity, 18
Pennant, 79
Pennon, 79
Pentomic division, 171
People sniffers, 188
People's war, 69. *See also* Small wars.
Percussion lock, 81, 83, 199
Percussion primer, 86
Pericles, 6
Perier, 52
Perimeter, 24
Perimeter Acquisition Radar (PAR), 178
Perimeter defense, 24
Peripheral operations, 118
Periscopes, 116
Permissions, 99
Pershing (missile), 175, 176
Pershing (tank), 185
Pershing II, 176, 178
Pershing, John, 95, 102
Persians, 29
Persimmon Regiment, 80
Personal staff, 78
Petersburg, 88
Petroleum, oil, and lubricants (POL), 189
Petty officers, 125
Phalangeal legion, 33, 34
Phalanxes, 29, 33, 196
"Phao-kich," 118
Phase lines, 161
Philip (of Macedon), 32
Philosophes, 59, 73
Photo reconnaissance, 163
Pick (artillery accessory), 66
Pickel, 114
Pickelhaube, 114
Pickets, 70
Piece, 47
Piecemeal attack, 21
Pierced steel plank (PSP), 181
Pig boats, 154
Pike, 34, 194
Pile on (form of tactics), 70
Pillbox, 115
Pilotless aircraft, 134
Pilots, 97–98
Pilum, 34
Pincer movement, 21
Pinched out, 159
Pioneer, 78
"Pip squeak," 158
Pipercub Grasshopper, 145
Pirates, 67, 170
Pistols, 48, 55, 56, 65, 81
Pitched battle, 23, 37
Platoon, 46, 159
Plattsburg Movement, 99
Plotting stations, 134
Plow tanks, 147
Plug bayonet, 47
Plunging fire, 55
Pneumatic cavalry, 102
Pocket battleship, 155
Poilu, 98
Point defense, 24
Point units, 188
Pointblank range, 55
Poitrails (poitrels), 38
Poitrels (poitrails), 38
POL (petroleum, oil, and lubricants), 189
Polaris, 178
Pole arms, 46, 47, 56
Polemarch, 32
Police power, 88
Poliorecetics, 55
Political commissars, 125
Political factors, 7, 8, 199
Pom-pom gun, 107
Pontonier, 78
Pontons (pontoons), 66
Pontoon battalions, 63
Pontoons (pontons) 66

Popular forces, 172
Poseidon, 178
Position defense, 24
Position warfare, 117
Positivism, 95
Post, 39
Poste de Command (P.C.), 64
Pot de fer, 49, 50, 51
Potato mashers, 108
Pottes, 55, 56
Pound Wonder, 108
Pour le Mérite, 99
POW (prisoner of war), 79, 170
POW cages, 79
POW camps, 79
PPSh-41 submachinegun, 130
Prague, Battle of, 18
Prague maneuver, 18
Precision bombing, 161
Predreadnoughts, 114
Pre-emptive attack, 23
Preparations, 118
Preparatory fires, 118
Preparedness, 193
Preps, 118
Presidential Unit Citation, 127
"Priest," the, 132
Primer, 48
Principes, 32
Principles of war, 8–11
Prisoner of war (POW), 79
Private (rank), 44
Privateers, 67
Probe, 23
Profession, 5, 200
Professional, 201
Professional institutions, 78, 95, 99, 194, 198
Professionalism, military, 5, 29, 32, 33, 95, 194, 198, 201
Projectiles. *See* Ammunition; Missile weapons.
Proletariat, 93
Prop, 137
Propeller, 87
Propeller plane, 137
Propelling screw, 87
Protecting, 68, 69
Protestant Reformation, 41
Protracted war, 123
Provender, 90
Provisions, 90
Provost marshal, 78
Proximity fuze, 135
Prussian Army, 63
Prussian needle gun, 81
Prussians, 93, 98, 198
Pseudocomitatenses, 32
PSP (pierced steel plank), 181
Psychological operations (PSYOPs), 69, 70, 172
Psychological warfare, 69, 70
PSYOPs (psychological operations), 69, 70, 172
PT (patrol torpedo) boat, 156
Pullman, G.M., 113
Punic War, First, 38
Punic War, Second, 6, 17
Punitive expedition, 159
Punji stakes, 172
Purple heart, 99
Pursuit, 17
Pursuit plane, 137
Pusan, 21
Pusher (aircraft), 110
Putsch, 2
Pyrotechnics, 88
PzKw I, 151, 152
PzKw II, 152
PzKw III, 152
PzKw IIIE, 152
PzKw IV, 152
PzKw V, 152

PzKw VI, **153**

Q-boats, **114**
Qua libra, 51
Quad 50s, **177**
Quarrel (of crossbow), **35**
Quartermaster, **44**
Quartermaster Corps, 11
Quartermaster Department, 80
Quartermaster general, **44**
Quartermeister, 56
Queen Victoria, 95, 99
Quick-firing artillery, **107**
Quick-firing weapons, **103**
Quillons, **35**
Quinine, **158**
Quinquereme, 37–38

R4D, **146**
R and D, **158**
Radar, **157**, 158, 200
Radio, **116**, 187
Radio detection stations, **157**
Radio direction finder, **158**
Radio Direction Finding (RDF), **157**
Radio net, **116**
Radio silence, **158**
Radio simulation, **158**
Radio traffic, **116**
Radioactive poisons, **135**
RAF (Royal Air Force), **101**, 125, 144, 145
Rafts, **87**
Raid, **17**
Rail junctions, **87**
Railheads, **113**
Railroad bridges, **87**
Railroads, 113, 197
Raise (a siege), **13**
Raking fire, **55**
Ram (ship), **87**
Ram (siege machine), **36**
Rammer (artillery accessory), **66**
Ramp tanks, **147**, 179
Rampart, **38**
Ramrod, **49**
Range (aircraft performance measure), **137**
Ranger battalion, **129**
Ranger company, **129**
Rangers, **126**
Rank (component of formation), **11**
Rank (hierarchy), 32, 61, 63, 77–78, 95,
    96–97
Ransomed, **79**
RAP (rear area protection), **162**
Rapid Engineer Deployment Heavy
    Operation Repair Squadron, Engineering
    (Red Horses), **171**
Rapier, **35**, 46
Rapprochement, **123**
Rate of supply, 120, **121**
Ration, **89–90**
Ravelin, **88**
RCT (regimental combat team), **129**
RDF (Radio Direction Finding), **157**
Reagan, Ronald, 178
Rear, **11**, 16, 23
Rear admiral, **78**
Rear area protection (RAP), **162**
Rear security forces, **16**
Rearguard, **16**
Rebellion, **3**
Rebels, **75**
"Rebs," **75**
Recoil, **132**
Recoil systems, **107**, 200
Recoilless rifles, **132**
Reconnaissance, **70**, 117
Reconnaissance bombers, **137**
Reconnaissance by fire, **86**
Reconnaissance in force, **17**
Reconnaissance patrol, **89**
Reconnaissance planes, **137**

Reconnoitering in force, **17**
Recruiting parties, **102**
Recruits, **61**, 63, 194
Red Baron (Baron Manfred von Richthofen),
    101, 110
Red Horses (Rapid Enginner Deployment
    Heavy Operation Repair Squadron,
    Engineering), **171**
Redan, **54**
Redeye guided missile, **174**
Redoubt, **54**
Reentry vehicle, **178**
Refugee, **77**
Refugee camp, **77**
Refuse a flank, **19**
Regiment, **32**, 64
Regimental band, **101**
Regimental combat team (RCT), **129**
Regimental dressing stations, **101**
Regimental eagle, **64**
Regional forces, **172**
Registered, **118**
Regular army, **31**
Regular Army (American), 78
Regular Army (Napoleonic era), 63
Regulars, **31**, 63, 64
Regulating stations, **121**
Reichsmarschalls, 125
Re-lay, **107**
Release point, **161**
Relief (from command), **79**
Relief columns, **188**
Relief expedition, **55**
Relief force, **55**
Relief in place, **23**
Relieve (a unit), **23**
Religious factors, 31, 32, 34, 35, 93, 197
Renaissance, 41–58, 196, 197
Renault FT, **113**
Renault tank, **149**
Rendezvous area, **161**
Reorganization Objectives Army Division
    (ROAD), **171**
Reorganization of the Current Infantry
    Division (ROCID), **171**
Reparations, **117**
Repeating rifles, **83**
Reprisal, **3**
Republic P-47 Thunderbolt, **139**
Required supply rate, **119–120**
Research and development, **158**
Reserve, **15**
Reserve area, **23**
Reserve Officers Training Corps
    (ROTC), **129**
Reserve trench line, **115**
Reservists, **95**
Resistance, **127**
Respirator masks, **109**
Retirement, **25**
Retreat, **24**
Retrograde operation, **12**, 24–25,
Revanche, 95
Reverse slope defense, **68**
Revetments, **87**
Review, **79**
Reviewing officer, **79**
Revitaillment, 121
Revolt, **3**
Revolution, **2**
Revolutionary war, 3
Revolutionary warfare, 2–3, 69–70. *See also*
    Small wars.
Revolver, **83**
Ribbon bridge, **179**
Richthofen, Baron Manfred von
    (Red Baron), 101, 110
Rifle, **81**, 84, 196, 199
Rifle artillery, **85**
Rifle company, **129**
Rifle musket, **81**

Rifle pits, **88**
Rifled guns, **85**
Rifled musket, **81**
Riflemen, **43**
Rifling, **81**, 200
Rim fire cartridge, **83**
Ritter, 31
River crossing operation, **26**
ROAD (Reorganization Objectives Army
    Division), **171**

Roaring Twenties, **123**
ROCID (Reorganization of the Current
    Infantry Division), **171**
Rocket ships, **156**
Rockets, **66**. *See also* Missile weapons.
ROE (rules of engagement), **158**
ROKs (troops of the Republic of Korea), **169**
Roll up a flank, **23**
Rolling barrages, **118**
Rolling stock, **87**
Roman confederation, 198
Roman Empire, 56
Romanovs, 41
Romans, 29, 31, 32, 33, 34, 35, 36–37, 38,
    194, 196, 199
Romanticism, **95**
Rome plows, **179**
Roosevelt, Franklin D., 6
Root, Elihu, 193
Rotary-wing aircraft, 172. *See also*
    Helicopters.
ROTC (Reserve Officers Training Corps),
    129
Roulement, 99
Round ships, **53**
Rounds, **83**
Rout, **23**
Route (direction of attack), **159**
Royal Air Force (RAF), **101**, 125, 144, 145
Royal Flying Corps, **101**
Royal Naval Air Service, **101**
Royal Tank Corps, **101**
RPVs (remotely piloted vehicles), **179**
Rules of engagement (ROE), **158**
Rushes (individual tactics), **55**
Russian Army, 98, 100, 101, 102
Russian saps, **115**

Saber, **35**, 63, 64, 102
Saber rattling, **116**
Saber strength, **98**
Sabot, **66**
Sabotage, **163**
SAC (Strategic Air Command), **170**, 182
Sacred Band, **32**
Saddle, **37**
Safeguard Antiballistic Missile (ABM), **178**
Sailors, **125**
Sails. *See* Seapower and sails.
St. Chamond, 112, **113**
Salients, **24**
Salons, **59**
SALT (Strategic Arms Limitations Talks),
    169, 176
Salvo fire, **55**
"Sammy," **98**
Sampans, **156**
Samurai, 125
Samurai swords, 130
Sanctuary areas, **189**
Sanitary service, **98**
Sanitary trains, **101**
Sapper company, **129**
Saps, **115**
Sarissa, 34
Satchel charges, **135**
SATS (Short Airfield for Tactical Support)
    181
Saturation bombing, **161**
"Sausages," **109**
SB-3 bomber, **144**

S-boats, **154**
Scandinavian Army, 132
Scarp, **54**
Schmeisser, Hugo, 130
Schmeisser submachinegun, **130**
School of the individual soldier, 78
Schools of practice, **78**
Schooners, 87
Schutzengrabenvernichtigungsautomobil, 113
Schutzstaffel (SS), 129
Schwarze Marias, 108
Schwerpunkt defense, 24
Scimitars, 35
Scorched earth policy, 117
Scott, Winfield, 77
Scouts, 89
Scrambler radios, 158
Screen destroyers, 156
Screening, 68, 69
Screening force, **16**
Screw, 87
Screw pickets, 115
Scrub typhus, 158
Scuttling, 162
Se débrouiller, 100
Sea line of communication (SLOC), **12**
Seabees, 171
Seamen, 125
Seaplane carriers, 155
Seaplanes, 137
Seapower, **56**
Seapower and sails, 56–57
Search and clear operations, 188
Search and destroy operations, 188
Secession, 88
Second contingent of the *Landsturm,* 102
Second line reserve, **102**
Second World War. *See* World War II.
Secondary attack, **15**
Second-in-command, 77
Sections, 102
Sector, **159**
Security (principle of war), 10
Security force, **16**
Selective service, 170
Self-propelled guns, 132
Semaphores, 88, 116
Semiarmored cruisers, 114
Semiautomatic weapon, 104
Semifixed ammunition (Napoleonic era), 66
Semifixed ammunition (World War II), 135
Senegalese, 98
Sentries, 89
Sergeant (missile), 175
Sergeant (rank), **44**
Sergeant York Division Air Defense (DIVAD) gun, **177–178**
Serpentine, 48
Service, **97**
Service and support troops, 127
Service ceiling, 137
Service chief, **97**
Service of Supply (SOS), 119
Service schools, 99
Service units, **64**
Servicemen, 123
Services of Supply, 127
Servire, 44
Setpiece battle, 23
Seven Years' War, 21
Seversky, 138
SGM heavy machine gun, **174**
Shakedown marches, 117
Shallow draught (draft) gunboats, 67
SHAPE (Supreme Headquarters, Allied Powers Europe), 169
Shaped charges, 135
Sharps, 83
Sharps carbine, **83**
Sharpshooters, 89
Shaw, Joshua, 81
Shell (ammunition), **86,** 108–109

Shell guns, 86
Shell shock, **99**
Shelling, 86
Sherman, William T., 80
Shermans, 147
Shields, **38**
Shillelagh missile, 185
Shinyo, 154
Ship of the line, 53–54
Shipyards, 162
Shock, 33
Shock action, **63,** 196
Shock armies, 128
Shock tactics, 188
Shock troops, 63
Shogun, 125
Shore parties, 119, 162
Short Airfield for Tactical Support (SATS), 181
Short arm inspections, 99
Short drop, 162
Short take off and landing (STOL), 180
"Shorts," 86
Shortwave radios, 116
Shot (ammunition), 53, 66, 86, 108
Shotgun, 51
Shoulder arm, **47**
Shoulder weapon, **47**
Show of force, 1
Shrapnel, **66**
Shrapnel, Lieutenant Henry, 66
Shrapnel bombs, **108**
Sick bays, 158
Sidearms, 81
Sidedoor Pullmans, 113
Side-wheeler, 87
Sidewinder, 177, 179
Siege artillery, 53, 66, 107
Siege engines, 36–37
Siege gun, 53
Siege line, 54
Siege train, 53
Siegecraft, 36–37
Siegfried Stellung, 116
Signal Corps, 11, **101**
Signal guns, 88
Silk hat, 97
Simonev SKS, 173
Simple line of operations, 68
Simplicity (principle of war), 11
Single line of operations, 68
Single phase searches, 161
Single-seater, 137
Single-shot weapon, 83
Six-shooter, 83
Skip bombing, 161
Skipper, 126
Skirmishers, 15–16, 63, 68, 89, 196
SLBM (submarine launched ballistics missile), 178
Slicks, 183
Sling, 35
SLOC (sea line of communication), 12
Sloops of war, 67
Small arm, 48
Small wars, 69–70, 167–168, **169,** 170–189, 196
Smart bomb, 135
Smoke, 109
Smoke pot, 135
Smokeless powder, 102, 103, 107
Snipers, 89
Sniperscope, 130
Snorkel, 154
Social Darwinism, 95
Social factors, 7, 8, 199
Socialism, 93, 95
Société Pour Aviation et ses Dérivés (SPAD) 13, 110
SOCOM (Special Operations Command), 172
Soda-bottle gun, 85

Soda-water bottle, 85
Soft information, 163
Soft targets, 174
Soften, 118
SOGs (Studies and Observation Groups), 171
Soixante-quinze, 107
Soldier of fortune, 31
Soldiers, 125
Soldiers of the line, 63
Solid shot, 86
Somme, Battle of, 113
Sonar, 154
Sopwith Camel, 109–110
Sopwith Tabloid, 109
Sortie (aerial action), **161**
Sortie (defensive action), 23
SOS (Service of Supply), 119
Sound locating sets, 158
South Vietnamese forces, 171–172
Soviet 12.7mm Degtyarevs, 132
Soviet Navy, 187
SPAD (*Société Pour Aviation et ses Dérivés*) 13, 110
Spanish Army, 45
Spanish Civil War, 139
Spanish square, 45
Spanish-American War, 99
Spartan (missile) 178
Spartans, 198
Spatha, 35
Spear, 34, 35
Spearhead, 23
Special forces units, 171, 172
Special operations, 26–27
Special Operations Command (SOCOM), 172
Special orders, 89
Special staff, 78
Spencer, Christopher, 83
Spencer Arms Company, 85
Spencer carbine, 83, 84
Spencer rifle, 83, 85
Spherical case-shot, 66
Spigot, 133
Spoiling attack, 23
Sponge (artillery accessory), 66
Spontoon, 46, 47
Springfield, Model 1903, **103–104**
Spruance (DD963) destroyers, 187
Spur, 53
Squadra, 32
Squadron, 32–33
  naval, 33
Squads, 102
Square, 55
Square division, 128
Srednii bombovos, 144
SS (*Schutzstaffel*), 129
SSBN (ballistic missile submarine), 187
SSNs, 187
SSs, 187
Stack arms, 79
Staff, 44
Staff duty officer, 127
Staff studies, **118–119**
Staging area, 162
Stahlhelm, 114
Stalemate, 116, 118, 196
Standard (flag), 79
Standard (officially accepted item), **81**
Standard gauge (gage) system, 113
Standard missiles, 187
Stand-down, 117
Standing army, 31, 194
Standing operating procedure, 68
Stand-to, 117
"Star Wars," **178**
START (Strategic Arms Reduction Talks), 169
Static defense, 24
Static division, 128
Stationed (troops), 39

Status quo ante, 116
Stavka, 100
Stay-behind agents, 163
Stealth bomber, 182
Steam power, 66, 113, 114, 196, 198
"Steel pots," 114
"Stellenbosched," 100
Stellung, 116
Stern-wheeler, 87
Stinger, 174
Stinger POST, 174
Stinger RMP, 174
Stirrup, 37
Stock (of gun), 47
Stockpiled, 115
Stokes, Sir Wilfred S., 108
Stokes mortar, 108
Stokes trench mortar, 108
STOL (short take off and landing), 180
Stoner, Eugene, 172
Storch, 145
Stores, 89
Storm troops, 98, 102
Strafing, 109
Straggling, 79
Straight leg infantry, 171
Stratagem, 5
Strategic Air Command (SAC), 170, 182
Strategic Arms Limitations Talks (SALT),
    169, 176
Strategic Arms Reduction Talks (START),
    169
Strategic attrition, 116
Strategic bombers, 137
Strategic bombing, 117, 178
Strategic Defense Initiative, 178
Strategic defensive, 68
Strategic envelopment, 20
Strategic exchange, 2
Strategic flank, 20
Strategic hamlets, 188
Strategic mobility, 86
Strategic points, 67–68
Strategic reserve forces, 16
Strategos, 5, 32
Strategy, 5, 6, 196–197, 198, 199
Strategy of annihilation, 6, 197
Strategy of attrition, 6
Strategy of evasion, 6
Strategy of exhaustion, 6, 196
Strategy of survival, 6
Stratified society, 199
Streltsi, 63
Strike force, 129
Strike the colors, 79
Striker, 86
Striking weapons, 33, 34–35, 36, 46–47,
    64–66, 81, 102, 130
Strongpoint defense, 24
Stuarts (tanks), 149, 150
Studies and Observation Groups (SOGs), 171
Stuka, 145
Sturmabteilung, 102
Sturmbattaillon, 102
Sturmpanzerkraftwagen, 113
Subaltern, 78
Subchasers, 154
Submachine guns, 106
Submarine launched ballistics missile
    (SLBM), 178
Submarine pens, 113
Submarines, 113, 114, 167, 187, 196
Subsistence Department, 80
Subsistence stores, 89
Subsistence supplies, 89
Substitutes, 75
Subversion, 163
Suicide boats, 154
Summary executions, 163
Summits, 169
Summons, 89
Suomi submachine gun, 132

Super bazooka, 132
Supercarrier, 170
Superfortress, 143
Superior lateral communications, 21
Supermarine Spitfire, 139
Superpowers, 169
Supersonic aircraft, 179, 180
Supply bases, 89
Supply line, 68, 197
Supply officer, 78
Supply trains, 89
Support, 11
Support command, 128
Support line trenches, 115
Supporting arms, 55
Supporting attack, 15
Supporting distance, 20
Supreme commander, 61
Supreme Headquarters, Allied Powers
    Europe (SHAPE), 169
Supreme War Council (SWC), 100
Surface effect systems, 187
Surgeon, 78
Surprise (principle of war), 10–11
Surrendered, 88
Sutherland, General, 201
Sutler, 56
Suzerainty, 158
SWC (Supreme War Council), 100
Swedish Army, 52, 198
Sweeps, 117
Swinton, Colonel E.D., 113
Swiss Army, 132
Swiss phalanx, 45
Swiss square, 194
Switch Line, 116
Switch positions, 157
Swordfish, 144
Swords, 34–35, 64
Sympathizers, 75
Synchronized, 110
Syntagma, 33
Système D, 100

T-26 light tank, 151
T-34, 151
T-34/85, 151
T-62 medium tank, 184, 185
T-72, 185
T-80, 185
Tables of Organization and Equipment
    (TOEs), 128
TAC (Tactical Air Command), 170
Tache d'huile, 188
Tactical Air Command (TAC), 170
Tactical aviation, 109
Tactical command post, 64
Tactical flank, 19
Tactical integrity, 118
Tactical interior lines, 21
Tactical manuals, 78
Tactical nuclear weapons, 171
Tactical operations, 69, 70
Tactical tailoring, 45
Tactics, 5, 33, 68–69, 70, 89, 117–118,
    159–161, 188, 194, 196, 198, 199
Tailored unit, 129
Takko units, 129
Taktos, 5
Tandem, 137
Tank destroyer, 153
Tank ditches, 157
Tank dozer, 153
Tank lighter, 156
Tankers, 156
Tanks, 64, 101, 112–113, 147–153, 179,
    184–186, 196
Tannenberg, 21
Taper ax, 35
Target ships, 156
Task force, 45
Tautog, 154

Taxis, 33
TBD-1 Devastator, 144
Teacher's Regiment, 80
Teams, 102
Teamster, 78
Tear gas grenades, 109
Technology, 8, 33–38, 46–55, 64–66, 81–88,
    102–116, 129–158, 172–188, 193, 199
Telecommunications, 88
Telegraph (Civil War), 88
Telegraph (Napoleonic era), 67
Telegraph key, 88
Television, 187, 188
Television-guided bombs, 179
Tenant, 44
Tenders, 156
Tercio, 45, 46
Terrain exercises without troops, 99
Terrain rides, 99
Territorial Army, 102
Territorial division, 80
Territorial organizations, 80
Territorials, 102
Terror attacks, 159
Terror bombing, 159
Terror raids, 159
Testudo, 36, 38
Tet Offensive of 1968, 17
Tethered artillery observation balloons, 109
Tetrahedrons, 157
Thanes (*thegns*) 31
Theater of war, 12
Thebans, 32
Thegns (thanes), 31
Theory and doctrine, military, 7, 13,
    197–198, 199
"Thin Man," 135, 143
Thin order, 68
Thinskinned, 155
Third line reserve, 102
Thirty Years' War, 59, 197, 198
Thompson submachinegun, 130
Threads of continuity. *See* Continuity,
    threads of.
Three-quarter armor, 54
Thrusting weapons, 33, 34–35, 64
Thucydides, 197
Thumbstall, 66
Tiger (tank), 153
Tiger Moth, 145
Time bombs, 135
"Tin hats," 114
Tinclad, 87
Tirailleurs, 63
Titan, 178
Titan II, 178
TNT (trinitrotulene), 108
Tochka, 157
Toehold, 119
(TOEs) Tables of Organization and
    Equipment, 128
Tomahawk (aircraft), 139
Tomahawk (striking weapon), 47
"Tommy," 98
Topographical engineers, 64
Torpedo boats, 114
Torpedo bombers, 137
Torpedo nets, 153
Torpedoes, 87
Torsion bar suspension, 152
Tortoise, 36
Total war, 2, 7, 59, 194, 197
Touchhole, 47, 51, 86
Toulon, siege of, 59
TOW (tube-launched, optically tracked,
    wire-guided missile), 175, 183, 186
TOW II, 175
TOWs (aircraft), 146
TOWs (ships), 114
TRADOC (Training and Doctrine
    Command), 171
Trail (artillery), 53

Train (supply unit), **64**, 197
Training and Doctrine Command (TRADOC), **171**
Training centers, **99**
Training teams, **99**
Trajectory, **52**, 55
Transport areas, **119**
Transports, **87**
Trap ships, **114**
Trappings, 34, **38**
Traverses, **115**
Trawlers, **114**
Treason, **69**
Treaty, **68**
Trebuchet, 37
Trench, **54**
Trench foot, **116**
Trench knife, **102**
Trench mortar, **108**
Trench stores, **115**
Trench tractor, **113**
Trench warfare, 93, 99, 108, 115, **117**, 196
Triangular divisions, **128**
Triarii, 32
Trigger, **48**
Trinitrotolulene (TNT), **108**
Triplane, **110**, 137
Triple Entente, **95**
Trireme, 37
Troop, **46**
Truce, **68**
Trucks, 34, **109**
Truman, Harry S., **167**
Trunnions, **53**
Tsutsugamushi disease, **158**
Tube-launched, optically tracked, wire-guided missile (TOW), **175**, 183, 186
Tugs, **87**
Tunnel-rat teams, **171**
Turma, 33
Turning movement, **20**
Turnpikes, **86**
Turret, 113
Turret defilade, **159**
Turreted gunpits, **116**
"Twice and one-half," **136**
Twin 40s, **177**
Two and one-half ton truck, **136**
"Two up and one back," **159**
Two-front war, **68**
Type H light machinegun, **130**
Typhoon barrages, **118**
Tyre, siege of, **38**

U-boats, **113**, 156
Uhlans, **98**
UH-1 Iroquois, **183**
UH-60 Blackhawk, **183**
Ultra, **163**
Umbrella talks, **169**
Unconditional surrender, **117**, 197
Uncover, **23**
Underground, **127**
Underwater demolition team, **126**
Unified command, **127**
Union, the, **88**
Union Army, 79, 88
Union Ordnance Department, **83**
Uniservice operation, **118**
United Nations, **167**
United States Air Force, 138, 170, 171, 178, 179, 180
United States Army, 5, 9, 11, 16, 24, 44, 45, 46, 55, 66, 69, 77, 79, 82, 86, 95, 96, 97, 98, 99, 100, 101, 102, 103, 106, 114, 118, 125, 127, 128, 130, 132, 138, 143, 170-171, 172, 174, 176, 177, 178, 185. *See also* American Army.
  organization of in Civil War, 80
  organization of in World War I, 101-102
  organization of in World War II, 127-128, 129

reorganization of, post-World-War-II, 170-171
United States Army Air Forces, 125
United States Army Air Service, 137
United States Army Special Forces, 129
United States gun motor carriage M7, **132**, 133
United States 105-mm howitzer, **132**, 133
United States Marine Corps, 118, 125
United States Military Academy, 78
United States Navy, 99, 118, 129, 132, 140, 143, 156, 170, 180, 182, 187
Units of fire, **135**
Unity of command (principle of war), 10
Universal military obligation, **61**
Unrestricted submarine warfare, **117**
Untersee boot, 113
Upton, Emory, 193
U.S. Pistol, M1911, **106**
U.S. Special, **83**
USS, **114**
USS *New Jersey,* 187
USS *Nimitz,* 187
USS *Washington,* 155
USS *Willard Keith,* 155
UTTAS (Utility Tactical Transport Aviation System), **183**

V-1, **134**
V-2, **134**
Val, **145**
Van, **16**
Vanguard, **16**
Variable time (VT) fuze, **135**
Vasi, 49, **50**
V.C. (Victoria Cross), **99**
VC (Viet Cong), **169**, 170
Vegetius, **197**
Vehicle and tank parks, **189**
Vent, **48**
Vergeltungswaffen, **134**
Versailles, Treaty of, 155
Vertical envelopment, **19**
Vertical take off and landing (VTOL), **180**
Very pistol, **116**
Veteran, **75**
Vickers machineguns, **109-110**
Vickers Supermarine Spitfire, 140
Vickers-Maxim 1-pounder automatic machine cannon, **107**
Victoria Cross (V.C.), **99**
Victorianism, 93, **94**
Victory Medal, **99**
Vielle, Paul, 103
Viet Cong (VC), **169**
Vietnam War, 17, 39, 64, 98, 135, 146, 156, 167-168, 169, 170, 171-172, 179, 180, 181-182, 183, 188, 189
Viking sword, **35**
Vikings, **38**
VOLAR (Volunteer Army), **170**
Volley fire, **55**
Voltigeurs, **63**, 64, 65
Voluntary withdrawal, **25**
Volunteer Army (VOLAR), **170**
Volunteers, **61**, 102, 194, 199
Von Dreyse, Johann, 81
Von Moltke, Helmuth, 198
Vought F4U Corsair, 141
VT (variable time) fuze, **135**
VTOL (vertical take off and landing), **180**
Vulcan, **177**
V-weapons, **134**

Waco CG-4A, 146
Wadding, 49, **81**
Waffe, **128**
Waffen SS divisions, 129
Wagon trains, **64**
Walleyes, **179**
War, **1**, 2, 3
  future of, 200-201

levels of, 1-2
nature of, 3
operational level of, 5
principles of, **8-11**
War area, **128**
War bonds, **117**
War cabinets, **100**
War chests, **90**
War colleges, **127**
War councils, **100**
War debt, **117**
War Department, 77, 78, 80, 170
War games, **99**
War gas, **109**
War Ministry, **61**
War neurosis, **158**
War of 1812, 66
War Office, **61**
War on multiple fronts, **68**
War orders, **88**
War plans, **100**
War powers resolution, **170**
War zones, **171**
Warhawks, **139**
Warhead, **135**
Warning order, **89**
Warrant, **95**
Warrant officer, **95-96**
Wars of German Unification, 93, **198**
Wars of national liberation, 167. *See also* Small wars.
Warsaw Pact nations, 172, 185
Washington, George, 39, 99, 198
Wastage, **98**
Watch, **126**
Watch officer, **127**
Waterloo, 65, 73, 81
Wave assaults, **159**
Waves, **56**
Weapons, 33. *See also* Artillery; Missile weapons; Striking weapons.
Weapons of mass destruction, **172**
Weapons platoon, **129**
Wedge block, **66**
"Weeps," **100**
Wehr, **128**
Wehrmacht, **128**
Wellington, Duke of, 68
West Point (NY), 78
West Point Iron and Cannon Foundry, 85
Western Front, 100, 101, 114, 116, 118
Western strategy, **116**
Wheeling movement, **56**
Wheellock, **48**
Whippet, **112**, 113
Whirly-bird, **182**
White forces, **102**
"Whiz Bang," **108**
WIA (wounded in action), **79**
Wildcats, **141**
William II, Emperor, 95, 98
William the Conqueror, 44
Wilson, Woodrow, 99
Winchester Arms Company, 85
Winchesters, **85**
Windward, **56**
Wings, **16**
Winter quarters, **39**
Winterization, **39**
"Wipers," **100**
"Wipers Express," **108**
Wireless interceptor teams, **102**
Wireless radio, **116**
Wiring parties, **102**
Withdrawal, **25**
Wolfpack, **129**
Wood, Leonard, 193
Wooden box mines, **157**
Wordsworth, William, 200
World War I, 8, 39, 44, 93, 94, 95-98, 99-121, 123-165, 172, 196, 197

World War II, 6, 8, 12, 21, 24, 66, 103, 104,
    114, 123–165, 167, 169, 170, 171, 172, 175,
    176, 178, 179, 181, 185, 187, 188, 196, 197
Wormer (artillery accessory), **66**
Wotan Stellung, **116**
Wounded, **79**
Wounded in action (WIA), **79**
Woyrsch Korps, **101**

X-Day, **158**
Xenophon, 197
XO, **77**

Yakovlev, Aleksander, 143
Yakovlev Yak-3, 142
Yaks, **143**
"Yankees," **98**
"Yanks," **98**
Yeomen, **43**
Yperite, **109**
Ypres, Belgium, 100, 108, 109

Z-day, **117**
Zeke, **138**
Zeppelin, Count Ferdinand von, 109

Zeppelins, **109**
Zero (Japanese plane), **138**, 141
Zero (marksmanship term), **89**
Zero day, **117**
Zero hour, **117**
Zero in, **89**
Zero their sights, **89**
Z-hour, **117**
Zone, **159**
Zone defense, **24**
Zone of action, **159**, 161

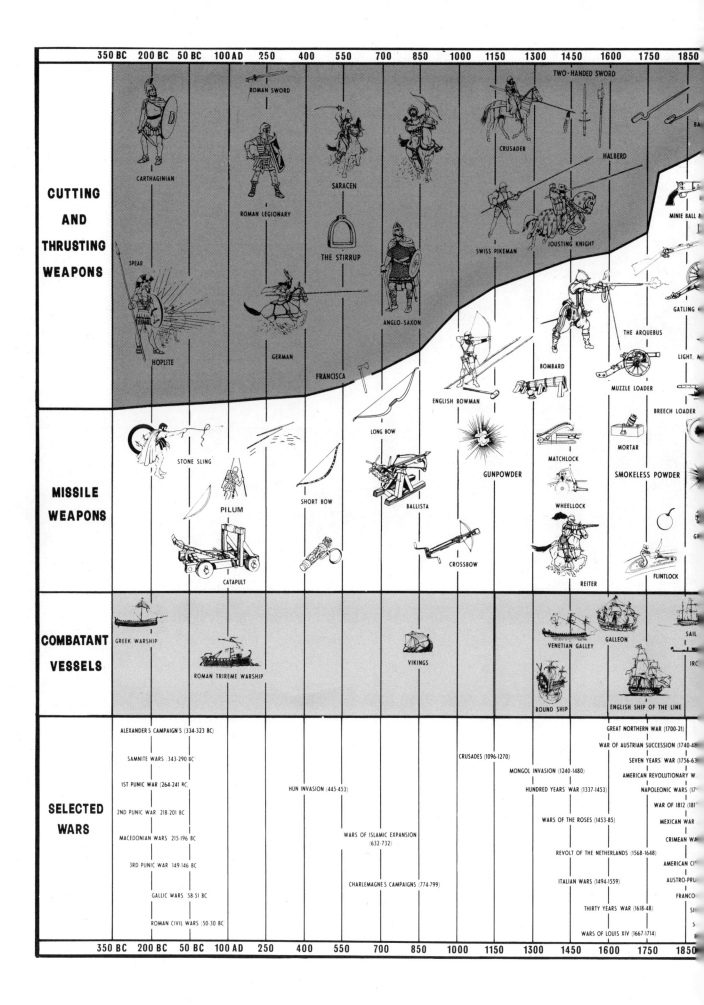

| | 350 BC | 200 BC | 50 BC | 100 AD | 250 | 400 | 550 | 700 | 850 | 1000 | 1150 | 1300 | 1450 | 1600 | 1750 | 1850 |

CUTTING AND THRUSTING WEAPONS

CARTHAGINIAN

ROMAN SWORD

TWO-HANDED SWORD

SARACEN

CRUSADER

HALBERD

SPEAR

ROMAN LEGIONARY

THE STIRRUP

SWISS PIKEMAN

JOUSTING KNIGHT

MINIE BALL &

HOPLITE

GERMAN

ANGLO-SAXON

THE ARQUEBUS

GATLING

FRANCISCA

ENGLISH BOWMAN

BOMBARD

MUZZLE LOADER

LIGHT,

MISSILE WEAPONS

STONE SLING

LONG BOW

GUNPOWDER

MATCHLOCK

BREECH LOADER

MORTAR

SMOKELESS POWDER

PILUM

SHORT BOW

BALLISTA

WHEELLOCK

CATAPULT

CROSSBOW

REITER

FLINTLOCK

COMBATANT VESSELS

GREEK WARSHIP

VENETIAN GALLEY

GALLEON

SAIL

ROMAN TRIREME WARSHIP

VIKINGS

ROUND SHIP

ENGLISH SHIP OF THE LINE

IRC

SELECTED WARS

ALEXANDER'S CAMPAIGNS (334-323 BC)

GREAT NORTHERN WAR (1700-21)

SAMNITE WARS (343-290 BC)

CRUSADES (1096-1270)

WAR OF AUSTRIAN SUCCESSION (1740-48

SEVEN YEARS' WAR (1756-63

1ST PUNIC WAR (264-241 BC)

MONGOL INVASION (1240-1480)

AMERICAN REVOLUTIONARY W

HUN INVASION (445-453)

HUNDRED YEARS WAR (1337-1453)

NAPOLEONIC WARS (17

2ND PUNIC WAR 218-201 BC

WAR OF 1812 (181

MACEDONIAN WARS 215-196 BC

WARS OF THE ROSES (1453-85)

MEXICAN WAR

WARS OF ISLAMIC EXPANSION (632-732)

CRIMEAN WA

3RD PUNIC WAR 149-146 BC

REVOLT OF THE NETHERLANDS (1568-1648)

AMERICAN CI

GALLIC WARS 58-51 BC

CHARLEMAGNE'S CAMPAIGNS (774-799)

ITALIAN WARS (1494-1559)

AUSTRO-PRU

FRANCO

ROMAN CIVIL WARS (50-30 BC)

THIRTY YEARS WAR (1618-48)

S

WARS OF LOUIS XIV (1667-1714)

S

| | 350 BC | 200 BC | 50 BC | 100 AD | 250 | 400 | 550 | 700 | 850 | 1000 | 1150 | 1300 | 1450 | 1600 | 1750 | 1850 |

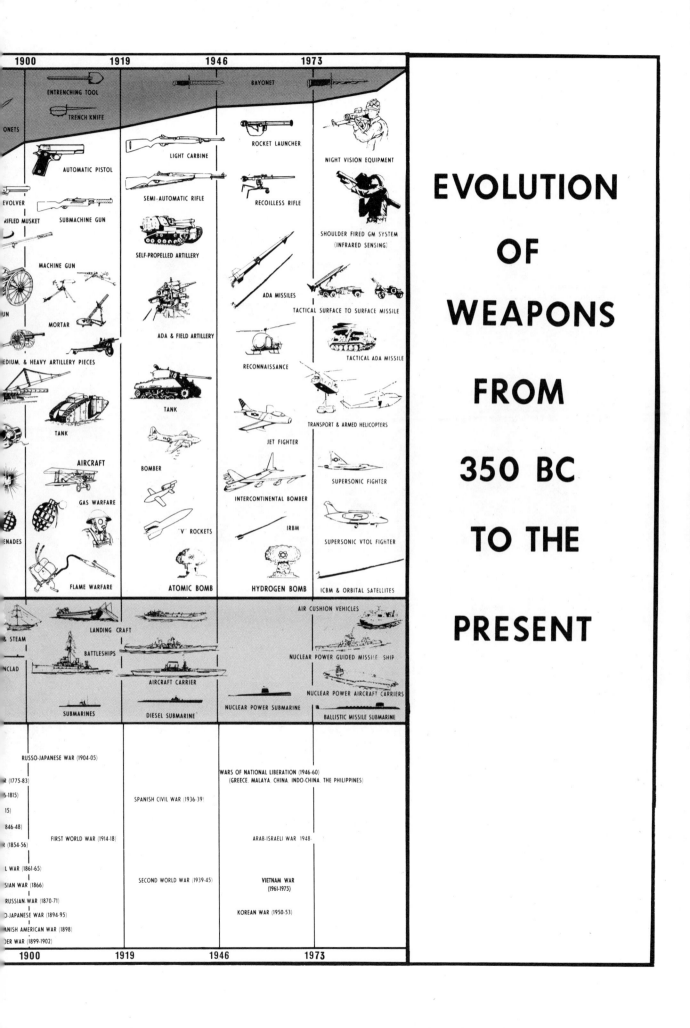

EVOLUTION OF WEAPONS FROM 350 BC TO THE PRESENT